0744228122 11

p.
S. Bacillus
E Coli

AF791230

UNIVERSITY OF BRADFORD
Library

Accession No. 028471701

This book should be returned not later than the last date stamped below.
The loan may be extended on request provided there is no waiting list.

Please note: telephone renewals are not allowed on QSC items.

FINES ARE CHARGED ON OVERDUE BOOKS

UNIVERSITY OF BRADFORD
WITHDRAWN AND DISCARDED
LIBRARY

L45 350A

L45 1/03/04 350

BD 028471701 8

ISOLATION AND IDENTIFICATION OF MICRO-ORGANISMS OF MEDICAL AND VETERINARY IMPORTANCE

A complete list of titles in the Society for Applied Bacteriology Technical Series appears at the end of this volume.

THE SOCIETY FOR APPLIED BACTERIOLOGY
TECHNICAL SERIES NO. 21

ISOLATION AND IDENTIFICATION OF MICRO-ORGANISMS OF MEDICAL AND VETERINARY IMPORTANCE

Edited by

C. H. COLLINS

Public Health Laboratory, Dulwich Hospital,
London, UK

AND

J. M. GRANGE

Department of Microbiology, Cardiothoracic Institute,
Brompton Hospital, London, UK

1985

ACADEMIC PRESS
(Harcourt Brace Jovanovich, Publishers)

LONDON · ORLANDO · SAN DIEGO · NEW YORK
TORONTO · MONTREAL · SYDNEY · TOKYO

COPYRIGHT © 1985, BY THE SOCIETY FOR APPLIED BACTERIOLOGY
ALL RIGHTS RESERVED.
NO PART OF THIS PUBLICATION MAY BE REPRODUCED OR
TRANSMITTED IN ANY FORM OR BY ANY MEANS, ELECTRONIC
OR MECHANICAL, INCLUDING PHOTOCOPY, RECORDING, OR
ANY INFORMATION STORAGE AND RETRIEVAL SYSTEM, WITHOUT
PERMISSION IN WRITING FROM THE PUBLISHER.

ACADEMIC PRESS INC. (LONDON) LTD.
24–28 Oval Road
LONDON NW1 7DX

United States Edition published by
ACADEMIC PRESS, INC.
Orlando, Florida 32887

British Library Cataloguing in Publication Data

Isolation and identification of micro-organisms
of medical and veterinary importance.–(The
Society for Applied Bacteriology technical
series; no. 21)
1. Medical microbiology 2. Micro-organisms,
Pathogenic 3. Veterinary medicine
I. Title II. Collins, C.H. III. Grange, John M.
616'.01 QR175

Library of Congress Cataloging in Publication Data
Main entry under title:

Isolation and identification of micro-organisms of
medical and veterinary importance.

(Technical series / Society for Applied Bacteriology;
no. 21)
Includes index.
1. Bacteriology, Medical--Cultures and culture media.
2. Veterinary bacteriology--Cultures and culture media.
3. Bacteria, Pathogenic--Identification. I. Collins,
C. H. (Christopher Herbert) II. Grange, John M.
III. Series: Technical series (Society for Applied
Bacteriology: no. 21) [DNLM: 1. Bacteria--isolation &
purification--congresses. 2. Bacterial Infections--con-
gresses. 3. Bacterial Infections--veterinary--congresses.
W1 S0851F no. 21 / QY 100 I85 1983]
QR66.I627 1985 616'.014 84-14535
ISBN 0-12-181460-2 (alk. paper)

PRINTED IN THE UNITED STATES OF AMERICA

85 86 87 88 9 8 7 6 5 4 3 2 1

UNIVERSITY OF BRADFORD
LIBRARY
14 JAN 1986
ACCESSION No.
0284741401
Q 576·8·06 COL
LOCATION

Contributors

A. S. BAILEY, *North Manchester Regional Virus Laboratory, New Holzel Laboratory, Booth Hall Children's Hospital, Manchester M9 2AA, UK*

JUDITH M. G. BAILEY, *North Manchester Regional Virus Laboratory, New Holzel Laboratory, Booth Hall Children's Hospital, Manchester M9 2AA, UK*

D. BEIGHTON, *Dental Research Unit, Royal College of Surgeons of England, West Hill, Downe, Near Orpington, Kent BR6 7JJ, UK*

J. S. BRAZIER, *Anaerobic Reference Unit, Public Health Laboratory, Luton and Dunstable Hospital, Luton LU4 0DZ, UK*

R. A. BREWER, *Ministry of Agriculture, Fisheries and Food, Central Veterinary Laboratory, New Haw, Weybridge, Surrey KT15 3NB, UK*

D. W. BURDON, *The General Hospital, Birmingham B4 6NH, UK*

C. K. CAMPBELL, *Public Health Laboratory Service, Mycological Reference Laboratory, London School of Hygiene and Tropical Medicine, London WC1E 7HT, UK*

D. C. A. CANDY, *Institute of Child Health, Birmingham B16 8ET, UK*

J. A. CARMAN, *Vaccine Research and Production Laboratory, Public Health Laboratory Service, Centre for Applied Microbiology and Research, Porton Down, Salisbury, Wiltshire SP4 0JG, UK*

C. H. COLLINS, *Regional Centre for Tuberculosis Bacteriology, Public Health Laboratory, Dulwich Hospital, East Dulwich Grove, London SE22 8QF, UK*

E. MARY COOKE,[1] *Department of Microbiology, University of Leeds, Leeds LS2 9JT, UK*

M. J. CORBEL, *Ministry of Agriculture, Fisheries and Food, Central Veterinary Laboratory, New Haw, Weybridge, Surrey KT15 3NB, UK*

P. E. CURTIS, *Sub-Department of Avian Medicine, University of Liverpool Veterinary Field Station, Leahurst, Neston, Wirral L64 7TE, UK*

C. DAVIS, *Public Health Laboratory Service, Mycological Reference Laboratory, London School of Hygiene and Tropical Medicine, London WC1E 7HT, UK*

P. J. DENNIS, *Environmental Microbiology and Safety Reference Laboratory, Public Health Laboratory Service, Centre for Applied Microbiology and Research, Porton Down, Salisbury, Wiltshire SP4 0JG, UK*

[1]Present address: Division of Hospital Infection, Central Public Health Laboratory, 175 Colindale Avenue, London NW9 5HT, UK.

T. J. Donovan,[2] *Public Health Laboratory, Preston Hall Hospital, Maidstone Kent ME20 7NH, UK*

L. R. Duddridge, *Mycobacterium Reference Unit, Public Health Laboratory, University Hospital of Wales, Heath Park, Cardiff CF4 4XW, UK*

C. S. F. Easmon, *Department of Medical Microbiology, Wright–Fleming Institute, St. Mary's Hospital Medical School, London W2 1PG, UK*

S. I. Egglestone,[3] *Gonococcus Reference Unit, Public Health Laboratory, Bristol BS2 8EL, UK*

P. G. Francis, *Ministry of Agriculture, Fisheries and Food, Central Veterinary Laboratory, Bacteriology Department, New Haw, Weybridge, Surrey KT15 3NB, UK*

G. N. Frerichs, *Institute of Aquaculture, University of Stirling, Stirling FK9 4LA, UK*

H. Gaya, *Microbiology Department, Brompton Hospital, London SW3 6HP, UK*

K. P. W. Gill, *Ministry of Agriculture, Fisheries and Food, Central Veterinary Laboratory, New Haw, Weybridge, Surrey KT15 3NB, UK*

M. Gitter, *Ministry of Agriculture, Fisheries and Food, Central Veterinary Laboratory, New Haw, Weybridge, Surrey KT15 3NB, UK*

P. Hambleton, *Vaccine Research and Production Laboratory, Public Health Laboratory Service, Centre for Applied Microbiology and Research, Porton Down, Salisbury, Wiltshire SP4 0JG, UK*

Margaret S. Hendrie, *Torry Research Station, Aberdeen AB9 8DG, UK*

D. McL. F. D. Hendry, *Ministry of Agriculture, Fisheries and Food, Central Veterinary Laboratory, New Haw, Weybridge, Surrey KT15 3NB, UK*

C. A. Ison, *Department of Medical Microbiology, Wright–Fleming Institute, St. Mary's Hospital Medical School, London W2 1PG, UK*

P. A. Jenkins, *Mycobacterium Reference Unit, Public Health Laboratory, University Hospital of Wales, Heath Park, Cardiff CF4 4XW, UK*

A. E. Jephcott, *Gonococcus Reference Unit, Public Health Laboratory, Bristol BS2 8EL, UK*

J. M. Ketley, *Department of Microbiology, University of Birmingham, Birmingham B15 2TT, UK*

Anna King, *Department of Microbiology, St. Thomas's Hospital Medical School, London SE1 7EH, UK*

K. P. Lander, *Ministry of Agriculture, Fisheries and Food, Central Veterinary Laboratory, New Haw, Weybridge, Surrey KT15 3NB, UK*

J. V. Lee, *Environmental Microbiology and Safety Reference Laboratory, Public*

[2]Present address: Public Health Laboratory, The William Harvey Hospital, Willesborough, Ashford, Kent TN24 0LZ, UK.

[3]Present address: Public Health Laboratory, Institute of Pathology, General Hospital, Westgate Road, Newcastle upon Tyne NE4 6BE, UK.

Health Laboratory Service, Centre for Applied Microbiology and Research, Porton Down, Salisbury, Wiltshire SP4 0JG, UK

P. N. LEVETT, *Anaerobic Reference Unit, Public Health Laboratory, Luton and Dunstable Hospital, Luton LU4 0DZ, UK*

D. W. R. MACKENZIE, *Public Health Laboratory Service, Mycological Reference Laboratory, London School of Hygiene and Tropical Medicine, London WC1E 7HT, UK*

GILL MEARNS, *North Manchester Regional Virus Laboratory, New Holzel Laboratory, Booth Hall Children's Hospital, Manchester M9 2AA, UK*

J. MELLING, *Vaccine Research and Production Laboratory, Public Health Laboratory Service, Centre for Applied Microbiology and Research, Porton Down, Salisbury, Wiltshire SP4 0JG, UK*

T. J. MITCHELL, *Department of Microbiology, University of Birmingham, Birmingham B15 2TT, UK*

I. PHILLIPS, *Department of Medical Microbiology, St. Thomas's Hospital Medical School, University of London, London SE1 7EH, UK*

K. D. PHILLIPS, *Anaerobic Reference Unit, Public Health Laboratory, Luton and Dunstable Hospital, Luton LU4 0DZ, UK*

SUSAN C. REDMOND, *Department of Microbiology, University of Birmingham, Birmingham B15 2TT, UK*

SHIRLEY J. RICHMOND, *North Manchester Regional Virus Laboratory, New Holzel Laboratory, Booth Hall Children's Hospital, Manchester M9 2AA, UK*

K. ROBINSON, *Division of Agricultural Bacteriology, School of Agriculture, Aberdeen AB9 1UD, UK*

ELAINE SAUVÉ,[4] *Division of Agricultural Bacteriology, School of Agriculture, Aberdeen AB9 1UD, UK*

RUTH SHINEBAUM,[5] *Department of Microbiology, University of Leeds, Leeds LS2 9JT, UK*

J. STEPHEN, *Department of Microbiology, University of Birmingham, Birmingham B15 2TT, UK*

P. TAYLOR, *Microbiology Department, Brompton Hospital, London SW3 6HP, UK*

SHEENA A. WAITKINS, *Public Health Laboratory Service, Leptospira Reference Unit, County Hospital, Hereford HR1 2ER, UK*

J. WATKINS,[6] *Yorkshire Water Authority, South Eastern Division, Divisional Laboratory, Green Lane, Scawthorpe, Doncaster, UK*

[4]Present address: Scotbec, 22 Great King Street, Edinburgh EH3 6QH, UK.

[5]Present address: Department of Bacteriology, University Medical School, Teviot Place, Edinburgh EH8 9AG, UK.

[6]Present address: Yorkshire Water Authority, Skeldergate Laboratory, Skeldergate, York YO1 1HL, UK.

A. T. WILLIS, *Anaerobic Reference Unit, Public Health Laboratory, Luton and Dunstable Hospital, Luton LU4 0DZ, UK*

A. E. WRIGHT,[7] *Environmental Microbiology and Safety Reference Laboratory, Public Health Laboratory Service, Centre for Applied Microbiology and Research, Porton Down, Salisbury, Wiltshire SP4 0JG, UK*

M. D. YATES, *Regional Centre for Tuberculosis Bacteriology, Public Health Laboratory, Dulwich Hospital, East Dulwich Grove, London SE22 8QF, UK*

[7]Present address: Public Health Laboratory, General Hospital, Westgate Road, Newcastle upon Tyne NE4 6BE, UK.

Preface

This is the twenty-first book in the Technical Series of the Society for Applied Bacteriology and reflects the contributions made at the Society's Demonstration Meeting, held at Birmingham University on 21 September 1983.

Although the study of micro-organisms has advanced and diversified on many fronts since the time when it was a handmaiden to human and veterinary medicine, the isolation and identification of microbes capable of causing disease is still a very important aid to the diagnosis, treatment and prevention of infection.

Progress in this field has been erratic. Some methods have hardly changed for half a century, while others have altered, some slowly, some dramatically.

Medical and veterinary bacteriologists have not always moved at the same pace nor in the same direction, and some evidence of this is apparent in the following chapters. Some organisms have always been difficult to culture and identify but new serological methods offer rapid and more certain assistance in diagnosis.

The contributors to the Demonstration Meeting brought their own particular expertise in the detection, isolation and identification of a variety of microbes and in the serological diagnosis of some diseases. The methods they present in this book have been tried and found to be successful, and it is hoped that other workers will find them useful.

We would like to thank the contributors for the speed with which they produced their manuscripts and the assistance they have given us. We also wish to thank Dr Alan Linton for his help in compiling the programme, and Dr Jack Hopton and Dr Janet Corry for the excellent local organisation which made the meeting successful.

December 1984

C. H. Collins
J. M. Grange

Contents

ISOLATION AND IDENTIFICATION OF MICRO-ORGANISMS OF MEDICAL AND VETERINARY IMPORTANCE

Pseudomonads and Related Bacteria

ANNA KING AND I. PHILLIPS

Department of Michrobiology, St. Thomas's Hospital Medical School, London, UK

The identification of pseudomonads and related bacteria has long been a task not happily undertaken by workers in most diagnostic laboratories. Our own interest in this area began in 1963 when it became clear that the identification of atypical strains of *Pseudomonas aeruginosa* posed problems (Phillips 1969) and was reinforced in 1970 when three patients were infected with *Ps. cepacia*, which was also isolated from heparinised saline solution (Phillips *et al.* 1971). This was followed by an outbreak of infection caused by contaminated autoclaved fluids (Phillips *et al.* 1972), and the isolation and identification of the causative organism of this outbreak led to a study of non-fermentative, aerobic Gram negative bacilli (King *et al.* 1979). Much of the impetus in the study of "unusual" Gram negative bacilli from clinical sources came from King (1964), and our investigations were based on her work and the subsequent revisions and extensions of it (Weaver *et al.* 1972; Tatum *et al.* 1974) and on the approach to identification in *Cowan and Steel's Manual for the Identification of Medical Bacteria* (Cowan 1974). Our aim was to simplify identification and to choose tests that were both easy and reliable for use in a diagnostic laboratory. We decided to follow the practice of the NCTC Computer Trials Laboratory, Colindale, and to use ammonium salt sugars (Snell & Lapage 1971) for the study of carbohydrate metabolism, rather than Hugh and Leifson's medium (Weaver *et al.* 1972) or the more arduous substrate-utilisation tests (Jessen 1965; Stanier *et al.* 1966). We recognize the inherent difficulties of the use of antibiotic sensitivity as a guide to identification, but, like Gilardi (1971), we have included those tests that seem useful. Finally, we have studied only those organisms that we have isolated in the course of the examination of clinical specimens, and from environmental samples examined in the search for sources in outbreaks of infection. We recognise that

Isolation and Identification of
Micro-organisms of Medical and
Veterinary Importance

Copyright © 1985 by the Society for Applied Bacteriology
All rights of reproduction in any form reserved
0-12-181460-2

there are therefore many omissions, but have found that the scheme is useful for the identification of most organisms isolated from both clinical and environmental samples examined in a medical diagnostic laboratory.

Tests Used

The temperature of incubation is 37°C unless otherwise stated, and tests are recorded as negative only after the recommended period of incubation as attempts at short cuts can lead to considerable confusion.

Oxidase test. This is done by the Kovács method (Cowan 1974).

Oxidation–fermentation tests. Hugh and Leifson medium (Cowan 1974) with glucose as the substrate is used. It is important to distinguish between alkali production and no reaction.

Catalase production. Hydrogen peroxide, 1 ml of 6% v/v in water, is run down an 18- to 24-h nutrient agar slope culture, and the presence of bubbles is recorded as positive.

Motility. Suspensions of 6-h peptone water cultures at room temperature and 37°C are examined on a slide under a coverslip by phase-contrast microscopy (King & Phillips 1978). It is important to appreciate that cultures of certain motile species contain only a small proportion of motile organisms.

Preparation of inoculum for biochemical tests. Colonies from a purity plate are suspended in normal saline to give approximately 10^6 organisms/ml.

Carbohydrate metabolism. Two methods are used: (1) ammonium salt "sugars" (Difco) (Cowan 1974, method 4), except that the final concentration of bromocresol purple is 0.004% w/v, and (2) peptone water "sugars" with Andrade's indicator (Cowan 1974, method 1).

Citrate utilisation. The medium of Simmons is used (Cowan 1974).

Arginine hydrolysis. The medium of Thornley (Cowan 1974, method 2) is used.

Decarboxylases. The medium of Møller (Cowan 1974, method 1) is used with lysine, arginine and ornithine as substrates. A simple change to an alkaline reaction is recorded as positive, provided that there is no pH change in the control.

Gluconate oxidation. Carpenter's modification of Shaw and Clark's method in 2-ml volumes of medium is used (Cowan 1974, method 2).

Urease activity. This is tested in Christensen's medium (Cowan 1974, method 1).

Indole production. Peptone water cultures, after incubation for 48 h, are tested with Kovács' reagent (Cowan 1974).

Hydrogen sulphide production. Triple sugar iron agar (Cowan 1974) is modified by the omission of sugars and indicator.
Nitrate and nitrite reduction. Nitrate and nitrite broth media (Difco) are used (Cowan 1974). It is necessary to use both media as some organisms are able to reduce only nitrite.
Temperature range for growth. A very light inoculum is placed in peptone water, which is heated to 42°C or cooled to 4°C before inoculation; this eliminates the necessity for serial sub-cultures.
Casein digestion. Skim milk agar is used (Lapage *et al.* 1970).
Gelatin hydrolysis. The agar method of Frazier is used (Cowan 1974).
Aesculin hydrolysis. This is observed in aesculin agar (Cowan 1974)
Tyrosine hydrolysis and pigment production on tyrosine agar. The method of Gordon and Smith (Cowan, 1974) is used.
Lipolysis of Tween 20 and Tween 80. The method of Sierra (Cowan 1974) is used.
Growth and inclusion formation on polyhydroxybutyrate. The mineral base agar of Owens & Keddie (1968) is modified by the addition of 0.5% DL-3-hydroxybutyric acid. Inclusion formation is confirmed by staining a thin film with Sudan black (Stanier *et al.* 1966).
Pigment production. Colony pigmentation on nutrient agar and the production of soluble pigments on King's medium A and B (Cowan 1974) are used.
Sensitivity testing. The antibiotic sensitivity is determined by a disc method on lysed-blood agar (King & Phillips 1978). The agents tested are carbenicillin (disc content 100 μg), cephaloridine (25 μg), cephalothin (30 μg), cephalexin (30 μg), colistin (100 units), gentamicin (10 μg) and sulphafurazole (500 μg). A zone diameter of 18 mm or more is taken to indicate sensitivity for all antibiotics except carbenicillin, for which a zone diameter of 12 mm or more indicates sensitivity.

Identification

The genera included in this scheme are *Pseudomonas*, *Acinetobacter* and *Alcaligenes*.

Three principal tests (oxidation–fermentation, oxidase and motility) are performed on all isolates. When tested in Hugh and Leifson's medium, these organisms are all either oxidative or produce alkali. The alkali production is sometimes very weak and difficult to detect, as for example with *Acinetobacter lwoffii*, which produces a thin surface pH change after 24 h. With some other organisms, particularly *Ps. maltophilia*, the initial strong alkaline reaction sometimes disappears after incubation for 48 h. In general the oxidative organisms produce a characteristic reaction after 24 h and thereafter do not change. Yellow-pigmented organisms such as *Ps. paucimobilis* may, however, soon resemble fermenters as the growth spreads throughout the medium.

Table 1 shows the most useful tests for oxidative, oxidase-positive, motile organisms. Production of a water-soluble pigment on King's medium A is characteristic of *Ps. aeruginosa*. Most isolates of *Ps. aeruginosa, Ps. fluorescens* and *Ps. putida* produce a fluorescent pigment on King's medium B. The presence or absence of arginine dihydrolase activity permits the division of non-pigmented variants of these three species plus a variety of other pseudomonads into two groups.

The supplementary tests listed in Tables 2–6 are for the identification of atypical strains, and in the majority of cases they are not required.

Table 2 shows the tests that are most useful for identifying fluorescent pseudomonads. It should be noted that all these species hydrolyse arginine. Strains of *Ps. aeruginosa* are rarely atypical in any of the first-line tests. *Pseudomonas fluorescens* produces typical reactions at room temperature and, apart from occasional strains that reduce nitrate to nitrite, atypical results are rare. The only atypical results seen with *Ps. putida* are that a few strains are ethanol negative. *Pseudomonas putida* produces a pale brown pigment on tyrosine and acid from glucose in peptone water, as do a few strains of *Ps. fluorescens*. *Pseudomonas aeruginosa* is never positive in these two tests. Highly

TABLE 1

Oxidative, oxidase-positive, motile organisms

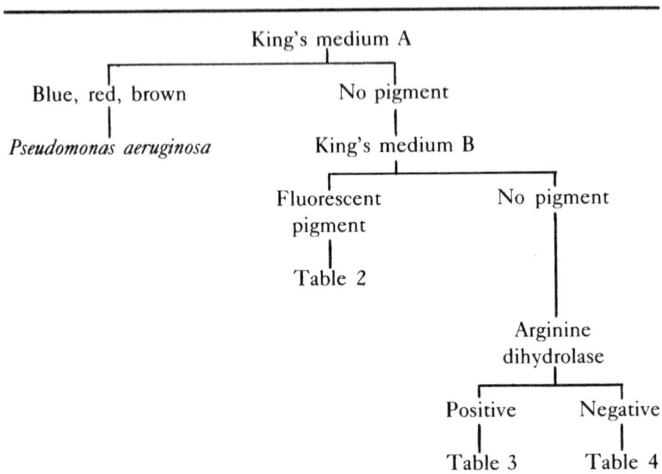

TABLE 2

Fluorescent, arginine-positive pseudomonads

Test	*Pseudomonas aeruginosa*	*Ps. fluorescens**	*Ps. putida*
First-line tests			
growth at 42°C	+	−	−
growth at 4°C	−	+	+
reduction of nitrate	+	+	−
reduction of nitrite	+	(−)†	−
ammonium salt "sugars"			
ethanol	+	−	+
glucose	+	+	+
mannitol	+	+	−
Supplementary tests			
gluconate	+	−	v
urease	+	−	−
casein digestion	+	v‡	−
gelatin hydrolysis	+	+	−
tyrosine hydrolysis	+	+	+
tyrosine pigment	−	(−)	+

* Typical reactions at room temperature.
†(−), usually negative.
‡ v, variable.

mucoid strains of *Ps. aeruginosa* are usually non-motile, slow to produce an oxidative reaction in Hugh and Leifson's medium and are sometimes non-pigmented.

Table 3 lists the characteristics of non-fluorescent, arginine-positive pseudomonads. It should be noted that results for *Ps. fluorescens*, *Ps. fragi* and *Ps. taetrolens* were obtained at room temperature. We have little experience with isolates of *Ps. pseudomallei*, but the colonies of the strains we have studied were of a distinctive pale-cream colour, and we had no difficulty in distinguishing them on the basis of the listed tests. *Pseudomonas stutzeri*, which is often arginine negative, has small yellow colonies that are shiny and convex when young and become flat, dry and wrinkled when older. It is not uncommon to see a mixture of both types of colony on 24-h cultures.

Table 4 shows characteristic results for oxidative, oxidase-positive, arginine-negative organisms. Colony pigmentation is a useful characteristic of certain organisms in this group: *Ps. paucimobilis* strains show bright yellow pigmentation and those of *Ps. vesiculare* are orange. *Pseudomonas cepacia* is usually, but weakly, ornithine decarboxylase positive. *Pseudomonas cepacia*, *Ps. thomasii* and *Ps. pickettii* grow readily on blood agar but die rapidly. Weaver *et al.* (1983) have designated *Ps. thomasii* as a biovar of *Ps. pickettii*.

TABLE 3

Oxidative, oxidase-positive, non-fluorescent, arginine-positive pseudomonads

Test	*Pseudomonas aeruginosa*	*Ps. fluorescens**	*Ps. putida*	*Ps. pseudomallei*	*Ps. ovalis*	*Ps. mendocina*	*Ps. fragi**	*Ps. taetrolens**	*Ps. stutzeri*†
First-line tests									
ammonium salt "sugars"									
dulcitol	−	−	−	+	−	−	−	−	−
ethanol	+	−	+	−	−	−	+	+	(+)
glucose	+	+	+	+	+	+	+	+	+
maltose	−	−	−	+	−	−	+	+	+
mannitol	+	v	−	+	−	−	−	+	+
reduction of NO_3 to NO_2	+	(−)	−	+	−	+	−	−	+
urease	+	−	−	−	−	−	−	−	−
gluconate	+	−	v	−	−	−	+	+	−
casein digestion	+	v	−	+	+	−	−	−	−
Supplementary tests									
growth at 42°C	+	−	−	−	−	−	−	−	v
growth at 4°C	−	+	+	−	−	−	+	+	+
reduction of nitrite	+	−	−	+	−	+	−	−	+
gelatin hydrolysis	+	+	−	+	+	−	−	+	−
tyrosine hydrolysis	+	+	+	+	+	+	+	+	+
tyrosine pigment	−	(−)	+	−	+	+	+	+	+

* No or poor growth at 37°C; tests at room temperature.
† Some isolates are arginine-negative.

TABLE 4

Oxidative, oxidase-positive, non-fluorescent, arginine-negative pseudomonads

Test	*Pseudomonas cepacia*	*Ps. thomasii**	*Ps. pickettii*	*Ps. stutzeri*	*Ps. vesiculare*	*Ps. paucimobilis*
First-line tests						
decarboxylases						
arginine	–	–	–	(–)	–	–
ornithine	(+)	–	–	–	–	–
lysine	+	–	–	–	–	–
ammonium salt "sugars"						
dulcitol	+	–	–	–	–	–
ethanol	–	–	–	(+)	–	–
glucose	+	+	+	+	–	(–)
maltose	+	+	–	+	+	+
mannitol	+	v	–	+	+	+
reduction of NO_3 to NO_2	–	v	+	+	–	–
urease	(–)	v	v	–	–	–
colony pigmentation	–	–	–	pale yellow	orange	yellow
Supplementary tests						
reduction of nitrite	–	v	+	+	–	–
casein digestion	+	–	–	–	–	–
gelatin hydrolysis	+	(+)	–	–	+	–
tyrosine hydrolysis	+	+	+	+	–	+
tyrosine pigment	–	–	–	+	–	–
growth at 42°C	–	+	–	v	–	–
growth at 4°C	–	–	–	+	+	–

* In some schemes *Ps. thomasii* is considered a biotype of *Ps. pickettii*.

Table 5 gives results for the two oxidative, oxidase-negative, non-motile organisms. The demonstration of acid production from glucose in peptone water by *Acinetobacter anitratus* is the only test necessary in most cases. *Pseudomonas mallei* can be further distinguished by its production of arginine dihydrolase, reduction of nitrate and failure to grow at 42°C.

Table 6 lists results for alkali-producing, oxidase-positive organisms. *Pseudomonas maltophilia* is not strictly oxidase positive as the blue colour develops only after 10 s, but its inclusion in this group seems appropriate. This organism is often confused with *Ps. aeruginosa* as it produces greenish colonies and has a strong ammoniacal odour. Despite its name it does not always produce acid from maltose, but is easily distinguishable by its positive reaction for lysine decarboxylase and aesculin hydrolysis. *Pseudomonas acidovorans* is distinguished by its constant ability to produce acid from mannitol and the occasional production of a yellow pigment in the test for indole. *Pseudomonas putrefaciens* is unmistakable in its unpleasant odour, ornithine decarboxylase activity and, usually, production of hydrogen sulphide. The only other easily identifiable organism in this group is *Alcaligenes odorans*, which has a highly characteristic fruity odour, is often alpha-haemolytic and reduces nitrite but not nitrate. The remainder of the pseudomonads in this group, along with *Alcaligenes* species which are not included in the tables, are difficult to identify with confidence. It is often necessary to look at the arrangement of flagella to distinguish the genera, but identification still remains a problem in this group of organisms as they are non-reactive in the majority of conventional biochemical tests. Weaver *et al.* (1983) have pro-

TABLE 5

Oxidative, oxidase-negative, non-motile organisms

Test	*Acinetobacter calcoaceticus* (var. *anitratus*)	*Pseudomonas mallei*
First-line tests		
peptone water "sugars"		
arabinose	+	−
glucose (acid)	+	−
Supplementary tests		
ammonium salt "sugars"		
arabinose	+	−
ethanol	(+)	−
glucose	+	+
lactose	+	−
rhamnose	+	−
xylose	+	−

TABLE 6

Alkali-producing, oxidase-positive organisms

Test	*Pseudomonas maltophilia**	*Ps. acidovorans*	*Ps. pseudoalcaligenes*	*Ps. testosteroni*	*Ps. alcaligenes*	*Ps. putrefaciens*	*Ps. diminuta*	*Alcaligenes odorans*
First-line tests								
ammonium salt "sugars"†								
ethanol	–	–	(–)	–	–	–	–	–
glucose	(+)	–	(+)	–	–	v	–	–
maltose	(+)	–	–	–	–	v	–	–
mannitol	–	+	–	–	–	–	–	–
decarboxylases								
arginine	–	–	+	–	–	–	–	–
lysine	+	–	–	–	–	–	–	–
ornithine	–	–	–	–	–	+	–	–
reduction of NO_3 to NO_2	v	+	+	+	v	+	–	–
reduction of nitrite	–	–	(–)	–	–	–	–	+
Tween 80 hydrolysis	+	+	–	+	+	+	–	(–)
Tween 20 hydrolysis	+	+	–	+	+	+	+	(–)
growth on β-hydroxybutyrate	–	+	+	+	–	–	–	+
citrate	–	+	+	+	(+)	–	–	+
aesculin hydrolysis	+	–	–	–	–	–	–	–
Supplementary tests								
growth at 42°C	–	–	–	–	–	v	–	v
growth at 4°C	–	+	v	–	–	v	v	+
tyrosine hydrolysis	+	+	+	v	+	v	v	+
tyrosine pigment	+	–	v	v	+	+	+	–
H_2S	–	–	–	–	–	(+)	–	–

* Slow oxidase positive (10 s).
† +, acidification of sugar.

TABLE 7

Antibiotic sensitivity of some common isolates of* Pseudomonas *species*

Antibiotic	*Pseudomonas aeruginosa*	*Ps. putida*	*Ps. fluorescens*	*Ps. capacia*	*Ps. thomasii*	*Ps. maltophilia*	*Ps. pseudoalcaligenes*	*Ps. acidovorans*	*Ps. stutzeri*
Carbenicillin	(S)	R	R	V	R	S	S	V	S
Cephaloridine	R	R	R	R	R	R	R	R	S
Cephalothin	R	R	R	R	I	R	R	R	R
Cephalexin	R	R	R	I	S	R	R	V	R
Colistin	S	S	S	R	R	S	V	(S)	S
Gentamicin	(S)	S	S	R	(R)	V	V	R	S
Sulphonamide	R	R	S	S	S	S	S	S	S

* S, sensitive; I, intermediate; R, resistant; V, variable; (R/S), usually resistant/sensitive.

duced identification tables containing as many as 40 tests, but there are no reactions that are absolute for any of these organisms, and identification remains one of elimination based on variable reactions.

The only alkali-producing, oxidase-negative, non-motile organism we have studied is *A. lwoffii*, which can be distinguished from *Bordetella parapertussis* by its inability to produce urease.

Table 7 lists the antibiotic sensitivity test results that are useful as an aid to identification. In addition to the results shown, there are some interesting observations with the newer compounds. Cefsulodin is active only against *Ps. aeruginosa*, imipenem is active against all the species except *Ps. maltophilia*, and the 4-quinolones norfloxacin and ciprofloxacin are active against all species, although *Ps. maltophilia* is less sensitive. We have not yet correlated disc sensitivity tests with minimum inhibitory concentrations for these compounds.

Identification Kits

The increase in interest in non-fermentative Gram negative bacilli over the past few years has led to the development of kits for the identification of this group of organisms. These include Oxi/Ferm Tube (Roche), Minitek (BBL) and the NF System (Flow Laboratories). We have limited experience with these kits but feel that they may have a use for those workers who are unfamiliar with many of the tests that assist in the identification of non-fermenters, or for those who wish to identify only occasional isolates. Because all the tests included in these kits are based on conventional biochemical tests, they also are inadequate for the identification of the non-reactive alkali-producing organisms. API 20E, with the addition of four extra tests to produce a nine-figure profile number, is also useful for the identification of the more biochemically active organisms.

References

COWAN, S. T. 1974 *Cowan and Steel's Manual for the Identification of Medical Bacteria* 2nd edn. London: Cambridge University Press.

GILARDI, G. L. 1971 Antimicrobial susceptibility as a diagnostic aid in the identification of non-fermenting Gram-negative bacteria. *Applied Microbiology* **22**, 821–823.

JESSEN, O. 1965 *Pseudomonas aeruginosa* and other green fluorescent Pseudomonads. A taxonomic study. Copenhagen: Munkgaard.

KING, A. & PHILLIPS, I. 1978 The identification of Pseudomonads and related bacteria in a clinical laboratory. *Journal of Medical Microbiology* **11**, 165–176.

KING, A., HOLMES, B., PHILLIPS, I. & LAPAGE, S. P. 1979 A taxonomic study of *Pseudomonas pickettii, P. thomasii* and Group IVd bacteria. *Journal of General Microbiology* **14**, 137–147.

KING, E. O. 1964 *The Identification of Unusual Pathogenic Gram-negative Bacteria*. Atlanta, Georgia: US Department of Health, Education and Welfare, Center for Disease Control.

LAPAGE, S. P., SHELTON, J. E. & MITCHELL, T. G. 1970 Media for the maintenance and preservation of bacteria. In *Methods in Microbiology*, ed. Norris, J. R. & Ribbons, D. W., Vol. 3A, pp. 1–33. London & New York: Academic Press.

OWENS, J. D. & KEDDIE, R. M. 1968 A note on the vitamin requirements of some coryneform bacteria from soil and herbage. *Journal of Applied Bacteriology* **31**, 344–348.

PHILLIPS, I. 1969 Identification of *Pseudomonas aeruginosa* in the clinical laboratory. *Journal of Medical Microbiology* **2**, 9–16.

PHILLIPS, I., EYKYN, S., CURTIS, M. A. & SNELL, J. J. S. 1971 *Pseudomonas cepacia* (multivorans) septicaemia in an intensive-care unit. *Lancet* **1**, 375–377.

PHILLIPS, I., EYKYN, S. & LAKER, M. 1972 Outbreak of hospital infection caused by contaminated autoclaved fluids. *Lancet* **1**, 1258–1260.

SNELL, J. J. S. & LAPAGE, S. P. 1971 Comparison of four methods for demonstrating glucose breakdown by bacteria. *Journal of General Microbiology* **68**, 221–225.

STANIER, R. Y., PALLERONI, N. J. & DOUDOROFF, M. 1966 The aerobic pseudomonads: a taxonomic study. *Journal of General Microbiology* **43**, 159–313.

TATUM, H. W., EWING, W. H. & WEAVER, R. E. 1974 Miscellaneous Gram-negative bacteria. In *Manual of Clinical Microbiology* 2nd edn, ed. Lennette, E. H., Spaulding, E. H. & Truant, J. P., pp. 270–294. Washington DC: American Society for Microbiology.

WEAVER, R. E., TATUM, H. W. & HOLLIS, D. G. 1972 *The Identification of Unusual Pathogenic Gram-negative Bacteria*. Atlanta, Georgia: US Department of Health, Education and Welfare, Center for Disease Control.

WEAVER, R. E., HOLLIS, D. G., CLARK, W. A. & RILEY, P. 1983 *Revised Tables from the Identification of Unusual Pathogenic Gram-negative Bacteria*. Atlanta, Georgia: US Department of Health and Human Services, Centers for Disease Control.

Vibrio, Aeromonas and Plesiomonas

J. V. Lee

Environmental Microbiology and Safety Reference Laboratory, Public Health Laboratory Service Centre for Applied Microbiology and Research, Porton Down, Salisbury, UK

T. J. Donovan*

Public Health Laboratory, Preston Hall Hospital, Maidstone, Kent, UK

In this chapter the isolation and identification of the pathogenic species of *Vibrio*, *Aeromonas* and *Plesiomonas* will be described. *Photobacterium* is the remaining genus within the Vibrionaceae, but it is not known to include any pathogenic strains and therefore will not be discussed. Vibrionaceae are usually aquatic organisms. Species of *Vibrio* are most common in brackish, estuarine and marine situations whereas *Aeromonas* and *Plesiomonas* are associated with fresh water. The methods applicable to the three genera differ so they will be discussed separately.

The Genus *Vibrio*

There has been considerable research into the taxonomy of the genus during the past decade, and it has expanded from 5 recognised species in 1974 to 29 validly published or in press. The genera *Beneckea* and *Lucibacterium* have been abolished and are now included within *Vibrio*. Our knowledge of some of the newly recognised species is limited, and more may prove to be pathogens than is indicated below. Many can be isolated by the methods used for the well-recognised pathogenic vibrios, but these may not prove to be ideal.

*Present address: Public Health Laboratory, William Harvey Hospital, Ashford, Kent, UK.

Isolation and Identification of
Micro-organisms of Medical and
Veterinary Importance

Copyright © 1985 by the Society for Applied Bacteriology
All rights of reproduction in any form reserved
0-12-181460-2

Diseases of man

Diarrhoea

The well-recognised enteropathogens are cholera toxin producing strains of *Vibrio cholerae* 0:1 (the cause of epidemic cholera) and *V. parahaemolyticus*. Non-0:1 *V. cholerae*, *V. mimicus* (Davis *et al.* 1981) and *V. fluvialis* biovar 1 (Lee *et al.* 1981) have more recently been shown to cause diarrhoea. *Vibrio furnissii* (formerly *V. fluvialis* biovar 2; Brenner *et al.* 1983) and *V. hollisae* (Hickman *et al.* 1982) have also been isolated from cases of diarrhoea and may be enteropathogenic. Apart from cholera epidemics, infection with these organisms is most commonly associated with the ingestion of inadequately cooked or raw seafoods.

Non-enteric infections

Vibrio vulnificus (Farmer 1979) deserves special mention. It causes primary septicaemia, particularly in patients with liver disease, and has caused several fatalities in the USA. Infection is probably via the gastrointestinal tract. *Vibrio vulnificus* can also cause wound infections which are usually more severe than those caused by other vibrios.

Non-enteric infections caused by other vibrios are reported increasingly. Species isolated include *V. cholerae*, *V. parahaemolyticus*, *V. alginolyticus*, *V. metschnikovii* and the newly described *V. damsela* (Love *et al.* 1981). Ears and wounds may become infected while bathing in the sea. Septicaemia is very rare and is usually a sequel to intestinal infection in patients with impaired hepatic function or other underlying disease.

Diseases of animals

"Vibriosis" among cultured and feral marine fish has been recognised for many years, but the identification of the organisms responsible has generally been inadequate. Studies of this economically important disease should be aided by the improved taxonomy of the genus. *Vibrio anguillarum*, *V. ordalii* (*V. anguillarum* biovar II), *V. vulnificus* biogroup 2 (Tison *et al.* 1982) and *V. damsela* (Love *et al.* 1983) all cause infections in marine fish, and *V. tubiashi* (Hada *et al.* 1984), *V. anguillarum*, *V. alginolyticus* and possibly *V. splendidus* (Jeffries 1982) cause bacillary necrosis of larval and juvenile bivalve molluscs. *Vibrio* species have been isolated from a wide range of animals and birds but have not been shown to cause disease in them.

Isolation

Selective media

Thiosulphate–citrate–bile-salt–sucrose agar (TCBS). The usual formula of TCBS is (g/litre): yeast extract, 5.0; peptone, 10.0; sodium thiosulphate,

10.0; sodium citrate, 10.0; ox bile, 8.0; sucrose, 20.0; sodium chloride, 10.0; ferric citrate, 1.0; bromothymol blue, 0.04; thymol blue, 0.04 and agar, 14.0. The pH is adjusted to 8.6.

This medium was developed for the isolation of *V. parahaemolyticus* (Kobayashi *et al.* 1963) but has proved to be useful for the isolation of other vibrios, in particular *V. cholerae*. The selective agents in the medium are bile salts, thiosulphate, citrate and a high pH of 8.6. It has achieved wide acceptance as the medium of first choice for the isolation of vibrios because it is highly selective, suitable for the isolation of most of the enteropathogenic species of vibrio, is readily available commercially, is easy to prepare (requiring no autoclaving), and permits the preliminary differentiation of the sucrose-fermenting species (yellow colonies) such as *V. cholerae* 0:1 from the non-sucrose-fermenters (green colonies) such as *V. parahaemolyticus*. This combination of factors also means it is suitable for use by workers who rarely encounter vibrios.

There is some variation in the efficacy of different brands of TCBS. McCormack *et al.* (1974) found that Eiken TCBS was the most effective for recovering *V. cholerae*. Nicholls *et al.* (1976) tested four brands with a wider range of *Vibrio* species and concluded that Eiken performed best followed closely by Oxoid TCBS. Subsequently we have found that Mast TCBS also performs well, and currently we recommend the use of Eiken (Tokyo, Japan), Oxoid or Mast TCBS. The medium needs careful preparation, and the manufacturers' instructions need some elaboration. The medium should be prepared in a flask at least three times larger than the required volume of the medium. The appropriate amount of dried powder is added to warm (35–40°C) distilled water and slowly heated with constant agitation or stirring. The medium is just brought to the boil, immediately removed from the heat, cooled to 50°C and poured into petri dishes. Before use the plates are dried at 37–45°C to remove the surface moisture.

A bench top steriliser, such as that produced by New Brunswick (Emerald St. London), will also produce satisfactory TCBS medium. The machine should be set to heat the medium to 100°C for 2 min and cool it to 60°C before pouring plates.

There may be variation between different batches of the same manufacturer's medium and even between different batches prepared in the laboratory from the same manufacturer's batch. It is therefore recommended that each batch should be quality controlled. The methods of Nicholls *et al.* (1976) and West *et al.* (1982) can be used. If *V. furnissii* (*V. fluvialis* biovar 2) NCTC 11218 is used as recommended by Taylor & Barrow (1981), a recovery of at least 50% should be achieved.

The ability of the various species of *Vibrio* to grow on TCBS is shown in Tables 2–4. Most strains produce colonies of 2–5 mm diameter after overnight incubation. Some species such as *V. hollisae* (Hickman *et al.* 1982) fail to grow or grow poorly on TCBS. New media may have to be developed for these species, such as that recently described for the isolation of *V. vulnificus* (Brayton *et al.* 1983). Some strains of *Aeromonas, Plesiomonas, Proteus* and *Streptococcus* spp. may grow, but usually the colonies are much smaller than those produced by *Vibrio* species.

Taurocholate–tellurite–gelatin agar (TTGA). Monsur (1963) developed this highly selective medium for the isolation of *V. cholerae*. After overnight incubation at 37°C, *V. cholerae* grows on this medium as small translucent colonies with a greyish centre and a faint turbid halo due to gelatinase production. *Vibrio alginolyticus* and *V. parahaemolyticus* produce larger semi opaque colonies, and most normal faecal flora is inhibited. This medium is not available commercially, needs regular quality control, particularly with regard to the potency of the tellurite, is relatively complicated to prepare and is not as selective as TCBS. Familiarity with the medium is essential to gain full advantage from its use, and it is therefore inappropriate for a laboratory only looking occasionally for vibrios.

Enrichment media

Alkaline peptone water (APW). The medium contains (g/litre): peptone 10.0 and sodium chloride 10.0. The pH is adjusted to 8.6; the medium dispensed as required and autoclaved at 121°C for 15 min.

This is the time-honoured enrichment broth which is suitable for the isolation of any of the vibrios. Enrichment occurs because of the ability of vibrios to grow very rapidly and at a high pH in contrast to most of the normal faecal flora. Volumes of not less than 20 ml should be used for the isolation of vibrios from faeces, as smaller volumes do not remain adequately alkaline. A short incubation period is very important if enrichment is to be achieved. Thus cultures of faeces grown at 37°C should be subcultured to TCBS or a secondary enrichment after only 5–8 h. Longer incubation (18–20 h) can be used at lower temperatures (18–22°C) when short incubation at 37°C is not practicable or appropriate.

Some workers have recommended various modifications of APW such as the addition of tellurite and bile salts (Monsur 1963), but in our experience these are not necessary. Many media have been developed specifically for the isolation of *V. parahaemolyticus*, and two of the most widely used are salt colistin broth and glucose salt teepol broth (Taylor *et al.* 1982) or variations of them (Joseph *et al.* 1982). These media are not suitable for the isolation of many other vibrios.

Diluents

T_1N_1. This is a simple tryptone water adjusted to pH 7.2 and containing (g/litre): Oxoid tryptone, 10 and NaCl, 10. Distribute as required and autoclave.

Vibrio diluent. This contains (g/litre): Bacto-peptone (Difco), 10.0; NaCl, 5.00 and $MgSO_4 \cdot 7H_20$, 4.94. The pH is adjusted to 8.0. The medium is distributed as required and autoclaved. In addition, APW may also be used as a diluent.

Isolation from faeces

Fresh specimens should be used, but when essential suitable transport media can be used such as those of Cary & Blair (1964) and Venkatraman & Ramakrishnan (1941).

A TCBS plate is inoculated with two or three loopsful of faeces and 20 ml APW with about 1 g faeces. The APW is incubated at 37°C for 6–8 h and subcultured from the top to a second TCBS plate. It has often been suggested that a second APW may be inoculated, incubated at 37°C overnight and subcultured to a third TCBS, but this step is not usually necessary. The TCBS plates are incubated overnight at 37 °C and examined for vibrio colonies immediately after removal from the incubator. If left on the bench, sucrose fermenting colonies (yellow) may revert and become green. Suspicious colonies should be subcultured to nutrient agar for the oxidase test and for the identification. The oxidase test cannot be done directly on growth taken off a TCBS plate because the acid produced by the fermentation of the sucrose inhibits the oxidase reaction.

Isolation from other clinical material

This should present no problem as all the pathogenic vibrios will grow well on commonly used non-selective media such as nutrient agar (provided it contains at least 0.5% NaCl) and blood agar. Growth on some other commonly used media such as MacConkey agar is variable, especially as they may contain insufficient sodium chloride. The problem is not isolation but identification of these organisms from such specimens.

Isolation from fish pathological material

Specimens are cultured directly onto TCBS and one of the following non-selective media: Marine Agar (Difco); seawater agar which contains (g/litre) peptone (Evans), 10, Lab. Lemco powder (Oxoid), 10, and is made up in a 3:1 mixture of filtered aged seawater and distilled water or J. agar which contains (g/litre) Nutrient Broth No. 2 (Oxoid) dried base, 25.0; NaCl, 17.5; KCl, 1.0; $MgCl_2 \cdot 6H_20$, 4.0; Davis agar, 15.0. The plates are incubated at 20°C for 24–48 h and colonies are subcultured for identification.

Isolation from food

Contaminated foods are frequently responsible for diarrhoea caused by vibrios. Not only seafoods need be involved. For example, *V. cholerae* can grow on or survive on a wide range of non-acidic foodstuffs (Kolvin & Roberts 1982).

Seafoods. Seafoods can contain a variety of species of *Vibrio*, so it may be necessary to adjust the techniques according to the species being sought. If only APW containing 1% NaCl is used, *V. cholerae* may not be detected because it can be overgrown by *V. alginolyticus*, which is extremely common in seafoods. The following technique is recommended for the routine examination of seafoods.

Flesh (10 g) is added to 100 ml APW containing 1% NaCl and to 100 ml salt-free APW. Both are homogenised, incubated at 37°C overnight and subcultured to TCBS.

Other foods. A 10-g portion of food is added to 100 ml APW (containing 1% NaCl), homogenised, incubated overnight at 37°C and subcultured to TCBS. In this case only one species of vibrio is likely to be present, and the use of salt-free APW would produce a food homogenate that could have too little salt for all species including *V. cholerae* to grow.

Isolation and enumeration from water

It may be necessary to examine various types of water for vibrios. Preliminary selection of the procedures to be followed can be made with the knowledge of the pH of the water, its salinity, temperature, organic content and chlorination if any. Brackish, static waters of high pH and temperature may contain large numbers of of vibrios, in particular *V. cholerae*, whereas water of low pH, low temperature, low electrolytic concentration and low organic content is unlikely to contain vibrios except in low numbers.

The methods described next assume that human pathogens are sought. If all incubation temperatures are kept to 20–25°C the methods can be used for any vibrio.

Direct (high numbers expected). Volumes (0.5ml) of water or dilutions of it in T_1N_1 or vibrio diluent are pipetted directly onto duplicate plates of TCBS (dried for 1 h at 40°C) and allowed to spread over the surface. After the fluid has absorbed, the plates are incubated at 25°C for 2–4 h followed by overnight at 37°C. Suspicious colonies are subcultured for further identification.

Direct enumeration by filtration through 0.45 or 0.2 μm pore size cellulose acetate membranes and incubation of the membranes on the surface of TCBS plates is not satisfactory, as the membranes become heavily overgrown by other bacteria.

Most probable number (MPN, moderate numbers expected). One 50-ml and 5 10-ml amounts of water are pipetted into equal volumes of double strength APW. If required, 5 1.0-ml and 5 0.1-ml amounts may also be pipetted into 20-ml volumes of single strength APW. The enrichment broths are incubated at 25°C for 2 h followed by 12–16 h at 37°C and then subcultured to

TCBS plates which are incubated at 37°C for 16–24 h. The plates are examined for vibrio-like colonies, and at least three of each colony type are subcultured for identification. Following identification the MPN of all *Vibrio* species or particular species can be estimated using the standard tables.

If *V. cholerae* is sought, waters containing >250 mmole Na^+/litre should be inoculated into salt free APW to prevent overgrowth by *V. alginolyticus*.

Concentration using filter aid. Sterile suspensions of filter aid (Hyflo-supercel, Hopkins and Williams) are added to a water sample to give a final concentration of 0.5 g/litre. The water with filter aid is shaken well and filtered through a sterile Whatman No. 17 grade filter paper. The filter paper with the filter aid retained on it is transferred to 100 ml APW, shaken well, and incubated at 25°C. Subcultures are made to TCBS after 8 and 22 h.

Concentration by membrane filtration. Filtration of clear water through a 0.45 or 0.2 μm pore size cellulose acetate membrane and incubation of the membrane in APW as for the filter aid above may also be a suitable means of concentration, but it has not been widely applied.

Concentration using cotton wool. This can easily be done in the field. Several litres of water may be passed through sterile absorbent cotton wool pushed into a sterile funnel. The cotton wool is then transferred into at least 200 ml APW, incubated at 25°C for 2–4 h and then at 37°C for 8–12 h. The enrichment is subcultured to TCBS.

Sewer (Moore) swabs. Swabs prepared from rolls of gauze (Moore 1948) have been used successfully to survey sewage for the presence of *V. cholerae* 0:1. The gauze swabs are suspended in the flowing sewage or other flowing water for 24–48 h and subsequently treated in the same manner as the cotton wool swab above.

Concentration by immunoabsorption. An immunological method has been described for the isolation of *V. cholerae* 0:1 from water using beads coated in 0:1 antiserum (Hranitzky *et al.* 1980). Despite its theoretical elegance other workers have not reported using it.

Isolation from other environmental specimens

If required, sediments, soil, and plant material can be examined by the methods described for foods.

The Genus *Aeromonas*

The taxonomy of *Aeromonas* is much less satisfactory than that of *Vibrio*. For the purposes of this chapter the genus is conveniently divided into two groups, *A. salmonicida* and the "Hydrophila group". The classification scheme of Popoff & Veron (1976) is accepted increasingly and divides the hydrophila group into *A. hydrophila* biovar *hydrophila*, *A. hydrophila* biovar *anaerogenes* and *A. sobria*. More recently it has been proposed that *A. hydrophila* biovar *anaerogenes* be renamed *A. caviae* (Popoff *et al.* 1981).

Aeromonas of the "hydrophila group" are extremely common in all freshwater environments and frequently may form as much as 50% of the population of aerobic heterotrophic bacteria in natural waters. In contrast *A. salmonicida* is only found in association with fish.

Diseases of man

Strains of the "hydrophila group" may cause bacteraemia and septicaemia in patients with immune deficiency and may infect wounds incurred in or near fresh water. They are increasingly incriminated as a cause of diarrhoea. In developed countries they are isolated from cases of diarrhoea at a much higher frequency than from healthy individuals, but which particular strains or species, if any, are enteropathogenic in man has not been established clearly yet.

Diseases of animals

Aeromonas salmonicida causes furunculosis of salmonid and other fish. Strains of the "hydrophila group" also cause haemorrhagic lesions in fish and reptiles.

Isolation

Selective media

A variety of selective media have been developed for the isolation of the "hydrophila group". They have been compared in our laboratories (unpublished results) and by others (von Graevenitz & Bucher 1983). Many of these were primarily developed for their isolation from water, for which they are of questionable value, and are not suitable for other purposes. In general *Aeromonas* do not grow on TCBS. There are no selective media for *A. salmonicida*, which will not grow on the media described below.

Xylose deoxycholate citrate agar (XDCA). This medium (Shread *et al.* 1981) contains (g/litre): Nutrient Broth No. 2 (Oxoid), 12.5; sodium citrate, 5.0; sodium thiosulphate, 5.0; ferric ammonium citrate, 1.0; sodium deoxycholate (Ward Blenkinsopp batch 19359), 2.5; agar No. 3 (Oxoid), 12.0; xylose, 10.0 and neutral red, 0.025. The ingredients are added to the water and gently heated to boiling with constant agitation. The mixture is allowed to simmer for 20 s, removed from the heat, cooled to 50°C, the pH adjusted to 7.0, and the plates poured. Strains of the "hydrophila group" and *Plesiomonas shigelloides* grow on XDCA as non-xylose fermenters. After overnight incubation at 37°C, the majority of aeromonads produce colonies of at least 1 mm diameter but some may be much smaller. Accordingly, non-xylose fermenting colonies of any size should be subcultured for identification. Many Enterobacteriaeceae are unable to grow on XDCA, and the majority of the remainder are xylose fermenters and readily distinguished from *Aeromonas* and *Plesiomonas*.

Ampicillin blood agar (ampicillin BA). Horse blood agar containing ampicillin (10 μg/ml) is also of value for the isolation of *Aeromonas* of the hydrophila group from faeces. (Goodwin *et al.* 1983). Strains of *Aeromonas* are usually strongly haemolytic and resistant to ampicillin. Theoretically the presence of ampicillin-resistant Enterobacteriaceae, particularly *Proteus*, could cause problems, but so far this has not been the case.

Enrichment media

The only enrichment broth developed specifically for *Aeromonas* is trypticase soy–ampicillin broth (von Graevenitz & Zinterhofer 1970), but this appears to have no advantage over APW, which is also suitable for their enrichment and can be used for vibrios.

Isolation from non-enteric infections in man

Aeromonas spp. of the "hydrophila group" grow readily on ordinary blood agars and nutrient agars at 37°C, and so there should be no problem isolating them from these clinical specimens.

Isolation from faeces

Faeces are inoculated directly onto XDCA and/or ampicillin BA, and the plates are incubated at 37°C overnight. In addition an APW should be inoculated, incubated at 20–25°C for 18 h, and subcultured to XDCA or ampicillin BA. Characteristic colonies are subcultured for identification.

Isolation from fish lesions

Pathological material is inoculated directly onto tryptone soya agar or other nutrient agars and incubated at 20–25°C for 48 h. *Aeromonas salmonicida* usually produces a brown diffusible pigment, as do some strains of other *Aeromonas* species. The pigmentation may take longer than 48 h to develop fully. In contrast to other *Aeromonas* spp., *A. salmonicida* is rather slow growing. Colonies growing on these non-selective media should be subjected to the oxidase test and, if oxidase positive, retained for identification.

Isolation from food and water

Should it be necessary to examine food for the presence of *Aeromonas*, the methods used for the examination of faeces are probably suitable. On occasions there may be very high numbers of *Aeromonas* in milk of poor quality.

Aeromonas may be present in untreated waters in extremely high numbers, and they may be isolated by simply subculturing the water or dilutions of it directly to nutrient agar. After overnight incubation at 25 or 37°C, oxidase-positive colonies are subcultured for identification.

If it is necessary to examine drinking water, enrichment in APW as for faeces above may be necessary in addition to direct subculturing.

Plesiomonas shigelloides

The genus *Plesiomonas* includes the single species *P. shigelloides*, which is now considered to be enteropathogenic to man. Strains have also been isolated from various diseased captive animals, in particular reptiles. This species is also common in the aquatic environment although not to the same degree as *Aeromonas*.

Isolation

Selective media

On XDCA, *P. shigelloides* grows as small non-xylose-fermenting colonies. Thus XDCA has the advantage over ampicillin BA that it can be used for both *Aeromonas* and *Plesiomonas*. *Plesiomonas* will also grow on normal desoxycholate citrate agar as a non-lactose-fermenting colony resembling *Shigella sonnei*. The colonies of *P. shigelloides* can be readily distinguished because they are oxidase positive and have an odour similar to *Haemophilus influenzae*.

Inositol brilliant green bile salt agar (IBB). This medium (Schubert 1982) was specifically derived for the isolation of *Plesiomonas* from faeces and has the following composition (g/litre): proteose peptone (Difco), 10.0; Lab Lemco powder (Oxoid), 5.0; sodium chloride, 5.0; bile salts no. 3 (Difco), 8.5; brilliant green (Merck), 0.00033; neutral red, 0.025; meso-inositol, 10.0 and agar, 15.0. The pH is adjusted to 7.2 and the medium autoclaved at 121°C for 15 min.

On IBB *P. shigelloides* grows as pink to red (inositol-fermenting) colonies that are larger than those formed on XDCA. Using artificially seeded faeces, von Graevenitz & Bucher (1983) attained a slightly higher isolation rate on IBB than on XDCA. It is likely that XDCA will prove more effective in practice since IBB is much less selective. This remains to be established by long term trials of the two media in parallel.

Enrichment media

No enrichment media have been developed specifically for the isolation of *P. shigeloides*. Although von Graevenitz & Bucher (1983) found APW to be satisfactory, we have been unable to show any enrichment of *Plesiomonas* in APW.

Isolation from faeces

Faeces are cultured directly onto XDCA, DCA, and possibly IBB, and the plates are incubated at 37°C for 24 h. Appropriate colonies (non-xylose-fermenters on XDCA, non-lactose-fermenters on DCA, inositol-fermenters on IBB) are subcultured to nutrient agar for the oxidase test and further identification.

Isolation from other materials

There have been very few studies of the incidence of *Plesiomonas* in food or the natural environment, and so it is not possible to recommend any particular method. Adaptations of the methods used for faeces are probably most appropriate.

Identification

There is not space in this chapter to describe fully the methods used for the identification of all of the species within the Vibrionaceae, but they have been well documented in several recent publications (Lee *et al.* 1979; Furniss *et al.* 1978; Baumann *et al.* 1984; West *et al.* 1983) and so only a few points of particular importance will be highlighted here.

The accurate routine identification of the Vibrionaceae depends on the reliable performance of a few tests: Kovács oxidase; Hugh and Leifson's O/F test; Møller's decarboxylases; sugar fermentations; the Voges–Proskauser (VP) test and salt tolerance. Sensitivity to the vibriostatic agent 2-4 di-amino-6, 7-di-isopropylpteridine phosphate, usually called 0/129, can also be useful. Media commonly used for the identification of the Enterobacteriaceae are usually satisfactory provided the NaCl concentration is increased to 1% (w/v).

Oxidase

The method of Kovács (1956) is used but for absolute reliability must be done on growth taken from a medium containing no fermentable carbohydrate, as the reaction is inhibited by acid thus causing false negatives.

Decarboxylases

The method of Møller (1955) is followed using Difco decarboxylase base with the NaCl concentration increased to 1% (w/v). It is essential that the medium is overlaid with sterile liquid paraffin. The tests are read after 1 and

2 day's incubation at 30–37°C. If it is necessary to use 20 or 25°C, 3 days incubation is required. Other methods of detecting decarboxylases, including those in "kits", are not reliable for the Vibrionaceae.

Sugar fermentations

Ordinary peptone water sugars can be used provided they contain 1% (w/v) NaCl. Bromocresol purple appears to give slightly more consistent results than other indicators. Some carbohydrates, in particular amygdalin, arbutin, inositol and salicin, may give delayed results, and so the fermentation tests should also be incubated for 5 days.

Voges–Proskauer (VP) test and motility

The method described by Lee *et al.* (1979) is used. The medium is a semisolid one which is stab inoculated, and so this test also provides a useful method of screening for motility. The medium is observed after incubation and before the addition of the reagents. Motile strains show diffuse growth throughout the medium, whereas non-motile strains grow only down the line

TABLE 1

Preliminary division of the genera Vibrio, Aeromonas *and* Plesiomonas*,†

Group‡	1	2	3	4	5	6	7	8
Oxidase	+	+	−	+	+	−	+	+
Nitrate	+	+	−	+	−	−	+	+
Thornley's or Møller's arginine	−	+	+	−	−	−	+	+
Møller's lysine	+	−	d	−	−	−	d	+
ornithine	d	−	−	−	−	−	−	+
O/129 150 μg disc:	S	S	S	S	S	NT	R§	d
growth in								
0% NaCl	d	d	d	−	−	−	+	+
3% NaCl	+	+	+	+	+	NT	+	+
6% NaCl	d	d	d	d	+	NT	−	−
Refer to Table	2	3	3	4	4	4	5	5

* Symbols/abbreviations: +, all strains positive; −, all strains negative; d, character differs from strain to strain; S, sensitive; R, resistant; NT, not tested.

† All strains ferment glucose.

‡ Group 1: *V. cholerae, V. mimicus, V. parahaemolyticus, V. alginolyticus, V. vulnificus, V. harveyi, V. campbellii, V. fischeri* and *V. marinus*. Group 2: *V. fluvialis, V. furnissii, V. anguillarum, V. diazotrophicus, V. splendidus* biovar 1, *V. tubiashii, V. costicola, V. nereis, V. damsela, V. proteolyticus, V. orientalis* and *V. aestuarianus*. Group 3: *V. metschnikovii*. Group 4: *V. hollisae, V. natriegens, V. splendidus* biovar II, *V. pelagius, V. nigripulchritudo* and *V. logei*. Group 5: *V. ordalii*. Group 6: *V. gazogenes*. Group 7: *Aeromonas hydrophila, A. caviae, A. sobria* and *A. salmonicida*. Group 8: *P. shigelloides*.

§ Some strains of *A. salmonicida* are sensitive.

TABLE 2

Identification of the Møller's lysine-positive, arginine-negative Vibrio *species**

Characteristics	*V. cholerae*	*V. mimicus*	*V. vulnificus*	*V. parahaemolyticus*	*V. alginolyticus*	*V. harveyi*	*V. campbellii*	*V. fischeri*	*V. marinus*
TCBS growth	Y	G	G/−	G	Y	G/Y	G	−	·
42°C growth	+	+	−	+	d	−	−	−	−
Growth in									
0% NaCl	+	+	−	−	−	−	−	−	−
6% NaCl	−	−	+	+	+	+	+	d	·
8% NaCl	−	−	−	+	+	d	+	−	·
10% NaCl	−	−	−	d	+	−	−	−	−
Voges–Proskauer	+†	−	−	−	+	−	−	−	−
Swarming on MA	−	−	−	d	+	−	−	−	−
Luminescence	−‡	−	−	−	−	d	−	d	−
Pigment	−	−	−	−	−	−	−	Y	−
Acid production from									
arabinose	−	−	−	d	−	−	−	−	−
arbutin	−	−	+	−	−	+	−	−	·
salicin	−	−	+	−	d	+	+	d	−
sucrose	+§	−	−	−	+	d	−	−	−
ONPG	+	+	+	−	−	d	−	d	−
Indole	+	+	+	+	+	+	+	−	−
Amylase	+	−	+	+	+	+	+	−	−
O/129 10 μg disc	S	S	S	R	R	S	S	S	S
Growth on									
cellobiose	−	−	+	−	d	+	d	−	−
gluconate	+	+	+	+	+	+	−	−	+
glucuronate	−	+	+	d	−	+	−	−	−
ethanol	−	d	−	+	d	−	−	−	−
glucosamine	+	·	d	+	d	−	−	−	·
serine	d	·	−	+	+	+	d	−	+
putrescine	−	−	−	+	d	−	−	−	−

* Symbols/abbreviations: +, ≥80% strains possess character; −, ≥80% strains do not possess character; d, 21–79% strains possess character; ·, data unavailable; S, sensitive; R, resistant; Y, yellow; G, green; MA, marine agar.

† *V. cholerae* 0:1 biovar classical is negative.

‡ Some strains of non-0:1 *V. cholerae* are luminous.

§ Some strains are sucrose-negative but distinguishable from *V. mimicus* by VP, amylase and growth on glucuronate.

TABLE 3

Identification of the Møller's or Thornley's arginine-positive Vibrio *species**

Characteristics	*V. fluvialis*	*V. furnissii*	*V. diazotrophicus*	*V. metshnikovii*	*V. anguillarum*	*V. splendidus* 1	*V. tubiashii*	*V. costicola*	*V. nereis*	*V. damsela*	*V. proteolyticus*	*V. orientalis*	*V. aestuarianus*
TCBS growth	Y	Y	Y	Y	Y	d	Y	–	Y/–	G	G	·	Y
Luminous	–	–	–	–	–	+	–	–	–	–	–	+	–
Swarming	–	–	–	–	–	–	–	–	–	–	+	–	–
Growth in 12% NaCl	–	–	–	–	–	–	–	+	–	–	–	·	–
Oxidase	+	+	+	–	+	+	+	+	+	+	+	+	+
$NO_3 \rightarrow NO_2$	+	+	+	–	+	+	+	d	+	+	+	+	·
Møller's lysine	–	–	–	–	–	–	–	–	–	d	+	·	–
Gas from glucose	–	+	–	–	–	–	–	–	–	+	–	–	–
Voges–Proskauer	–	–	–	+	+	–	–	–	–	+	+	–	–
Growth at 4°C	–	–	+	–	+	–	–	–	–	·	–	+	+
Acid production from L-arabinose	+	+	+	–	d	–	–	–	–	–	–	–	–

arbutin	+	−	+	d	−	−	−	·	−	−	−	·	·
cellobiose	d	−	+	−	d	+	d	·	−	−	+	+	+
lactose	−	−	+	d	−	−	−	·	−	−	−	−	+
sorbitol	−	−	−	d	d	−	−	·	−	−	+	·	+
sucrose	+	+	+	+	+	d	+	+	+	−	−	+	+
xylose	−	−	+	−	−	−	−	−	−	−	−	·	−
Extracellular breakdown of													
alginate	−	−	−	−	−	d	−	·	−	−	−	−	·
chitin	+	+	−	+	+	+	+	·	d	+	+	+	·
elastin	−	−	−	+	+	−	−	·	−	−	+	·	−
lecithin	+	+	−	+	d	d	+	·	−	d	+	·	·
Tween 80	+	+	−	+	+	+	+	·	−	−	+	+	·
urea	−	−	d	−	−	−	−	·	−	+	−	−	·
tyrosine	−	−	−	−	−	−	+	·	+	−	−	−	·
xanthine	−	−	−	−	−	−	+	·	−	−	−	−	·
Growth on													
DL-3-hydroxybutyrate	+	+	+	−	−	−	d	·	+	−	−	+	·
galacturonate	+	+	+	−	−	−	−	·	−	−	−	·	·
glucuronate	+	−	+	−	−	+	d	·	−	−	−	−	·
glutarate	−	+	−	−	−	−	−	·	d	−	−	·	·
leucine	−	−	−	−	−	−	−	·	+	−	−	·	·

*Symbols/abbreviations as in Table 2.

of the inoculum. The motility of those strains that appear non-motile should be checked by the hanging drop method using the tube of salt tolerance medium showing maximum growth (discussed next).

Salt tolerance

Tryptone water (1% w/v Oxoid tryptone) is used containing 0, 3, 6, 8, 10 and 12% (w/v) NaCl as described by Lee *et al.* (1979). The results are more consistent if the medium is distributed in 20 ml rather than the 4-ml amounts recommended previously. Growth is recorded after 24- and 48-h incubation.

0/129 sensitivity

The method of Lee *et al.* (1979) is used, but discs are now available from Oxoid.

ONPG test

The method of Cowan (1974) is used, but the ONPG broth is made up with sodium potassium magnesium (SPM) broth substituted for peptone water. The SPM broth contains (g/litre): tryptone (Oxoid), 10.0; NaCl, 10.0; KCl, 1.0 and $MgCl_2{\cdot}6H_2O$, 4.0. The hydrolysis of the ONPG by galactosidase is recorded after 24-h incubation.

Aesculin hydrolysis

The medium contains (g/litre): tryptone (Oxoid), 10.0; aesculin, 1.0; ferric citrate, 0.5; NaCl, 10.0 and agar no. 3 (Oxoid), 15.0. The agar, peptone and NaCl are dissolved by heating the water, and then the aesculin and ferric citrate are added and dissolved. The medium is autoclaved at 115°C for 10 min and poured in plates.

Up to 12 strains are inoculated onto the plates as spots, and the test is incubated for up to 5 days. Blackening indicates aesculin hydrolysis. Liquid media do not give such consistent results.

Nutritional screening

The ability of species of *Vibrio* to utilise compounds as the sole organic carbon source is determined using the methods of Lee *et al.* (1981). The basal method is a modification of the minimal medium of Clowes & Hayes (1968) and contains (g/litre): NH_4Cl, 5.0; NH_4NO_3, 1.0: Na_2SO_4, 2.0; K_2HPO_4, 3.0: KH_2PO_4, 1.0; $MgSO_4{\cdot}7H_2O$,0.1; NaCl, 10.0; $MgCl_2{\cdot}6H_2O$, 4.0 and purified agar (Oxoid), 10.0. The medium is prepared as two separate solutions, one containing the agar $MgCl_2{\cdot}6H_2O$ and NaCl and the other containing the remaining ingredients. The two solutions are combined aseptically after autoclaving at 121°C for 15 min.

For *Aeromonas* and *Plesiomonas* the same medium is used with the NaCl and $MgCl_2 \cdot 6H_2O$ left out. The terrestrial basal medium (TBM) of Baumann *et al.* (1984) can also be used for *Aeromonas* and *Plesiomonas* as well as *V. cholerae* and *V. metschnikovii.* The basal medium (BM) of Baumann *et al.* (1984) can be used for the remaining *Vibrio* species.

Choice of tests for identification

The reactions of the various species of *Vibrio, Aeromonas* and *Plesiomonas* in the tests most useful for their identification are shown in Tables 1–5. Tables 1–4 have been compiled from some original data but mostly from that included in various publications (Baumann *et al.* 1984; West *et al.* 1983; Guerinot *et al.* 1982; Hada *et al.* 1984; Tison & Seidler 1983; Yang *et al.* 1983;

TABLE 4

Identification of the Møller's decarboxylase-negative Vibrio *species**

Characteristics	*V. hollisae*	*V. natriegens*	*V. splendidus II*	*V. pelagius I*	*V. pelagius II*	*V. nigripulchritudo*	*V. ordalii*	*V. logei*	*V. gazogenes*
TCBS	−	Y/−	Y/−	Y/−	d	G/−	−	·	·
Pigment	−	−	−	−	−	B	−	Y/O	R
Oxidase	+	+	+	+	+	+	+	+	−
$NO_3 \rightarrow NO_2$	+	+	+	+	+	+	−	|	−
Luminous	−	−	−	−	−	−	−	+	−
Gas from glucose	−	−	−	−	−	−	−	−	+
Growth at 30°C	+	+	+	+	+	+	+	−	+
Extracellular breakdown of									
alginate	−	−	−	+	+	−	−	−	−
chitin	+	−	+	d	+	+	d	+	−
gelatin	−	d	+	−	+	+	+	−	+
starch	−	d	+	−	+	+	−	−	+
Tween 80	−	+	+	+	+	+	−	d	+
Growth on									
L-arabinose	+	+	−	−	−	−	−	−	+
gluconate	+	+	−	+	+	d	−	+	−
glucuronate	−	d	−	−	−	+	−	−	−
inositol	−	d	−	−	−	+	−	−	−
rhamnose	−	+	−	−	−	−	−	−	−
sucrose	−	+	−	+	d	−	+	−	+
xylose	−	−	−	−	−	−	−	−	+
putrescine	−	+	−	+	+	−	−	−	−

*Symbols/abbreviations: B, blue black; O, orange; R, red; others as in Table 2.

TABLE 5

Identification of Aeromonas *and* Plesiomonas *species**

	A. hydrophila	*A. sobria*	*A. cavaie*	*A. salmonicida*	*P. shigelloides*
TCBS growth	Y/−	−	Y/−	−	G/−
Gas from glucose	d	+	−	d	−
Voges–Proskauer	+	+	−	−	−
Gluconate oxidation	+	+	−	−	−
Møller's					
lysine	+	+	−	+	+
ornithine	−	−	−	−	+
Acid from					
L-arabinose	+	−	+	+	−
arbutin	+	−	+	d	−
inositol	−	−	−	−	+
salicin	+	d	+	−	−
sucrose	+	+	+	d	−
Aesculin hydrolysis	+	−	+	d	−
Elastin breakdown	+	−	−	d	−
Motility	+	+	+	−	+
Growth					
at 37°C	+	+	+	−	+
at 42°C	−	+	−	−	+
Pigment	−	−	−	+†	−

* Symbols/abbreviations as in Table 2.
† Brown diffusible.

Love *et al.* 1981; Hickman *et al.* 1982; Lee *et al.* 1981). Table 5 is based on original data (Lee, Jeffries, & Bryant, unpublished observations) but is generally in agreement with the data of Popoff & Veron (1976).

The tables include all of the species of *Vibrio*, *Aeromonas* and *Plesiomonas* currently validly published or in press. Leaving some out could have caused misidentification of pathogens as non-pathogens or vice versa. The tables are quite comprehensive, but if difficulties are still encountered in differentiating some species further tests could be selected from the appropriate references just listed.

For routine identifications, a selection of the tests shown in the tables can be made according to the organisms sought. A good set of tests for routinely identifying a wide range of commonly occurring Vibrionaceae includes the O/F test of Hugh and Leifson; oxidase, Vogev–Proskauer, motility (in semi-solid VP medium), nitrate reduction, Møller's arginine, lysine and ornithine, 0/129 sensitivity (10 and 150 μg discs), salt tolerance (0, 3, 6, 8 and 10% NaCl), gas production from glucose and acid production from arabinose, arbutin, inositol, salicin and sucrose.

References

BAUMANN, P., FURNISS, A. L. & LEE, J. V. 1984 Genus *Vibrio*. In *Bergey's Manual of Systematic Bacteriology* 9th edn, ed. Krieg, N. R., Baltimore, Maryland: Williams & Wilkins, pp. 518–538.

BRAYTON, P. R., WEST, P. A., RUSSEK, E. & COLWELL, R. R. 1983 New selective plating medium for isolation of *Vibrio vulnificus* biogroup 1. *Journal of Clinical Microbiology* **17**, 1039–1044.

BRENNER, D. J., HICKMANN-BRENNER, F. W., LEE, J. V., STEIGERWALT, A. G., FANNING, G. R., HOLLIS, D. G., FARMER, J. J., III, WEAVER, R. E., JOSEPH, S. W. & SEIDLER, R. J. 1983 *Vibrio furnissii* (formerly aerogenic biogroup of *V. fluvialis*), a new species isolated from human faeces and from the environment. *Journal of Clinical Microbiology* **18**, 816–824.

CARY, S. G. & BLAIR, E. B. 1964 A new transport medium for shipment of clinical specimens. *Journal of Bacteriology* **88**, 96–98.

CLOWES, R. C. & HAYES, W. 1968 *Experiments in Microbial Genetics*. Oxford: Blackwell Scientific Publications.

COWAN, S. T. 1974 *Cowan & Steel's Manual for the Identification of Medical Bacteria* 2nd edn. London: Cambridge University Press.

DAVIS, B. R., FANNING, G. R., MADDEN, J. M., STEIGERWALT, A. G., BRADFORD, H. B., SMITH, H. L. & BRENNER, D. J. 1981 Characterization of biochemically atypical *Vibrio cholerae* strains and designation of a new pathogenic species, *Vibrio mimicus*. *Journal of Clinical Microbiology* **14**, 631–639.

FARMER, J. J. III 1979 *Vibrio ("Beneckea") vulnificus*, the bacterium associated with sepsis, septicaemia and the sea. *Lancet* **ii**, 903.

FURNISS, A. L., LEE, J. V. & DONOVAN, T. J. 1978 *The Vibrios*. Public Health Laboratory Service Monograph Series No. 11. London: Public Health Laboratory Service.

GOODWIN, C. S., HARPER, W. E. S., STEWART, J. K., GRACEY, M., BURKE, V. & ROBINSON, J. 1983 Enterotoxigenic *Aeromonas hydrophila* and diarrhoea in adults. *Medical Journal of Australia* **70**, 25–26.

GUERINOT, M. L., WEST, P. A., LEE, J. V. & COLWELL, R. R. 1982 *Vibrio diazotrophicus* sp. nov., a marine nitrogen-fixing bacterium. *International Journal of Systematic Bacteriology* **32**, 350–357.

HADA, H. S., WEST, P. A., LEE, J. V., STEMMLER, J. & COLWELL, R. R. 1984 *Vibrio tubiashii* sp. nov., a pathogen of bivalve mollusks. *International Journal of Systematic Bacteriology* **34**, 1–4.

HICKMAN, F. W., FARMER, J. J., III, HOLLIS, D. G., FANNING, G. R., STEIGERWALT, A. G., WEAVER, R. E. & BRENNER, D. J. 1982 Identification of *Vibrio hollisae* sp. nov. from patients with diarrhoea. *Journal of Clinical Microbiology* **15**, 395–401.

HRANITZKY, K. W., LARSON, A. D., RAGSDALE, D. W. & SIEBELING, R. J. 1980 Isolation of Ol serovars of *Vibrio cholerae* by serologically specific method. *Science* (*Washington, D.C.*) **210**, 1025–1026.

JEFFRIES, V. E. 1982 Three *Vibrio* strains pathogenic to larvae of *Crassostrea gigas* and *Ostrea edulis*. *Aquaculture* **29**, 201–226.

JOSEPH, S. W., COLWELL, R. R. & KAPER, J. B. 1982 *Vibrio parahaemolyticus* and related halophilic vibrios. *CRC Critical Reviews in Microbiology* **10**, 77–124.

KOBAYASHI, T., ENOMOTO, S., SAKAZAKI, R. & KUWAHARA, S. 1963 A new selective medium for pathogenic vibrios, TCBS agar (modified Nakanishi's agar). *Nippon Saikingaku Zasshi* **18**, 387–391.

KOLVIN, J. L. & ROBERTS, D. 1982 Studies on the growth of *Vibrio cholerae* biotype eltor and biotype classical in foods. *Journal of Hygiene* **89**, 243–252.

KOVÁCS, N. 1956. Identification of *Pseudomonas pyocyanea* by the oxidase reaction. *Nature (London)* **178**, 703.

LEE, J. V., HENDRIE, M. S. & SHEWAN, J. M. 1979 Identification of *Aeromonas, Vibrio*, and related organisms. In *Identification Methods for Microbiologists* 2nd edn, ed. Skinner, F. A. & Lovelock, D. W., pp. 151–161. Society for Applied Bacteriology Technical Series No. 14. London & New York: Academic Press.

LEE, J. V., SHREAD, P., FURNISS, A. L. & BRYANT, T. N. 1981 Taxonomy and description of *Vibrio fluvialis* sp. nov. (synonym group F Vibrios, group EF6). *Journal of Applied Bacteriology* **50**, 73–94.

LOVE, M., TEEBKEN-FISHER, D., HOSE, J. E., FARMER, J. J., III, HICKAMN, F. W. & FANNING, G. R. 1981 *Vibrio damsela*, a marine bacterium, causes skin ulcers on the Damselfish *Chromic punctipinnis*. *Science (Washington, D.C.)* **214**, 1139–1140.

MCCORMACK, W. M., DEWITT, W. E., BAILEY, P. E., MORRIS, G. K., SOEHARJONO, P., & GANGAROSA, E. J. 1974 Evaluation of thiosulfate–citrate–bile salts–sucrose agar, a selective medium for the isolation of *Vibrio cholerae* and other pathogenic vibrios. *Journal of Infectious Diseases* **129**, 497–500.

MØLLER, V. 1955 Simplified tests for some amino acid decarboxylases and for the arginine dihydrolase system. *Acta Pathologica et Microbiologica Scandinavica* **36**, 158–172.

MONSUR, K. A. 1963 Bacteriological diagnosis of cholera under field conditions. *Bulletin of the World Health Organization* **28**, 387–389.

MOORE, B. 1948 The detection of paratyphoid carriers in towns by means of sewage examination. *Monthly Bulletin of the Ministry of Health* **7**, 241–248.

NICHOLLS, K. M., LEE, J. V. & DONOVAN, T. J. 1976 An evaluation of commercial thiosulphate citrate bile salt sucrose agar (TCBS). *Journal of Applied Bacteriology* **41**, 265–269.

POPOFF, M. Y. & VERON, M. 1976 A taxonomic study of the *Aeromonas hydrophila–Aeromonas punctata* group. *Journal of General Microbiology* **94**, 11–22.

POPOFF, M. Y., COYNAULT, C., KIREDJIAN, M. & LEMELIN, M. 1981 Polynucleotide sequence relatedness among motile *Aeromonas* species. *Current Microbiology* **5**, 109–114.

SCHUBERT, R. H. W. 1982 On the problems of isolation of *Plesiomonas shiggelloides* and on a new elective medium: Inositol brilliant green bile salts agar. In *Quality Assurance and Quality Control of Microbiological Culture Media*, ed. Corry, J. E. L., pp. 163–174. Darmstadt: G-I-T Verlag Ernst Giebeler.

SHREAD, P., DONOVAN, T. J. & LEE, J. V. 1981 A survey of the incidence of *Aeromonas* in human faeces. *Society for General Microbiology Quarterly* **8**, 184.

TAYLOR, J. A. & BARROW, G. I. 1981 A non-pathogenic vibrio for routine quality control of TCBS cholera medium. *Journal of Clinical Pathology* **34**, 208–212.

TAYLOR, J. A., MILLER, D. C., BARROW, G. I., CANN, D. C. & TAYLOR, L. Y. 1982 The isolation and identification of *Vibrio parahaemolyticus*. In *Isolation and Identification Methods for Food Poisoning Organisms*, ed. Corry, J. E. L., Roberts, D. & Skinner, F. A., pp. 287–301. Society for Applied Bacteriology Technical Series No. 17. London & New York: Academic Press.

TISON, D. L. & SEIDLER, R. J. 1983 *Vibrio aestuarianus:* A new species from estuarine waters and shellfish. *International Journal of Systematic Bacteriology* **33**, 699–702.

TISON, D. L., NISHIBUCHI, M., GREENWOOD, J. D. & SEIDLER, R. J. 1982 *Vibrio vulnificus* biogroup 2: new biogroup pathogenic for eels. *Applied and Environmental Microbiology* **44**, 640–646.

VENKATRAMAN, K. V. & RAMAKRISHNAN, C. S. 1941 A preserving medium for the transmission of specimens for the isolation of *Vibrio cholerae. Indian Journal of Medical Research* **29**, 681–684.

VON GRAEVENITZ, A. & BUCHER, C. 1983 Evaluation of differential and selective media for the isolation of *Aeromonas* and *Plesiomonas* spp. from human faeces. *Journal of Clinical Microbiology* **17**, 16–21.

VON GRAEVENITZ, A. & ZINTERHOFER, L. 1970 The detection of *Aeromonas hydrophila* in stool specimens. *Health Laboratory Science* **7**, 124–126.

WEST, P. A., RUSSEK, E., BRAYTON, P. R. & COLWELL, R. R. 1982 Statistical evaluation of a quality control method for isolation of pathogenic *Vibrio* species on selected thiosulphate citrate bile salts sucrose agars. *Journal of Clinical Microbiology* **16**, 1110–1116.

WEST, P. A., LEE, J. V. & BRYANT, T. N. 1983 A numerical taxonomic study of species of *Vibrio* isolated from the aquatic environment and birds in Kent, England. *Journal of Applied Bacteriology* **55**, 263–282.

YANG, Y., YEH, L., CAY, Y., BAUMANN, L., BAUMANN, P., TANG, J. S. & BEAMAN, B. 1983 Characterization of marine luminous bacteria isolated off the coast of China and description of *Vibrio orientalis* sp. nov. *Current Microbiology* **8**, 95–100.

Klebsiellas

RUTH SHINEBAUM* AND E. MARY COOKE†

Department of Microbiology, University of Leeds, Leeds, UK

Klebsiellas are an important cause of sporadic infections in hospitals (Cooke *et al.* 1979) as well as being associated with outbreaks of infection which may be serious, prolonged, and difficult to treat because of multiple antibiotic resistance (Price & Sleigh 1970; Martin *et al.* 1971; Curie *et al.* 1978). The routes of transmission of this organism are not completely elucidated although hand transmission (Casewell & Phillips 1977; Shinebaum *et al.* 1979) and faecal carriage (Selden *et al.* 1971; Cooke *et al.* 1979) may both be important.

This organism is also of veterinary significance and can be a major cause of coliform mastitis in both cows and sows (Blood *et al.* 1979). A possible method of transmission in these cases is from contaminated bedding materials, such as sawdust (Newman & Kowalski 1973).

For these reasons microbiologists may wish to isolate klebsiellas from a variety of sources, to identify them to species level and to type some of the strains isolated. We report here the methods that we have found to be useful in the epidemiological investigations we have carried out.

Isolation

In the isolation of klebsiellas from mixed cultures the colonial appearance, although useful, cannot be relied upon to differentiate these organisms from other enterobacteria. The method described below is a modification of that

*Present address: Department of Bacteriology, University Medical School, Teriot Place, Edinburgh EH8 9Ag, UK.

†Present address: Division of Hospital Infection, Central Public Health Laboratory, 175 Colindale Avenue, London NW9 5HT, UK.

Isolation and Identification of Micro-organisms of Medical and Veterinary Importance

Copyright © 1985 by the Society for Applied Bacteriology
All rights of reproduction in any form reserved
0-12-181460-2

chosen after faecal specimens had been examined by a variety of different cultural methods (Cooke *et al.* 1979).

For the examination of faeces a sterile swab is rubbed into the sample. To examine food, sawdust and other environmental materials, 1 g of material is emulsified in 5 ml of quarter-strength Ringer's solution, preferably using a stomacher, and a swab is dipped into this. This method may also be used for faeces too solid to be picked up by a swab. To sample environmental surfaces, a broth-moistened swab may be used.

The swab is inoculated onto MacConkey, MacConkey–inositol–carbenicillin (MIC), and Simmons citrate agar plates, and into Koser's citrate broth and MIC broth. The plates are incubated overnight and the broths for 48 h, both at 37°C. The broths are then subcultured to both MIC and citrate plates.

MIC medium is a modification of that described by Thom (1970). It contains in distilled water (g/litre): sodium taurocholate, 5 (or concentrated bile salts, e.g., Oxoid L56, 1.5); peptone, 20; inositol, 10; agar (for solid medium), 12. The bile salts are dissolved in 50 ml of the water. The other ingredients are dissolved in the remaining water by steaming and the bile salt solution is added. The pH is adjusted to 7.6, 5 ml of 1% (w/v) neutral red is added, and the medium is autoclaved at 121°C for 15 min. The medium is cooled to 56°C and carbenicillin (10 μg/ml) added before pouring or dispensing.

On this medium, klebsiella colonies are large and pink. A few other organisms may be cultured, but only rarely do they ferment inositol.

Quantitative assessment of klebsiella can be achieved by emulsification of the sample in Ringer's solution, as above, and plating 0.1 ml of suitable dilutions onto MacConkey, MIC, and citrate plates.

Samples of water can be tested by culturing 0.1 ml, or dilutions in Ringer's solution if growth is expected to be heavy, according to the above schedule. It is often useful to incubate plates/broths from water samples at a lower temperature (25°C), particularly for initial isolation of klebsiellas.

If time or media are at a premium, a shortened scheme using only MIC plates, and MIC broths subcultured to MIC plates can be used. This will result in a failure to isolate about 3% of klebsiellas when compared with the complete method. It is not recommended, however, for samples where initial numbers of klebsiellas are expected to be low.

Identification

Colonies from the MIC and citrate plates, and any showing appropriate morphology from the MacConkey plate, are inoculated onto Christensen's urea, Simmons citrate, and as a stab into Donovan's medium (Donovan 1966) and are incubated overnight at 37°C.

Donovan's medium contains (g/litre): tryptone, 10; sodium chloride, 5; agar, 3; 2,3,5-tetrazolium chloride, 0.05; ferrous ammonium sulphate, 0.3; sodium thiosulphate, 0.2; dipotassium hydrogen phosphate, 0.3; inositol, 10. After steaming to dissolve these, 3 ml of 1% bromothymol blue is added, and the medium is tubed in 3- to 5-ml amounts (stab culture) and autoclaved at 115°C for 20 min. Hydrogen sulphide production turns the medium black; motility is shown by spreading of the red growth indicator. Non-inositol fermenters tend to turn the medium deep blue. Klebsiellas grow as a thin red line in a yellow medium. The medium is read after 24–30 h.

All organisms positive in Donovan's, urea and citrate tests can be considered *Klebsiella* spp. (Table 1). Any organism which is non-motile, hydrogen sulphide negative and positive for two of citrate, urea, and inositol fermentations could be *Klebsiella* spp. and should be subjected to further identification by the methods of Cowan (1974).

The identification of *Klebsiella* to species level has been made confusing by the use of two systems of nomenclature, and by controversy regarding the species status of *K. oxytoca*. Cowan (1974) recognises seven species: *K. aerogenes, K. oxytoca, K. pneumoniae, K. edwardsii, K. atlantae, K. ozaenae* and *K. rhinoscleromatis*. In comparison, Edwards & Ewing (1972) recognise only three species, *K. pneumoniae*, which encompasses five species of the Cowan classification, *K. ozaenae* and *K. rhinoscleromatis*. The recognition of *K. oxytoca* as a separate species depends on the importance assigned to the properties of indole production and gelatin liquefaction in *Klebsiella*. Cowan (1974) requires both properties in *K. oxytoca*. Jain *et al.* (1974) consider that indole positivity alone gives sufficient differentiation to warrant the species *K. oxytoca*, a view reflected by the manufacturers of many commercial identification kits. Edwards & Ewing (1972) consider that indole-positive strains

TABLE 1

Short system to identify Klebsiella *spp.*

Species	Donovan's medium: H_2S	Donovan's medium: Motility	Donovan's medium: Acid from inositol	Citrate	Urea
Klebsiella spp.	−	−	+	+	+
Enterobacter spp.	−	+	+	+	−
Citrobacter spp.	+	+	−	+	−
Escherichia coli	−	+	−	−	−

belong in *K. pneumoniae* and that the ability to liquefy gelatin is insufficient to place some of these strains in a separate species.

In practice, the vast majority of strains are speciated as *K. aerogenes* (Britain) or *K. pneumoniae* (America, most commercial kits, increasingly common elsewhere). The preponderance of *K. oxytoca* depends on the identification system used and is common in Britain and in most commercial kits. These differences must be borne in mind when comparing strains from different sources or with those in the literature. We have always included all indole-positive strains, up to 15% of our isolates, in *K. aerogenes*, particularly as less than half of such strains liquefy gelatin.

Typing

Capsular typing

This typing method has been in use for many years (Kauffmann 1949; Ørskov 1955; Edwards & Fife 1955). Seventy-seven serotypes are recognised. Antisera are prepared by the injection of formalinised capsulate cultures into rabbits. A variety of immunisation schedules have been used, and these have been compared (Edmondson & Cooke 1979*a*). Titres of antisera are usually between 16 and 256, and many of the antisera produced will cross-react with a small number of other serotypes. The antisera may be pooled according to their cross-reactions, but because of the low titres of some of them the pools will generally not be large.

To encourage the production of capsules, the organisms are grown on Worfel–Ferguson agar (Difco) for 48 h at 30°C. They are then typed by emulsifying a small portion of the growth in 10% formal-buffered saline to give 5–10 organisms per high-power field. Equal volumes of antiserum and suspension are mixed on a slide and after 2–3 min are examined under a cover slip by oil immersion phase-contrast microscopy. A positive reaction is indicated by the presence of a dark line round the edge of the capsule (the quellung reaction). At lower dilutions clumping of organisms may occur.

Because capsular serotyping of klebsiellas requires the preparation of 77 antisera, it has been carried out in only a small number of laboratories. In our experience, however, 17 antisera against the most commonly found serotypes (7, 8, 9, 11, 12, 15, 16, 18, 21, 27, 33, 35, 38, 46, 47, 52, 62) account for 40% of all strains and 46% of all typable strains. It would therefore be possible for a small laboratory to set up a simplified typing scheme, particularly if undertaking epidemiological work.

As an alternative to the quellung method, indirect immunofluorescence (Riser *et al.* 1976) or counter-current immunoelectrophoresis may be used

(Palfreyman 1978), although this latter method may give problems with a high proportion of false negatives (Shinebaum & Cooke, unpublished observations).

Bacteriocin typing

Although bacteriocin typing had been described earlier (Hall 1971; Buffenmyer *et al.* 1976), we developed, in 1979, a klebecin typing method using 15 producer strains (Edmondson & Cooke 1979*b*).

Klebecin production

Klebecin producers are grown in a small volume of proteose peptone no. 3 broth (Difco PP3) and incubated at 37°C for 18 h. One millilitre of each culture is added to 8 ml PP3 and incubated in a 32°C water bath for 1.5 h. Mitomycin C is then added to each culture to give a final concentration of 1 μg/ml, and the cultures are incubated at 32°C for a further 18 h. The remaining viable cells are killed by the addition of 0.5 ml chloroform, and the cultures are clarified by centrifugation at 3000 rev/min for 15 min. The caps of the tubes are loosened and cultures are left at room temperature for 30 min to allow evaporation of traces of chloroform. The supernatant fluids are then transferred to sterile containers. One drop of each klebecin preparation and one drop of a control containing uninoculated PP3 and mitomycin C are applied to lawns of test strains. A multiple inoculator of the Ridgway–Watt system (Denley Instruments, Bolney, Sussex) is used.

Preparation of lawns

The test strains are grown in a small volume of PP3 for 18 h at 37°C. A 1-in-10 dilution is made of each culture in phosphate-buffered saline, and 0.1 ml of each diluted suspension added to 3 ml of 0.7% agar which has been melted and cooled to 50°C. After mixing the molten agar is poured over the surface of a well-dried proteose peptone agar plate to give an even overlay. The lids are removed and the plates left at room temperature for 1.5–2 h. The dried plates can then be tested with the klebecin preparations.

Recording results

Results are recorded as follows: clear inhibiting zone, +; some resistant growth but a discernible inhibition, ±; and no discernible inhibition, −. The results are recorded by the method devised by Farmer (1970), as follows: + + + = 1; + + − = 2; + − + = 3; − + + = 4; + − − = 5; − + − = 6; − − + = 7; − − − = 8. The klebecin type is therefore a five-digit number; "±" reactions are considered positive when recording the results. However, in distinguishing between strains, only the difference between a "+" and a "−" reaction is considered.

Reproducibility and discrimination

Laboratory tests show that 67% reproducibility is obtained if patterns differing by only one strong reaction are regarded as distinct. If two reactions are regarded as necessary to distinguish between strains, the reproducibility rises to 89% but the discriminating power of the method falls. This is more fully discussed by Edmondson & Cooke (1979*b*).

Observations

Apart from the aforementioned methods, other methods of typing klebsiellas have been described, including biotyping (Duncan & Rennie 1974) and phagetyping (Slopek *et al.* 1967). The choice of typing method will depend on both the facilities available and the degree of discrimination required. Obviously, a combination of methods will result in a finer subdivision of the organisms. In particular, we have found that bacteriocin typing will subdivide capsular serotypes that might otherwise have been regarded as homogeneous.

With the exception of capsular serotyping, the methods described here can readily be used in most laboratories to study the epidemiology of sporadic and epidemic klebsiella infections, in both man and animals.

Outbreaks of klebsiella infections in hospitals have tended to be associated with gentamicin resistance in a small number of serotypes (Curie *et al.* 1978; Casewell & Talsania 1979) and strains which cause infection differ from environmental strains in their temperature growth range (Edmondson *et al.* 1980) and in their ability to survive on the hands (Cooke *et al.* 1981). The application of isolation, identification and typing procedures together with studies of other properties should make it possible to gain information about the epidemiology, transmissibility and pathogenicity of these organisms.

References

BLOOD, D. C., HENDERSON, J. A. & RODOSTITS, O. M. 1979 *Veterinary Medicine* 5th edn, pp. 363–405. London: Ballière Tindall.

BUFFENMYER, C. L., RYCHECK, R. R. & YEE, R. B. 1976 Bacteriocin (klebocin) sensitivity typing of *Klebsiella*. *Journal of Clinical Microbiology* **4**, 239–244.

CASEWELL, M. & PHILLIPS, I. 1977 Hands as a route of transmission for *Klebsiella* species. *British Medical Journal* **ii**, 1315–1317.

CASEWELL, M., & TALSANIA, H. G. 1979 Predominance of certain klebsiella capsular types in hospitals in the United Kingdom. *Journal of Infection* **1**, 77–79.

COOKE, E. M., BRAYSON, J. C., EDMONDSON, A. S. & HALL, D. 1979 An investigation into the incidence and sources of klebsiella infections in hospital patients. *Journal of Hygiene* **82**, 473–480.

COOKE, E. M., EDMONDSON, A. S. & STARKEY, W. 1981 The ability of strains of *Klebsiella aerogenes* to survive on the hands. *Journal of Medical Microbiology* **14**, 443–450.

COWAN, S. T. 1974 *Cowan and Steel's Manual for the Identification of Medical Bacteria* 2nd edn. London: Cambridge University Press.

CURIE, K., SPELLER, D. C. E., SIMPSON, R. A., STEPHENS, M. & COOKE, D. I. 1978 A hospital epidemic caused by a gentamicin-resistant *Klebsiella aerogenes*. *Journal of Hygiene* **80,** 115–123.

DONOVAN, T. J. 1966 A *Klebsiella* screening medium. *Journal of Medical Laboratory Technology* **23,** 194–197.

DUNCAN, I. B. R. & RENNIE, R. P. 1974 Klebsiella in hospital-acquired infections *Lancet* **ii,** 157.

EDMONDSON, A. S. & COOKE, E. M. 1979*a* The production of antisera to the klebsiella capsular antigens. *Journal of Applied Bacteriology* **46,** 579–584.

EDMONDSON, A. S. & COOKE, E. M. 1979*b* The development and assessment of a bacteriocin typing method for klebsiella. *Journal of Hygiene* **82,** 207–223.

EDMONDSON, A. S., COOKE, E. M., WILCOCK, A. P. D. & SHINEBAUM, R. 1980 A comparision of the properties of klebsiella strains isolated from different sources. *Journal of Medical Microbiology* **13,** 541–550.

EDWARDS P. R. & EWING, W. H. 1972 *Identification of Enterobacteriaceae* 3rd edn. Minneapolis: Burgess Publishing Company.

EDWARDS, P. R. & FIFE, M. A. 1955 Studies of the Klebsiella–Aerobacter group of bacteria. *Journal of Bacteriology* **70,** 382–390.

FARMER, J. J., III 1970 Mnemonic for reporting bacteriocin and bacteriophage types. *Lancet* **ii,** 96.

HALL, F. A. 1971 Bacteriocin typing of *Klebsiella* spp. *Journal of Clinical Pathology* **24,** 712–716.

JAIN, K., RADSAK, K. & MANNHEIM, W. 1974 Differentiation of the oxytocum group from klebsiella by deoxyribonucleic acid–deoxyribonucleic acid hybridization. *International Journal of Systematic Bacteriology* **24,** 402–407.

KAUFFMANN, F. 1949 On the serology of the *Klebsiella* group. *Acta Pathologica et Microbiologica Scandinavica* **26,** 381–406.

MARTIN, C. M., IKARI, N. S., ZIMMERMAN, J. & WAITZ, J. A. 1971 A virulent nosocomial *Klebsiella* with a transferable R factor for gentamicin: emergence and suppression. *Journal of Infectious Diseases* **124,** Suppl., 24–29.

NEWMAN, L. E. & KOWALSKI, J. J. 1973 Fresh sawdust bedding—a possible source of *Klebsiella* organisms. *American Journal of Veterinary Research* **34,** 979–980.

ØRSKOV, I. 1955 Serological investigations in the *Klebsiella* group. 1. New capsule types. *Acta Pathologica et Microbiologica Scandinavica* **36,** 449–453.

PALFREYMAN, J. M. 1978 Klebsiella serotyping by counter-current immunoelectropheresis. *Journal of Hygiene* **81,** 219–225.

PRICE, D. J. E. & SLEIGH, J. D. 1970 Control of infection due to *Klebsiella aerogenes* in a neurosurgical unit by withdrawal of all antibiotics. *Lancet* **ii** 1213–1215.

RISER, E., NOONE, P. & POULTON, T. A. 1976 A new serotyping method for *Klebsiella* species: development of the technique. *Journal of Clinical Pathology* **29,** 296–304.

SELDEN, R., LEE, S., WANG, W. L. L., BENNETT, J. V. & EICKHOFF, T. C. 1971 Nosocomial klebsiella infections: intestinal colonization as a reservoir. *Annals of Internal Medicine* **74,** 657–664.

SHINEBAUM, R., COOKE, E. M. & BRAYSON, J. C. 1979 Acquisition of *Klebsiella aerogenes* by neonates. *Journal of Medical Microbiology* **12,** 201–205.

SLOPEK, S., PRZONDO-HESSEK, A., MILCH, H. & DEAK, S. 1967 A working scheme for bacteriophage typing of *Klebsiella* bacilli. *Archivum Immunologiae et Therapiae Experimentalis* **15,** 589–599.

THOM, B. T. 1970 *Klebsiella* in faeces. *Lancet* **ii,** 1033.

Pasteurella multocida

P. E. Curtis

Sub-department of Avian Medicine, University of Liverpool Veterinary Field Station, Leahurst, Neston, Wirral, UK

Pasteurella multocida is involved in disease processes in poultry, cattle, other ruminants, pigs, rabbits and many other species (Buxton & Fraser 1977). It is often present as a commensal in the throat or upper respiratory tract of cats, dogs, other carnivores, and many other species including wild rats. Man acquires infection from bites and scratches of domestic pets and also by other routes and from other sources. The virulence of strains for individual animal species varies enormously, as does the ease with which the organism may be recovered.

Sampling Methods

The sampling methods to be adopted will depend on the animal species involved, the presence or absence of disease, and whether the objective is diagnosis, screening, or to prove the absence of infection.

Plain swabs or calcium alginate swabs of appropriate length and flexibility may be employed for nasal sampling. Nasal washings have also been collected from pigs as an alternative to swabbing.

Impression smears may be prepared from the liver and other sites and examined for Gram negative organisms resembling *P. multocida*.

Poultry

In acute disease in live birds, swabs should be taken from the choanal slit or oropharynx, the nose and cloaca. Material from abscesses and lesions and abnormal faeces should also be examined. The following material should be

Isolation and Identification of
Micro-organisms of Medical and
Veterinary Importance

Copyright © 1985 by the Society for Applied Bacteriology
All rights of reproduction in any form reserved
0-12-181460-2

collected from dead birds: heart blood, liver, spleen and lungs. Other material, such as air sacs, peritoneal surfaces, meninges and bone marrow, may be sampled.

In subacute or chronic disease swabs should be taken from the choanal slit or oropharynx and from abscesses or lesions. Lesions visible at autopsy should be examined. These may occur in the liver, spleen, lungs, peritoneal surfaces, joints, wattles, middle ear, eye and brain, but in the more chronic forms of the disease it may be difficult to isolate the organisms.

Symptomless carriers may occur among survivors of outbreaks or among birds exposed to strains of low virulence. Swabs should be taken from the choanal slit or oropharynx, and in dead birds portions of the nasal turbinates should be collected.

Rats

Wild rats (*Rattus norvegicus*) are not susceptible to infection by *P. multocida*, but they may be carriers (Curtis *et al.* 1980). The swabbing of live rats is not recommended. To examine dead rats, the skin over the face should be reflected and the snout snipped with sterile bone forceps just posterior to the insertion of the incisors to expose the turbinate bones, portions of which are placed in nutrient broth for overnight incubation. Oropharyngeal swabs should be taken.

Cats, dogs and other carnivores

Many of these animals are healthy carriers of *P. multocida*. Cats and dogs may also be carriers of group EF-4 bacteria, which resemble *P. multocida* but are indole negative (Holmes & Ahmed 1981).

Oropharyngeal swabs and material from abscesses and other lesions should be collected from live and dead animals.

Cattle, pigs and rabbits

These animals may be carriers of low-virulence strains but may also acquire highly virulent infections leading to outbreaks of acute disease.

Nasal swabs or washings and material from abscesses and other lesions should be collected from live and dead animals.

Man

Local abscesses, throat swabs or suspicious lesions may be swabbed or otherwise sampled, depending on the features of the infection.

Isolation

Cultural, or biological followed by cultural methods may be used for the isolation of *P. multocida*, but although cultural methods are adequate when there is recent active infection: where the animals are possibly symptomless carriers biological methods may be expected to give a higher recovery rate.

Tables 1 and 2 indicate recommended media and procedures for the various kinds of specimens.

TABLE 1

*Pasteurella multocida infections in birds: recommended and optional specimens and procedures**

Infection	Blood agar	MacConkey agar	Selective media: Direct	Selective media: Via nutrient broth	Nutrient broth and mouse
Acute disease					
Live birds					
Choanal slit or oropharynx	R	R	R	R	O
Nasal swab	R	R	R	R	O
Lesions	R	R	O	O	O
Cloacal swab	O	O	R	R	R
Abnormal faeces	O	O	R	R	R
Dead birds					
Heart blood, Liver, Spleen, Lungs	R	R			
Airsacs, Peritoneum, Meninges, Bone marrow	O	O			
Subacute or chronic disease					
Live birds					
Choanal slit, etc.	O	O	R	R	R
Lesions	R	R	O	O	O
Dead birds					
Lesions	R	R	R	R	R
Symptomless carriers					
Live birds					
Choanal slit			R	R	R
Dead birds					
Choanal slit, Nasal turbinate			O	R	R

* R, recommended; O, optional procedures.

TABLE 2

*Pasteurella multocida infections in mammals: recommended and optional specimens and procedures**

Mammal	Blood agar	MacConkey agar	Selective media: Direct	Selective media: Via nutrient broth	Nutrient broth and mouse
Rats (dead)					
Nasal turbinates; Oropharyngeal swabs				R	R
Cats, dogs, other carnivores (live or dead)					
Oropharyngeal swabs	R	R	R	R	O
Lesions	R	R	O	O	O
Cattle, pigs, rabbits (live or dead)					
Nasal swabs or washings; Lesions	R	R	R	R	O

* R, recommended; O, optional procedures.

Culture media

Useful nonselective media include horse or bovine blood agar, MacConkey agar, nutrient broth and dextrose starch agar (Difco).

Morris's medium no. 2

This selective medium (Morris 1958) uses a tryptic digest of beef agar containing Fildes' sheep blood digest, 0.5 ml/litre; neomycin, 2.5 mg/litre; tyrothricin, 10 mg/litre; Actidione (cycloheximide), 100 mg/litre (or Mycostatin (nystatin), 200,000 units/litre); potassium tellurite, 2.5 mg/litre.

Morris's medium no. 2 modified

This differs from the original in using a tryptone agar (Oxoid) base instead of the beef digest.

Clindamycin medium

For this medium (Garlinghouse *et al.* 1981), blood agar base is used, with the addition of bovine, horse or sheep blood, 5%, and clindamycin, 2 mg/litre.

Clindamycin–gentamicin–tellurite (CGT) medium

This is heart infusion agar (Difco) with defibrinated horse blood, 5 ml/litre; clindamycin phosphate, 5 mg/litre; gentamicin sulphate, 0.75 mg/litre; potassium tellurite, 2.5 mg/litre (Knight *et al.* 1983).

Cultural methods

Swabs are cultured on solid media and also placed in nutrient broth. Both are incubated at 37°C. Blood and MacConkey agar cultures are examined after 24 h, and selective media cultures after 72 h. Broth cultures are plated on selective media after 24 h.

Biological methods

Overnight broth cultures may be employed for mouse inoculation to detect the organism, using mice from a colony with no history of *P. multocida* infection. The intramuscular route may be used; the mouse is killed, if not already dead, after 24 h and cultures are prepared from liver, spleen and inoculation site (Curtis & Ollerhead 1981). Pedersen & Barfod (1981) used porcine nasal swabs rinsed in 1 ml of 0.9% saline; they injected 0.2 ml of the washing intraperitoneally into mice and observed them for 1 week. Cultures were made from the livers of animals that died within this time and the cultures were incubated aerobically for 48 h.

Identification

Colonies on blood agar may be smooth or mucoid, are always non-haemolytic, greyish in colour, and are 1–4 mm in diameter; mucoid colonies are relatively large. There is usually no growth on MacConkey agar, but very occasionally small (0.5 mm) colonies may be found in the inoculation area. If colonies on dextrose starch agar are examined with a stereomicroscope and oblique illumination, they may show varying degrees of iridescence: orange-yellow or yellow-green, as well as non-iridescent grey colonies may be seen.

Cultures in nutrient broth are moderately turbid after 18–24 h. Colonies on Morris agar no. 2 after 18–72 h are colorless or grey and 0.5–2.0 mm in diameter. On clindamycin blood agar and CGT media, they are similar to those on blood agar, but some strains may grow poorly on CGT.

The cells are non-motile, Gram negative, oval, coccobacilli or short rods, usually single and 0.2–0.4 × 0.5–0.7 μm. Indian ink preparations may show heavily capsulated organisms. Electron microscopy shows coccobacilli, with capsules present in 0 to 97% of cells.

Biochemical and other tests

The indole test, with tryptone water and Kovac's reagent, is positive. The urease test on Christensen's medium, incubated at 37°C for 48 h, is negative. In triple sugar iron agar (Oxoid) a yellow (acid) slope and butt with no gas production is seen after 18–24 h at 37°C. No H_2S is produced on this medium by *P. multocida*. The oxidase and catalase tests are positive.

Sugar fermentation tests are done in 1% solutions of the carbohydrates in peptone water with Andrade's indicator. Acid is produced from glucose and sucrose, usually from mannitol, but not usually from lactose, maltose, arabinose or dulcitol.

Other identification systems

The API 20E system may be used to identify *P. multocida* providing the initial inoculum is prepared in distilled water containing 0.1% yeast extract and not in distilled water alone, as recommended for Enterobacteriaceae.

Serotyping

There have been several attempts to devise a serotyping system for *P. multocida,* and these have been reviewed by Brogden & Packer (1979). Carter & Chengappa (1981) suggested that a combination of methods be employed, involving the Heddleston system and capsular typing systems.

The system of Heddleston *et al.* (1972) has been found to be useful for epidemiological purposes. Using type-specific strains, of which 16 have been described, sera can be produced in specific pathogen-free (SPF) chickens. Killed antigens emulsified in light oil are injected subcutaneously and sera are collected approximately 3 weeks later.

Antigens for the test are prepared from field isolates which have been grown overnight on dextrose starch agar at 37°C, removed and emulsified with 1 ml of phosphate-buffered saline containing 8.5% NaCl and 0.3% formalin, and heated in a waterbath at 100°C for 1 h. After cooling, the antigen is centrifuged at 4000 rev/min for 30 min and the supernatant employed as the antigen in a gel diffusion plate test. This is done in 50-mm-diameter petri dishes containing 4 ml of 1% agar with 8.5% NaCl and 0.1% Trypan blue. The antigen is placed in the central well and specific sera in the six surrounding wells. The wells are 4 mm in diameter and their centres 6 mm apart. The test is read after 24 and 48 h incubation in a moist atmosphere (sandwich box) at 37°C. Precipitin lines are recorded on charts, to show the location of any lines and also their strength (+, ++, +++).

Virulence

It is important to formulate an estimate of the virulence of a field strain either from clinical observation or by laboratory tests or both. The virulence range is considerable.

Virulence for poultry

Field isolates may be tested for virulence by the following method. The organism is grown overnight on dextrose starch agar, washed off with normal saline, and the density is adjusted to Wellcome opacity tube no. 1 to give approximately 10^9 organisms/ml. This is diluted with broth to give an estimated final dose of 10^6 organisms in a 0.1-ml dose. Viable counts may be performed. The inoculation is made into the right pectoral muscle of three 6- to 8-week-old SPF chickens or turkey poults. If SPF poultry are not available, commercial stock free of major mycoplasmas should be used. Virulence may be rated on the basis of lesions and mortality over a 6-day period.

Three chickens or three turkey poults are used, and the results are assessed 6 days after inoculation:

Virulence rating	Lesions and mortality rate
0	Nil
1	Small (i.e. less than 5 mm in diameter) lesions in muscle or associated tissues at the inoculation site in one or two birds
2	Small lesions in muscle or associated tissues at the inoculation site in three birds, or more extensive lesions at the inoculation site in one or two birds
3	More extensive lesions at the inoculation site or associated tissues in three birds
4	Death of one to three birds
5	Death of three birds in 72 h

Similar results may be obtained by examining the birds after 3 rather than 6 days and applying a similar interpretation.

Swabbing the choanal slit of susceptible turkeys or chickens with a fresh culture of the organism to be tested is an alternative method of challenge which has been used to test for virulence (Heddleston 1975).

Virulence for mice

This may be assessed by examining mice 24 h after an intramuscular inoculation of mice (6- to 12-week-old BALB "C" or N.M.R.I. strains) with approximately 10^6 organisms prepared as for poultry. Virulence can be rated as follows:

Virulence rating	Symptoms and mortality	Gross lesions
0	Nil	Nil
1	Nil	Small area of congestion at inoculation site

continued

Virulence rating	Symptoms and mortality	Gross lesions
2	Slight lameness on injected limb; otherwise normal	Slight haemorrhagic swelling of muscle at inoculation site
3	Lameness on injected limb; also reduced activity and other signs of ill health	Marked haemorrhagic swelling of muscle and subcutaneous congestion at the inoculation site
4	Immobile unless stimulated, lame; huddling; eyes closed; coat raised; very sick or moribund	Variable appearance of inoculation site; haemorrhagic swelling or congestion in some cases, but usually no marked local lesions; upper intestine distended with glistening appearance and mucoid contents
5	Dead	Lesions as for 4 above

Other workers have used intraperitoneal inoculation of an overnight broth culture at various dilutions to assess the LD50 from the accumulated deaths on the day 7 (Chandrasekaran & Yeap 1982).

Atrophic rhinitis in pigs

Toxigenic strains of *P. multocida* are said to be involved in causing this disease, and tests have been devised to recognize these strains. De Jong and others (1980) described intracutaneous tests on guinea pigs or rabbits using 0.2 ml of an overnight broth culture or a filtrate of it. The inoculation site was examined daily for 48–72 h, and the size and intensity of any reaction were noted. The same authors used mouse lethality tests, injecting 0.5 ml of a bacteria-free filtrate of a broth culture intraperitoneally and recording mortality in 1–10 days as an indication of the presence of thermolabile toxin. They also infected SPF piglets by instilling 1 ml of an overnight broth culture in each nostril for 5 sucessive days. A filtrate of such a culture was similarly administered to other SPF piglets. Cultures or culture filtrates causing positive reactions in guinea pigs were found to be capable of causing atrophic rhinitis in SPF piglets which were killed and examined up to 4 weeks after instillation.

Rutter & Rojas (1982) instilled 0.5 ml of a suspension of *P. multocida* into each nostril of 7- to 16-day-old gnotobiotic piglets to assess the atrophic rhinitis pathogenicity of strains. Rutter (1983) used a 1-ml dose of a suspension of *P. multocida* (mean c.f.u. 10^7–10^8/ml) given to 12-day-old germ-free piglets following earlier instillation of *Bordetella bronchiseptica* for the same purpose. A filtrate of a culture sonicate of *P. multocida* has been used to test for toxigenicity in a mouse lethality test in which 0.5-ml volumes of the filtrate were inoculated intraperitoneally into hybrid mice weighing about 17 g. BALB/C mice are particularly sensitive, and deaths within 72 h were recorded. This preparation was also used in guinea pig dermonecrotic tests in which a 0.1- to 0.2-ml dose was used and the reaction was read after 72 h.

A cell culture assay using embryonic bovine lung cells has been developed (Rutter & Luther 1984) for the detection of toxigenic strains of *P. multocida* from pigs. It has proved highly sensitive and is claimed to have several advantages over the mouse assay.

References

Brogden, K. A. & Packer, R. A. 1979 Comparison of *Pasteurella multocida* serotyping systems. *American Journal of Veterinary Research* **40**, 1332–1335.

Buxton, A. & Fraser, G. 1977 *Pasteurella, Yersinia* and *Francisella*. In *Animal Microbiology* Vol. 1. Oxford: Blackwell Scientific Publications.

Carter, G. R. & Chengappa, M. M. 1981 Recommendations for a standard system of designating serotypes of *Pasteurella multocida*. *Proceedings of Annual Meeting—American Association of Veterinary Laboratory Diagnosticians* **24**, 37–42.

Chandrasekaran, S. & Yeap, P. C. 1982 *Pasteurella multocida* in pigs: The serotypes and the assessment of their virulence in mice. *British Veterinary Journal* **138**, 332–336.

Curtis, P. E. & Ollerhead, G. E. 1981 Investigation to determine whether healthy chickens and turkeys are oral carriers of *Pasteurella multocida*. *Veterinary Record* **108**, 206–207.

Curtis, P. E., Ollerhead, G. E. & Ellis, C. E. 1980 *Pasteurella multocida* infection of poultry farm rats. *Veterinary Record* **107**, 326–327.

de Jong, M. F., Oei, H. L. & Tetenburg, G. J. 1980 AR-pathogenicity tests for *Pasteurella* isolates. In *Proceedings of the International Pig Veterinary Society*, ed. Nielsen, N. C., Høgh, P., Bille, N., p. 211. Copenhagen.

Garlinghouse, L. E., Di Giacomo, R. F., Van Hoosier, G. L. & Condon, J. 1981 Selective media for *Pasteurella multocida* and *Bordetella bronchiseptica*. *Laboratory Animal Science* **31**, 39–42.

Heddleston, K. L. 1975 Pasteurellosis. In *Isolation and Identification of Avian Pathogens*, ed. Hitchner, S. B., Domermuth, C. H., Purchase, G. H. & Williams, J. E., pp. 38–51. American Association of Avian Pathologists. Ithaca, New York: Arnold Printing Corporation.

Heddleston, K. L., Gallagher, J. E. & Rebers, P. A. 1972 Fowl cholera: gel diffusion precipitin test for serotyping *Pasteurella multocida* from avian species. *Avian Diseases* **16**, 925–936.

Holmes, B. & Ahmed, M. S. 1981 Group EF-4: A *Pasteurella*-like organism. In *Haemophilus, Pasteurella and Actinobacillus*, ed. Kilian, M., Frederiksen, W. & Biberstein, E. L., pp. 161–174. London & New York: Academic Press.

Knight, D. P., Paine, J. E. & Speller, D. C. E. 1983 A selective medium for *Pasteurella multocida* and its use with animal and human specimens. *Journal of Clinical Pathology* **36**, 591–594.

MORRIS, E. J. 1958 Selective media for some *Pasteurella* species. *Journal of General Microbiology* **19**, 305–311.

PEDERSON, K. B. & BARFOD, K. 1981 The aetiological significance of *Bordetella bronchiseptica* and *Pasteurella multocida* in atrophic rhinitis of swine. *Nordisk Veterinaermedicin* **33**, 513–522.

RUTTER, J. M. 1983 Virulence of *Pasteurella multocida* in atrophic rhinitis of gnotobiotic pigs infected with *Bordetella bronchiseptica*. *Research in Veterinary Science* **34**, 287–295.

RUTTER, J. M., & LUTHER, P. D. 1984 Cell culture assay for toxigenic *Pasteurella multocida* from atrophic rhinitis of pigs. *Veterinary Record* **114**, 393–396.

RUTTER, J. M. & ROJAS, X. 1982 Atrophic rhinitis in gnotobiotic piglets: Differences in the pathogenicity of *Pasteurella multocida* in combined infections with *Bordetella bronchiseptica*. *Veterinary Record* **110**, 531–535.

Brucellas

M. J. CORBEL AND D. McL. F. D. HENDRY

Ministry of Agriculture, Fisheries and Food, Central Veterinary Laboratory, Weybridge, Surrey, UK

The genus *Brucella* contains a small group of closely related Gram negative bacteria. At present they are classified into six species identified on the basis of the preferred natural host: *B. abortus*, *B. canis*, *B. melitensis*, *B. neotomae*, *B. ovis* and *B. suis*.

Brucella abortus primarily infects cattle but can also be transmitted to buffalo, camels, deer, dogs, horses, sheep, goats, yaks and man. *Brucella canis* is essentially an infection of dogs but is occasionally transmitted to man. *Brucella melitensis* primarily infects sheep and goats but can also infect cattle and man. *Brucella neotomae* is not known to infect any species other than the desert wood rat (*Neotoma lepida* Thomas) under natural conditions. *Brucella ovis* primarily infects sheep and has not been implicated in natural infections in any other species, including man. *Brucella suis* displays a wider range of host specificity than the other species, and this also varies with its biovars. Thus *B. suis* biovars 1 and 3 have the pig as their natural host but may also be transmitted to dogs, horses, rodents and man; biovar 2 occurs naturally in both hares and swine but is less pathogenic for man than biovars 1 and 3; biovar 4 has the reindeer (*Rangifer tarandus*) as its primary natural host but can also infect dogs, wild carnivores, musk oxen and man; and biovar 5 occurs naturally in murine and cricetine rodents and can be transmitted to man. Under normal circumstances, *Brucella* species show little tendency for lateral transmission in secondary hosts, the natural reservoir of infection being maintained in the primary host species.

Typically, the organisms produce infections which are localized in the lymphoid system and the reproductive organs. Thus in the female the uterus and mammary glands usually become infected, with extension to the placenta and fetal tissues often producing abortion in the pregnant animal. In the

Isolation and Identification of Micro-organisms of Medical and Veterinary Importance

Copyright © 1985 by the Society for Applied Bacteriology
All rights of reproduction in any form reserved
0-12-181460-2

male, the testes, epididymides and accessory glands are the main sites for localisation. In both primary and secondary hosts, however, almost any tissue can become infected, and the clinical manifestations of the disease can cover a wide spectrum.

The identification of typical brucella isolates is usually straightforward. Nevertheless, difficulties can arise, particularly in dealing with isolates which vary in one or more properties from those recognized for the genus. This is particularly the case with cultures recovered in the non-smooth phase. These may be very difficult to identify correctly without recourse to specialized tests. Application of the methods described in this chapter should, nevertheless, permit identification of all isolates of *Brucella*.

The Genus *Brucella* and Its Species

The genus and the six species of *Brucella* currently recognised are defined as follows.

Brucella Meyer and Shaw 1920

Small, non-motile, non-sporing, Gram negative cocci, coccobacilli or short rods, 0.5–0.7 μm by 0.5–1.5 μm, occurring singly, in pairs or short chains. Aerobic, no growth under strictly anaerobic conditions. Carboxyphilic. Metabolism is mainly oxidative. Usually show little fermentative action on carbohydrates in conventional media. Multiple amino acids, thiamin, biotin and nicotinamide are required. Growth of many strains is improved by calcium pantothenate and *meso*-erythritol. Haemin (X factor) and nicotinamide adenine dinucleotide (V factor) are not required. Catalase positive. Usually oxidase positive. Usually reduce nitrates. Produce H_2S and hydrolyse urea to a variable extent. Do not produce indole. Do not liquefy gelatin or lyse blood. Do not produce acetylmethyl carbinol [Voges–Proskauer (V–P) test]. Do not utilise citrate. Give a negative methyl red reaction. Do not usually release *o*-nitrophenol from *o*-nitrophenol β-D-galactoside. Do not change litmus milk or may render it alkaline. Optimum growth temperature 37°C, range 20–40°C. Optimum pH for growth between pH 6.6 and 7.4. Contain cytochrome C. Guanine plus cytosine content of DNA is 55–58 mole %. Show greater than 90% homology in DNA hybridisation tests. Possess characteristic internal antigens. Intracellular parasites producing characteristic infections in animals, transmissible to man.

Brucella abortus (Schmidt and Weiss 1901) Meyer and Shaw

Catalase and oxidase positive. Usually requires supplementary CO_2 for growth, especially on primary isolation. Usually produces H_2S from sulphur-containing amino acids or proteins. Usually hydrolyses urea (occasional strains may not). Usually grows in the presence of basic fuchsin, methyl violet, pyronin and safranin O but not thionin, at standard concentrations. Reduces nitrates to nitrites and may also reduce nitrites. Smooth strains may have A, M, or A and M surface antigens reactive in tests with monospecific antisera, depending upon biovar. Oxidises L-alanine, D-alanine, L-asparagine, L-glutamic acid, D-galactose, D-glucose, D-ribose and *meso*-erythritol. Does not oxidise D-xylose, L-arginine, DL-citrulline, DL-ornithine or L-lysine. Cultures in the smooth or smooth-intermediate phase are lysed by brucella phages of the Tbilisi (Tb), Weybridge (Wb), Firenze (Fi), M51-S708, Berkeley (Bk_2), MC/75 and D groups at routine test dilution (RTD) and 10^4 RTD. Non-smooth cultures are lysed by brucella phage R at RTD. Usually pathogenic for cattle, causing abortion, but may also infect other species including dogs, horses, sheep and man.

Eight biovars are currently recognised. Reference/neotype strains: *B. abortus* 544 (biovar 1); *B. abortus* 86/8/59 (biovar 2); *B. abortus* Tulya (biovar 3); *B. abortus* 292 (biovar 4); *B. abortus* B3196 (biovar 5); *B. abortus* 870 (biovar 6); *B. abortus* 63/75 (biovar 7); *B. abortus* C68 (biovar 9). *Brucella abortus* biovar 8 is no longer recognised.

Brucella melitensis (Hughes 1893) Meyer and Shaw 1920

Catalase and oxidase positive. Does not require supplementary CO_2 for growth. Does not produce H_2S, or no more than a trace, when grown on recommended media. Usually hydrolyses urea but some strains may not. Usually grows in the presence of basic fuchsin, thionin, methyl violet, pyronin and thionin blue at the standard concentrations. Reduces nitrates to nitrites and may also reduce nitrites. Smooth cultures may have the A, M, or both A and M surface antigens reactive in tests with monospecific sera. Oxidises D-glucose, *meso*-erythritol, L-alanine, D-alanine, L-asparagine and L-glutamic acid. Does not oxidise L-arabinose, D-galactose, D-ribose, L-xylose, L-arginine, DL-citrulline, DL-ornithine or L-lysine. Cultures in the smooth phase are usually susceptible to lysis by Bk_2 phage at RTD and 10^4 RTD, but are not lysed by brucella phages of the Tb, Wb, Fi, M51-S708, MC/75, D or R groups at RTD or 10^4 RTD. Non-smooth cultures are resistant to lysis by all of these phages. Usually pathogenic for sheep and goats but may also infect other species, including man and cattle.

Reference strains are *B. melitensis* 16M (biovar 1); *B. melitensis* 63/9 (biovar 2); and *B. melitensis* Ether (biovar 3).

Brucella suis Huddleson 1929

Catalase and oxidase positive. Supplementary CO_2 is not required for growth. Produces large amounts of H_2S (biovar 1) or none at all (other biovars). Usually hydrolyses urea rapidly. Usually grows in the presence of thionin, but most strains are inhibited by basic fuchsin, methyl violet, pyronin, safranin O and malachite green at standard concentrations. Smooth cultures have the A surface antigen predominant except for those of biovar 4, which react equally with anti-sera monospecific for the A and M surface antigens, and biovar 5, which has the M antigen dominant. Usually oxidises D-ribose, D-glucose, *meso*-erythritol, D-xylose, L-arginine, DL-citrulline, DL-ornithine and L-lysine. Oxidation of L-asparagine, L-glutamic acid, L-arabinose and D-galactose varies with the biovar. Does not oxidise L-alanine or D-alanine. Smooth or smooth-intermediate cultures are lysed by brucella phages of the Wb, M51-S708, Bk_2, MC/75 and D groups at RTD and 10^4 RTD. Phages of the Fi group produce partial lysis at RTD; those of the Tb group do not produce lysis at RTD but are lytic at 10^4 RTD. Non-smooth cultures are not lysed by any of these phages or by phage R at RTD or 10^4 RTD.

Usually pathogenic for pigs, except biovar 4, which is usually pathogenic for reindeer. Other species may be infected, however, including dogs, hares and rodents.

Five biovars are currently recognised. Reference strains: *B. suis* 1330 (biovar 1); *B. suis* Thomsen (biovar 2); *B. suis* 686 (biovar 3); *B. suis* 40 (biovar 4); and *B. suis* 514 (biovar 5).

Brucella neotomae Stoenner and Lackman 1957

Catalase positive, oxidase negative. Does not require supplementary CO_2 for growth. Usually produces H_2S in large amounts and hydrolyses urea rapidly. Does not grow in the presence of basic fuchsin even at 1/150,000, nor in the presence of safranin O at 1/10,000 or thionin blue at 1/150,000, but will grow in the presence of thionin at 1/150,000. Reduces nitrates to nitrites. In smooth cultures the A surface antigen is predominant. Oxidises L-asparagine, L-glutamic acid, L-arabinose, D-galactose, D-glucose, *meso*-erythritol and D-xylose. Does not oxidise L-alanine, D-alanine, L-arginine, DL-citrulline, DL-ornithine or L-lysine. Oxidation of D-ribose may be variable. In special media, ferments D-glucose, D-galactose, L-arabinose and D-xylose. Smooth or smooth-intermediate cultures are lysed by brucella phages of the Wb, M51-S708, Fi, Bk_2, MC/75 and D groups at RTD. Phages of the Tb group produce partial lysis with few very small plaques at RTD, complete lysis at 10^4 RTD. Rough or mucoid cultures are not lysed by these phages or by brucella phage R, at RTD or 10^4 RTD.

Pathogenic for the desert wood rat (*Neotoma lepida* Thomas). Authenticated natural infections of other species have not been reported.

No biovars are recognised. Reference/neotype strain, *B. neotomae* 5K33.

Brucella ovis Buddle 1956

Catalase positive, oxidase negative. Requires supplementary CO_2 for growth. Does not produce H_2S or usually hydrolyse urea. Inhibited by methyl violet but usually grows in the presence of basic fuchsin and thionin at standard concentrations. Does not produce nitrite from nitrate. A true smooth phase does not occur. Although the colonial morphology may resemble the smooth form, cultures are always in the rough phase on primary isolation. Rough-specific surface antigens cross-reacting with other non-smooth brucellas are predominant. Oxidises L-alanine, D-alanine, L-asparagine, D-asparagine, L-glutamic acid, DL-serine and adonitol. Does not oxidise L-arabinose, D-galactose, D-glucose, D-ribose, *meso*-erythritol, D-xylose, L-arginine, DL-citrulline, DL-ornithine or L-lysine. The cultures are not lysed by brucella phages of the Tb, Wb, M51-S708, Fi, Bk_2, MC/75, D or R groups at RTD or 10^4 RTD but are lysed by phages R/O and R/C at RTD and 10^4 RTD.

Pathogenic for sheep, causing epididymitis in rams and sometimes abortion in ewes. Natural infections in other species have not been reported but goats are susceptible to experimental inoculation.

No biovars are recognised. Reference/neotype strain, *B. ovis* 63/290.

Brucella canis Carmichael and Bruner 1968

Catalase positive, oxidase positive. Does not require supplementary CO_2 for growth. Does not produce H_2S. Hydrolyses urea very rapidly. Reduces nitrates to nitrites. Growth is inhibited by basic fuchsin but not by thionin at the standard concentration. A smooth phase is not known. Cultures are always in the rough or mucoid phase on primary isolation. Forms a mucoid sediment in liquid media. Rough-specific surface antigens cross-reacting with other brucellas are predominant. Oxidises D-ribose, D-glucose, L-arginine, DL-citrulline, DL-ornithine, L-lysine. Oxidation of *meso*-erythritol is variable. Does not oxidise L-alanine, D-alanine, L-asparagine, L-glutamic acid, L-arabinose, D-galactose. Is not lysed by brucella phages of the Tb, M51-S708, Wb, Fi, Bk_2, MC/75, D, R or R/O groups at RTD or 10^4RTD but are lysed by phage R/C at RTD.

Pathogenic for dogs, causing abortion in pregnant females and epididymo-orchitis in males, occasionally transmitted to man.

No biovars are recognised. Reference/neotype strain, *B. canis* RM6/66.

Biovars

The biovars within species are differentiated on the basis of specific combinations of characteristics. These include CO_2 requirement, H_2S produc-

TABLE 1

Characteristics used in the differentiation of Brucella *species and biovars*

Species	Biovar	CO_2 requirement	H_2S production	Growth on media containing: Thionin*	Growth on media containing: Basic fuchsin*	Agglutination with monospecific anti-sera: A	M	R	Lysis by phage at RTD†: Tb	Wb	Bk_2	Fi	Substrates metabolised oxidatively**: L-Alanine	L-Asparagine	L-Glutamic acid	L-Arabinose	D-Galactose	D-Ribose	D-Glucose	*meso*-Erythritol	D-Xylose	L-Arginine	DL-Ornithine	L-Citrulline	L-Lysine
B. abortus	1	(+)	+	−	+	+	−	−	L	L	L	L													
	2	(+)	+	−	−	+	−	−	L	L	L	L													
	3§	(+)	+	+	+	+	−	−	L	L	L	L													
	4	(+)	+	−	+‖	−	+	−	L	L	L	L	+	+	+	+	+	+	+	+	−	−	−	−	−
	5	−	−	+	+	−	+	−	L	L	L	L													
	6§	−	(+)‡	+	+	+	−	−	L	L	L	L													
	7	−	(+)	+	+	+	+	−	L	L	L	L													
	9	−	+	+	+	−	+	−	L	L	L	L													
B. suis	1	−	+	+	−#	+	−	−	NL	L	L	PL	−	−	+	+	+	+	+	+	+	+	+	+	+
	2	−	−	+	−	+	−	−	NL	L	L	PL	−	−	+	+	+	+	+	+	−	+	+	+	+
	3	−	−	+	+	+	−	−	NL	L	L	PL	−	−	+	−	−	+	+	+	−	+	+	+	−
	4	−	−	+	(−)	+	+	−	NL	L	L	PL	−	−	+	−	−	+	+	+	−	+	+	+	−
	5	−	−	+	−	−	+	−	NL	L	L	PL	±	+	+	−	−	+	+	+	+	+	+	+	+
B. melitensis	1	−	−	+	+	−	+	−	NL	NL	L	NL													
	2	−	−	+	+	+	−	−	NL	NL	L	NL	+	+	+	−	−	−	+	+	−	−	−	−	−
	3	−	−	+	+	+	+	−	NL	NL	L	NL													
B. ovis		+	−	+	(+)	−	−	+	NL	NL	NL	NL	−	+	+	−	−	−	−	−	−	−	−	−	−
B. canis		−	−	+	−	−	−	+	NL	NL	NL	NL	−	−	+	−	−	+	+	+	−	+	+	+	−
B. neotomae		−	+	−	−	+	−	−	NL or PL	L	L	L	−	−	+	+	+	+	+	+	−	−	−	−	−

* Concentration = 1/50,000 w/v.
† Phage R will lyse non-smooth *Brucella abortus* at RTD. Phage R/O will lyse *B. ovis* at RTD and Phage R/C will also lyse *B. canis* at RTD. L, confluent lysis; PL, partial lysis; NL, no lysis.
‡ (+), Most strains positive; (−), most strains negative.
§ For more certain differentiation of *B. abortus* biovars 3 and 6, thionin at 1/25,000 w/v is used in addition. Biovar 3 = +, biovar 6 = −.
‖ Some strains of this biovar are inhibited by basic fuchsin.
\# Some isolates may be resistant to basic fuchsin, pyronin and safranin O.
** + = $QO_2N > 100$; − = $QO_2N < 100$; ± = QO_2N varies between strains.

tion, serological specificity, tolerance to dyes and in the case of *B. suis* strains, oxidative metabolic profiles with standard substrates. The current biotyping scheme is summarised in Table 1.

Safety Precautions

In terms of the risk of causing laboratory-acquired infections, *Brucella* strains are among the most dangerous groups of bacteria to work with. Cultures of these organisms and materials containing them, including clinical specimens, should be handled only with proper precautions by persons who are adequately trained and experienced in microbiological techniques. The more hazardous techniques, such as oxidative metabolism tests, are best left to reference laboratories specializing in these.

It is strongly recommended that, whenever possible, all manipulations involving live brucella cultures should be done in microbiological safety cabinets of a design which enables the operator to be physically separated from any sources of contamination (class III). Some procedures, such as manometric measurements of oxidative metabolic activity, are very difficult to perform under completely enclosed conditions and may have to be done on the open bench in a room set aside for the purpose. In these cases, the operator should wear full protective clothing and a respirator fitted with a microbiological filter.

Culture suspensions should not be centrifuged in ordinary centrifuges unless these are fitted with sealable buckets or aerosol-proof rotors. The tubes or bottles should be filled and emptied only in a safety cabinet.

Whenever possible, use should be made of techniques which minimise aerosol production. Thus, pipettes should be filled and emptied using rubber bulbs or other pipetting aids, care being taken to avoid contaminating these and to prevent frothing or bubbling during the procedure. Catalase tests should not be done by flooding exposed cultures with hydrogen peroxide. Instead, the culture suspension should be contained in a test tube and the diluted hydrogen peroxide added by means of a Pasteur pipette inserted through the cotton-wool plug.

It is advisable to substitute disposable plastic apparatus (e.g. inoculating loops, swabs, pipettes, syringes, petri dishes) for non-disposable materials which require heating or are fragile. All contaminated materials should be discarded into leak-proof disposable containers, such as heavy-duty polythene bags, and then incinerated, or rendered safe by autoclaving. Facilities should also be available for sterilising protective clothing, the working area and the exhaust-protective cabinets by fumigation with either formaldehyde or ethylene oxide.

Work with virulent brucella cultures, especially those of *B. melitensis* or *B. suis*, should be kept to a minimum; wherever permissible, avirulent or attenuated strains should be substituted (e.g. *B. abortus* strain 19 or 99, *B. melitensis* Rev 1) for antigen production, phage propagation etc.

General recommendations on the safety precautions to be observed when working with pathogenic organisms are given by Collins (1983).

Transport of cultures

Brucella cultures that are to be sent through the post should preferably be lyophilised and sealed in ampoules. Alternatively, cultures on SDA slopes or blood agar slopes in bijou bottles may be used. The ampoule or bottle should be well wrapped in absorbent packing material, sealed into a leak-proof plastic bag and then packed into a rigid wooden or metal container, which is finally packed in a robust cardboard outer container. This should be addressed and labelled according to the national and international regulations relating to the transport of infectious perishable biological substances (IPBS). These vary between countries. A registered IPBS mail service is operated by the post office to most of the major countries. For others, the individual regulations have to be followed. Guidance on the transport of cultures across national boundaries is given by Collins (1983).

Isolation

Culture media, general

Serum dextrose agar (SDA)

This medium is suitable for making sub-cultures, for maintenance and for primary isolation of *Brucella* from uncontaminated sources. It is not recommended for primary isolation from grossly contaminated material, e.g. milk, placenta, semen.

Its composition (g or ml/litre) is as follows: blood agar base No. 2 (Oxoid), 40; equine serum, inactivated at 56°C for 30 min, 50 or 100; 25% w/v D-(+)-glucose autoclaved at 109°C for 15 min, 40.

The Oxoid base is added to the water, and the mixture is allowed to stand for 10 min to prevent powder caking. Gentle heat is applied to aid solution before autoclaving at 121°C for 15 min. The base is cooled to 56°C before adding the serum and glucose solution. Plates or slopes are then poured immediately.

Glycerol dextrose agar (GDA)

This medium is used for examining cultures for dissociation. It consists of (1) 40 g of blood agar base (Oxoid) in 1 litre of distilled water, autoclaved at 121°C for 20 min; (2) a solution of 25 g D-(+)-glucose and 50 ml glycerol in 50 ml of distilled water, sterilised by autoclaving at 105°C for 15 min. Solution (2) is added to 2500 ml of the melted blood agar base.

Albimi brucella broth

This medium is used in the propagation of brucella phages. The original formulation is no longer available, but equivalent preparations are produced by GIBCO and BBL. For use, 25 g are suspended in 1 litre of distilled water and sterilised by autoclaving at 121°C for 15 min.

Albimi brucella agar may also be used as a basal medium or as a serum-free medium for many strains.

Culture media, selective

Farrell's modified SDA

SDA medium can be made selective for brucella by the addition of antibiotics and antimicrobial agents. The formulation developed by Farrell (1974) is recommended for the isolation of brucella from material likely to be contaminated, e.g. milk, placenta, semen and vaginal discharge.

It is prepared from SDA basal medium by the addition of bacitracin, cycloheximide, nalidixic acid, nystatin and polymyxin B. Stock solutions of these are prepared as follows.

Bacitracin. (Wellcome Research, London). The sterile powder is dissolved in sterile distilled water to a concentration of 2000 units/ml and stored at 4°.

Cycloheximide. (Actidione; Upjohn and Co., Michigan, USA). One gram of actidione is dissolved in 5 ml of acetone and then diluted 1:20 with distilled water to give 10,000 μg/ml. This may be stored at 4°C.

Nalidixic acid. (Winthrop Laboratories, Edgefield Avenue, Newcastle). A 5% w/v stock solution is prepared in 0.5 mole/litre NaOH and stored at 4°C. Immediately before use, this is diluted 1/10 in distilled water.

Nystatin. (E. R. Squibb, New York, USA). Immediately before use the sterile powder is suspended in sufficient sterile distilled water to give a concentration of 50,000 units/ml.

Polymyxin B. (Wellcome Research, London). The sterile powder is dissolved in sterile distilled water to give a concentration of 5000 units/ml and stored frozen at −20°C. Unused solution should be discarded and not re-frozen.

Vancomycin. (Lilly and Co., Indianapolis, USA). A stock solution containing 50 mg/ml is made by dissolving the sterile powder in sterile distilled water immediately before use.

The volumes of each of these stock solutions to be added to a 1-litre volume of melted SDA base are given in Table 2. A ready-mixed antibiotic supplement based on Farrell's formulation is available in freeze-dried form from Oxoid Ltd. This is reconstituted in methanol according to the manufacturer's instructions and added to the basal medium.

Brodie and Sinton's liquid medium

This medium (Brodie & Sinton 1975) is useful for fluid enrichment of brucellas when attempting to isolate the organism from contaminated materials. Its composition (amount/litre, in distilled water) is as follows: Tryptone Soya Broth (Oxoid), 30 g; equine serum, inactivated at 56°C for 30 min, 50 ml; bacitracin, 20,000 units; polymyxin B sulphate, 5000 units; nalidixic acid, 5 mg; vancomycin, 20 mg; nystatin, 100,000 units; cycloheximide, 100 mg; amphotericin B, 4 mg; and D-cycloserine, 312.5 mg.

The medium is prepared by dissolving the tryptone soya broth powder in the distilled water and autoclaving at 121°C for 15 min. After cooling the broth, the sterile horse serum is added aseptically, followed by the antibiotics. Most of these are prepared as described for Farrell's modified SDA medium. The amphotericin B (Fungizone; E. R. Squibb and Co., New York, USA) is reconstituted in sterile distilled water to give a stock suspension containing 10 mg/ml. This is used immediately. The D-cycloserine (Sigma, London) is reconstituted in distilled water to a concentration of 50 mg/ml, sterilised by membrane filtration, and 6.25 ml is added to each litre volume of the medium.

Castañeda blood culture bottles

These are recommended (Castañeda 1947) for attempts to isolate *Brucella* spp. from blood or other liquid samples. They consist of bottles containing both a solid and a liquid medium. The purpose of this is to permit enrich-

TABLE 2

Antibiotic composition of Farrell's modified SDA medium

Antibiotic	Millilitres stock solution/litre basal medium	Approximate final concentration/ml
Bacitracin	12.5	25 units
Polymyxin B	1.0	5 units
Cycloheximide	10.0	100 μg
Vancomycin	0.4	20 μg
Nalidixic acid	1.0	5 μg
Nystatin	2.0	100 units

ment of small inocula by growth in the liquid phase, followed by transfer to the solid medium for formation of discrete colonies without the need for frequent sub-cultures. The method is not suitable for isolation of brucellas from grossly contaminated materials. Medical flat screw-capped bottles with a nominal capacity of 125 ml are suitable containers for the medium. For the culture of non-human blood samples, Farrell's modified SDA supplemented with 15 g/litre of agar (Oxoid no. 1) is recommended for the solid phase. About 20 ml of the molten medium is poured aseptically into each bottle, which is laid flat until the agar has solidified. About 15 ml of the liquid phase is then added to each bottle. This is prepared by dissolving 23 g Dextrose Broth (Oxoid) and 20 g of tri-sodium citrate in 1 litre of distilled water and autoclaving at 121°C for 15 min. On cooling, 50 ml of sterile horse serum, inactivated by heating at 56°C for 30 min, is added aseptically, followed by the antibiotics contained in Farrell's modified SDA medium to the concentrations given in Table 2. For human blood cultures, the antibiotics may be omitted. It is always advisable to pre-incubate the bottles before use, as a sterility check.

Modified vancmycin–colistimethate–nystatin (VCN) medium

This medium incorporates the antibiotic formulation used by Brown *et al.* (1971) for the isolation of *B. ovis* from ram semen. It is also suitable for *B. canis* and other *Brucella* species. It is less inhibitory than other selective media and is recommended for the isolation of fastidious or antibiotic-sensitive brucella strains from contaminated material.

The recommended basal medium is SDA (which should contain 10% horse serum for the isolation of *B. ovis*) or Bordet–Gengou agar. The VCN inhibitor is obtainable from BBL, Division of Becton-Dickinson, Cockeysville, Maryland, USA, in vials containing in each: vancomycin, 3000 μg; colistimethate, 7500 μg; and nystatin, 12,500 units.

The medium is prepared by adding the contents of one vial of VCN inhibitor, dissolved in 10 ml of sterile distilled water, and 1 ml of 1% furadantin in 0.1 mole/litre NaOH, to 1 litre of the basal medium, melted and cooled to 50°C. It is poured in petri dishes to form layers about 5 mm deep. If contamination by swarming organisms is a problem, the agar concentration in the basal medium may be increased to 3%.

Dye sensitivity test media

These are prepared by adding the relevant dye to a basal medium. The inhibitory activity of each dye to be used varies with its source and batch and also with the basal medium employed. Dyes equivalent to those formerly available from the National Aniline Division, Allied Chemical and Dye Company, New York, New York, USA are recommended for this purpose.

TABLE 3

Concentrations used in dye/erythritol sensitivity test media

Inhibitor	Stock solution, % (w/v)	Volume (ml) added to 1 litre	Final concentration
Basic fuchsin	0.20	10	1/50,000
Thionin	0.20	10	1/50,000
Safranin O	0.50	20	1/10,000
Thionin blue	0.01	20	1/500,000
Methyl violet	0.0125	10	1/800,000
Pyronin	0.01	20	1/500,000
Malachite green	0.01	20	1/500,000
meso-Erythritol	10.0	10	1 mg/ml
meso-Erythritol	10.0	20	2 mg/ml

Each new batch must be tested to determine the optimum dye concentration required to inhibit brucella reference strains growing on the basal medium used. SDA is recommended for this purpose as it will support the growth of all *Brucella* species and biovars. To prepare the media, the dyes are dissolved in distilled water to produce stock solutions of the concentration shown in Table 3. These are sterilised by steaming for 1 h. The volume of dye solution to produce the required final concentration is then added to the SDA basal medium. Plates or slopes are poured from this immediately. A similar procedure is used to prepare plates containing *meso*-erythritol.

The concentrations of these reagents which will usually give satisfactory results in inhibition tests are shown in Table 3.

Hydrogen sulphide test strips

Sheets of filter paper are soaked in a 10% w/v solution of lead acetate and hung up to dry at 37°C in air free of H_2S. When dry, the sheets are cut into strips of 8 × 100 mm, wrapped in grease-proof paper in bundles of 50, and vacuum autoclaved at 125°C for 20 min. They should be stored away from cultures or other sources of H_2S.

Oxidative metabolism tests

Brucellas are oxidative rather than fermentative. The specific pattern of activity produced by an isolate is usually characteristic of the species and in some cases may also identify the biovar. By determing the pattern of oxidation of a selected range of substrates, the identity of a strain can usually be established. The methods currently used for the determination of oxidative metabolic patterns include (1) manometric procedures using either the Warburg constant volume respirometer or the Gilson differential respirometer; (2) thin-layer chromatography of selected amino acid and carbohydrate substrates.

The manometric procedures are tedious and hazardous to perform. Details of the methods for these and the chromatographic technique are given by Corbel *et al.* (1979*a*). In general, these procedures are best suited to the specialist laboratory and are not recommended for routine identification. Most cultures can be identified successfully with the conventional cultural and serological tests and by phage typing. Oxidative metabolism tests are only essential for the identification of isolates which are not amenable to these procedures, for example non-smooth strains of *B. melitensis* and *B. suis*.

Preparation of Anti-sera

Anti-sera to smooth and rough brucella strains are prepared in healthy 6-month-old male or female rabbits of a large breed. Before use the animals must be shown by serum agglutination tests to be free of non-specific agglutinins and anti-brucella antibodies.

Smooth brucella anti-serum

(1) A suspension of smooth *B. abortus* 544 or *B. melitensis* 16M containing about 5×10^9 organisms/ml in 0.15 mole/litre NaCl is prepared.
(2) One-millilitre volumes of this are injected intravenously into rabbits.
(3) The rabbits are bled after 12–14 days and the serum is tested for brucella agglutinins. By this time most animals will have titres of at least 1:640 in the standard serum agglutination test. These animals are then exsanguinated and the sera pooled.
(4) The pooled serum is diluted 1:10 in 0.5% w/v phenol in 0.15 mole/litre NaCl and checked for its ability to agglutinate brucella strains in slide agglutination tests. The diluted serum should agglutinate suspensions of smooth cultures of *B. abortus*, *B. melitensis* and *B. suis* almost immediately. The phenolised serum may be stored at 4°C until required but must not be frozen.

Rough brucella anti-serum

(1) A suspension of *B. ovis* 63/290 containing about 10^{11} organisms/ml in 0.15 mole/litre NaCl plus 0.01% w/v thiomersal is prepared. This is

homogenised with an equal volume of Freund's incomplete adjuvant until a stable emulsion is formed.

(2) Rabbits are injected with four doses of 0.5 ml of this emulsion distributed over separate intramuscular sites.

(3) The animals are bled about 4 weeks later and the serum tested for agglutinins to *B. ovis*. A dilution of 1:20 in normal rabbit serum (using the pre-inoculation serum) should produce almost immediate agglutination of *B. ovis*, *B. canis* and *B. abortus* 45/20 cultures in slide agglutination tests. The pre-inoculation serum for dilution should give completely negative results in slide tests. The anti-serum may also agglutinate smooth brucella strains. This will not matter if it is only to be used for presumptive identification of brucella isolates.

(4) All of the animals giving adequate titers to *B. ovis* are exsanguinated and their sera pooled. This is diluted 1:10 in brucella-negative rabbit serum and thiomersal (0.01% w/v) is added as a preservative. The serum may be stored frozen in small volumes until required.

A and M monospecific anti-sera

(1) Suspensions of smooth *B. abortus* biovar 1 strain 544 and *B. melitensis* biovar 1 strain 16M containing about 10^9 viable organisms/ml in phosphate-buffered saline (PBS) are prepared.

(2) Single doses of 0.5 ml of these are injected intravenously into rabbits.

(3) The animals are bled daily from day 5 onwards and the sera titrated for brucella agglutinins. By day 9 or 10 most animals will have serum agglutination titres of at least 1:1280.

(4) Those animals with high agglutinin titres are exsanguinated and separate pools of anti-sera against *B. abortus* and *B. melitensis* are prepared.

(5) Each serum pool is titrated against standard *B. abortus* and *B. melitensis* suspensions. A titre of at least 1:1280 against the homologous antigen is desirable.

(6) The quantity of packed brucella cells of the heterologous species required to remove cross reacting agglutinins from each batch of serum is determined. This should be done by trial runs with 5-ml volumes. Usually about 1 ml of packed cells will be needed for each 10 ml of antiserum.

(7) Each batch of serum is absorbed by mixing it with the appropriate volume of packed cells and incubating at 37°C for 2 h. The anti-serum is recovered by centrifugation and again titrated against the homologous and heterologous brucella antigens. After satisfactory absorption it will agglutinate the homologous antigen at a dilution of at least 1:160 but will not agglutinate the heterologous antigen even at a 1:10 dilution. If nec-

essary the absorption process must be repeated until these conditions are met.

(8) The absorbed serum is diluted 1:5 in phenol–saline (0.5% w/v phenol in 0.15 mole/litre NaCl), and its reaction to *B. abortus* and *B. melitensis* is checked by slide agglutination tests. Satisfactory monospecific serum will rapidly agglutinate the homologous organism but have no visible effect on the heterologous species.

(9) The diluted serum should be stored at 4°C until required but allowed to warm to room temperature before use. It must not be frozen.

Anti-R specific serum

Anti-serum reacting only with rough strains of brucellas is prepared using the R antigen of *B. ovis* as antigen.

(1) *Brucella ovis* strain 63/290 is grown on SDA slopes incubated at 37°C in air plus 10% (v/v) CO_2 for 4 days.

(2) The slopes are checked for purity and the growth is resuspended in 0.15 mole/litre NaCl to give a concentration of about 10^{10} organisms/ml. Roux flasks are inoculated with 3-ml volumes of this.

(3) The Roux flasks are incubated at 37°C in air plus 10% (v/v) CO_2 for 4 days.

(4) The cultures are checked for purity and the growth is harvested from each in 10–15 ml of 0.15 mole/litre NaCl by shaking with a few glass beads.

(5) The suspension is strained through two layers of muslin and centrifuged at 3000**g** for 20 min. The supernatant is discarded and the deposited cells are washed twice by centrifugation in 0.15 mole/litre NaCl.

(6) The organisms are resuspended in 0.15 mole/litre NaCl and heated at 80°C for 2 h.

(7) The suspension is centrifuged at 3000 **g** for 30 min to remove the cells.

(8) The supernatant is centrifuged at 60,000 **g** for 6 h to sediment the antigen. The sediment is then discarded.

(9) The deposit is resuspended in 20 volumes of 0.15 mole/litre NaCl. This can be assisted by repeated freezing and thawing. The R antigen solution is stored at 4°C until required.

(10) The R antigen solution is homogenised with an equal volume of Freund's incomplete adjuvant.

(11) Six rabbits (more if a larger volume of serum is required) are each injected with 2 ml of antigen/adjuvant homogenate spread over four subcutaneous sites on the left flank. The injections are repeated 7 days later on the right flank.

(12) Four weeks later the rabbits are exsanguinated and the sera are collected individually.
(13) Each serum is titrated for agglutinins to rough brucella cells using a modification of the standard serum agglutination test. For this, serum dilutions are made in Menzel's buffer (0.5% w/v NaCl + 0.5% w/v phenol in a mixture of 1.5 volumes of 0.05 mole/litre Na_2CO_3 and 8.5 volumes of 0.1 mole/litre $NaHCO_3$; final pH 8.9). Equal volumes of antigen consisting of a suspension of *B. ovis* 63/290 suspended in the same buffer to an optical density (O.D.) of 3.30 at 450 nm are added to each tube. Incubation and reading of the test is as for the standard procedure using smooth antigen. (For further details, see Corbel *et al.* 1979*b*.) The sera are titrated against smooth *B. abortus* antigen in the standard test. Satisfactory sera should have a titre of at least 1:640 against the rough antigen and 1:20 or less against smooth antigen.
(14) Smooth agglutinins are removed by absorbing the sera with smooth *B. abortus* cells. This is done by adding 1 volume of packed *B. abortus* 544 or S99 cells to 20 volumes of serum and incubating at room temperature overnight. The serum is recovered by centrifugation.
(15) Each batch of serum is tested again for agglutinins for rough and smooth brucella strains in the tube and slide agglutination tests. Satisfactory samples will agglutinate rough organisms to high titre but produce no reaction with smooth brucella cells even at the lowest dilution. Batches of sera with similar titres are pooled and freeze-dried in 0.5 ml volumes, sealed under nitrogen in ampoules, and stored at 4°C. For use the serum is reconstituted in 0.5-ml volumes of distilled water and diluted 1:5 in brucella-negative rabbit serum. The diluent is used in parallel as a negative control.

Preparation of Phages

The brucella phages currently in routine use for the identification of cultures are the Tb, Wb, Fi, Bk_2 and R strains referred to in the section *The Genus* Brucella *and Its Species*. Seed stocks of these phages are kept at this laboratory.

Propagation

The host strain recommended for propagation of phages Tb and Fi is smooth *B. abortus* strain 19. For phage Wb smooth *B. suis* 1330 is recommended, and for phage Bk_2 smooth *B. melitensis* Isfahan. The propagating strain for phage

R is rough *B. abortus* 45/20 or B1119R, for phage R/O *B. ovis* 63/290, and for phage R/C *B. canis* Mex 51.

(1) The propagating strains are grown on SDA slopes incubated at 37°C in air for 24–36 h. The organisms are then suspended in Albimi brucella broth to a concentration of about 10^{10} cells/ml. The phage suspensions are diluted to slightly below RTD (see *Titration of phage stocks* for full definition), i.e. to between 10^4 and 10^5 p.f.u./ml.
(2) A 0.1-ml volume of the propagating strain suspension and an equal volume of phage stock are plated in the centre of a 9-cm petri dish containing SDA, spread uniformly over the agar surface, using a glass or stainless steel L-shaped spreader.
(3) The plates are inverted and incubated in air at 37°C for 24–36 h.
(4) Those plates showing confluent or near-confluent lysis of the host strain, with no bacterial contamination, are selected. Each plate is then flooded with about 3 ml of sterile Albimi brucella broth and left undisturbed on a level surface for 20 min at room temperature.
(5) The liquid from each plate is decanted into a sterile container to form the crude phage stock.
(6) The crude stock is clarified by centrifugation at 500 **g** for 10–15 min and filtered through a 0.45 μm pore diameter membrane filter. The filtrate is stored at 4°C. Toluene may be added as a preservative except for the R series of phages.
(7) Before use the titre of the phage stock and its host range are determined.

Titration of phage stocks

For culture typing purposes, brucella phages are normally used at the routine test dilution (RTD). This is defined as the lowest phage concentration which will produce confluent lysis on a lawn of the propagating strain. To determine this:

(1) Serial phage dilutions in the range 10^{-1} to 10^{-9} are prepared in sterile Albimi brucella broth, using a separate pipette for each stage.
(2) Then 0.1 ml of the stock suspension of the propagating strain is placed on an SDA plate and spread evenly over the surface using a glass or stainless steel L-shaped spreader. The plate is allowed to dry at room temperature for 1 h.
(3) Standard drops of the phage dilutions, e.g. from a 50-dropper pipette, are pipetted onto the agar surface. The drops should be discrete and well separated, hence too many should not be placed on one plate. The position and dilution of each drop should be noted.

(4) The plates are allowed to remain undisturbed at room temperature for 1 h, then inverted and incubated at 37°C in air for 48 h (or 72 h for Bk_2 and R phages).
(5) The plates are examined and the highest dilution at which confluent lysis is produced is noted. This is the RTD. The absolute phage titre (p.f.u./ml) can be determined by counting the number of plaques produced by a unit volume at limiting dilution. The counts should be limited to dilutions producing discrete, well-separated plaques.

Host range tests

Before use, the phage should be tested at RTD, RTD $\times 10^2$, and RTD $\times 10^4$ against authentic rough and smooth strains of *B. abortus*, *B. suis* and *B. melitensis* and preferably also, strains of *B. neotomae*, *B. canis* and *B. ovis*. Such a test will determine whether any host-range mutants are present in the phage stock. This is particularly important in the case of the R group phages, which have a high rate of mutation to variants having a host range similar to Wb phage. To minimise the effects of these mutants, the RTD of the rough specific phages should not contain more than 10 p.f.u./ml of the smooth variant. The procedure for doing the host-range determinations with the three dilutions of phage specified is as described in detail by Corbel & Thomas (1980).

Isolation

Choice and preparation of samples

Almost any tissue or secretion may be cultured in an attempt to isolate brucellas. Samples from farm animals most likely to yield positive cultures include abortion material (placental cotyledon, amniotic fluid, vaginal or uterine discharge, fetal gastric contents, fetal lung and fetal liver), bone marrow, lymph nodes, mammary glands, uterus, seminal vesicles and accessory glands, testes and epididymides or other organs with macroscopic lesions, hygroma fluid, milk, colostrum, semen and blood.

The best lymph nodes to examine are the mandibular, parotid, retropharyngeal, mesenteric, supramammary/external inguinal and internal iliac nodes of cattle, sheep and goats and the mandibular, parotid, supra-pharyngeal, gastric, hepatic, splenic and internal iliac nodes of swine.

For the diagnosis of brucellosis in man and also *B. canis* infection in dogs, blood culture is generally used, but other body fluids and tissue samples collected from almost any site may be examined. Blood samples are best inoculated directly into blood culture media. Otherwise they may be collected into 2% w/v sodium citrate solution. Preservative-free heparin may also be used as an anticoagulant, but others, particularly EDTA, are best avoided. Freezing and thawing of citrated blood samples may facilitate recovery of intracellular bacilli.

Samples of solid tissues should be homogenized with two or three volumes of Albimi brucella broth (or equivalent), either in a rotating blade homogenizer or, preferably, in a Colworth Stomacher. Large samples of tissue which are likely to be contaminated should be trimmed free of fat and the external surface disinfected by immersion in ethanol followed by flaming. The tissue is then sliced with a sterile blade and the freshly cut surface is applied directly to solid culture medium.

Direct culture

Samples of blood and tissue homogenate may be inoculated directly onto plates of SDA or other suitable medium. When small numbers of organisms are likely to be present or if antibiotics have been administered it is best to use a richer medium such as heated blood agar or Bordet–Gengou medium.

When plates of non-selective medium are used for isolation, the inoculum should be spread thickly over one-third of the plate surface and then streaked out over the remaining two-thirds. This facilitates growth of single colonies and their separation from contaminants.

When contaminating organisms are likely to be present a selective medium should be used. The SDA antibiotic medium of Farrell (1974) is suitable for most smooth brucella strains but may be too inhibitory for *B. canis* and *B. ovis*, for which modified VCN medium is preferable.

When selective media are used, approximately 0.1 ml of inoculum should be spread over the whole surface of the medium. For the culture of milk samples it is recommended that the cream layer be used, or if the sample is centrifuged, the deposit and cream layers should be combined.

For the culture of semen samples, particularly on the less selective modified VCN medium, it is recommended that the sample be diluted 1:5 or 1:10 in Albimi brucella broth or sterile 0.15 mole/litre NaCl and centrifuged at 500 *g* for 10 min. The supernatant is then forced through 0.8 μm pore size membrane filters in Swinney adaptors, and the filtrate is plated directly onto the selective medium.

Where possible, a minimum of six plates per sample should be inoculated.

Unless culture of CO_2-independent brucella strains is being attempted, all cultures should be incubated in air plus 10% v/v CO_2 at 37°C for up to 7 days. If brucellas are present, colonies will be visible within this time. Longer incubation is not likely to be profitable because of overgrowth by contaminants.

Enrichment culture

Grossly contaminated samples are probably best enriched in selective liquid medium before plating onto solid medium, or by inoculation of guinea pigs or embryonated eggs.

Enrichment of blood samples or tissue homogenates is conveniently performed by the Castañeda method. This is facilitated if the central part of the metal cap of the culture bottle is cut out with a cork borer. The sample is then injected directly into the bottle, through the rubber liner after swabbing this with 70% v/v iso-propanol. After thorough mixing, the bottles are incubated in a vertical position at 37°C in air plus 10% v/v CO_2. The cultures are inspected every 2 or 3 days and the liquid phase tipped over the agar if no colonies are visible. The bottles are then re-incubated in an upright position for 4–6 weeks if necessary. Any colonies which appear on the solid phase are sub-cultured for identification.

Milk or other contaminated material may be enriched for brucellas by inoculation into the liquid medium of Brodie & Sinton (1975). Usually, 0.2 ml of sample is added to each 5-ml volume of medium, which is then incubated with the cap loosened at 37°C for 5 days, in air plus 10% v/v CO_2. Volumes of 0.2 ml are then sub-cultured onto Farrell's SDA or other selective medium and then incubated under the same conditions.

Homogenates of tissue, milk, cream, cheese etc. may be injected by the intramuscular route into pairs of guinea pigs. Up to 1 ml of material may be injected into each animal, but the volume should be reduced for grossly contaminated or toxic samples (some cheeses and semen samples fall into this category). The animals may be protected against clostridial infection by prior injection with polyvalent clostridial anti-toxin. One of the animals is killed after 4 weeks and the other after 6 weeks. The spleen, liver, testes, epididymides and lymph nodes are examined carefully and sub-cultured onto SDA. A blood sample should also be tested for antibodies to brucella. This will sometimes disclose the presence of brucellas in the inoculum even when no organisms are recovered.

Yolk sac inoculation is sometimes practised for the isolation of *B. canis* from canine blood. Replicate volumes of 0.2 ml of citrated or heparinized blood are injected into the yolk sac of 6- to 8-day-old chick embryos. The eggs are re-incubated and inspected daily. Those with dead embryos are opened and the liver cultured on SDA or modified VCN medium. Embryos surviving on day 18 are killed by chilling and the yolk and liver are cultured.

Preservation of Cultures and Organisms

Brucella strains may be stored for short periods by streaking on SDA slopes, incubating for 72 h in air plus 10% v/v CO_2 (if required), then sealing the slopes and storing at 4°C. The procedure should be repeated every 6–8 weeks. This technique is not very satisfactory, however, as each sub-culture increases the chance of changes in the characteristics of the strain, and of contamination of the slope.

Vacuum drying is the method most commonly used at this laboratory. It requires relatively little apparatus other than a vacuum pump, compressor, and a manifold to which tubes may be attached for sealing. The organisms are grown on SDA slopes at 37°C in air plus 10% v/v CO_2 (if required) for 72 h. The growth is then washed off the slopes and suspended in sterile rabbit serum to give a concentration of about 10^{10} organisms/ml. (For preserving non-smooth strains, horse serum may be substituted for rabbit serum.) The suspension is distributed in 0.1-ml amounts into sterile 63 × 8-mm tubes which are sealed with cotton-wool plugs.

The tubes are placed in a desiccator containing calcium chloride or phosphorus pentoxide, and air is evacuated until small bubbles appear in the liquid. The desiccator is then sealed and stored at 4°C overnight. The following day, the air in the desiccator is evacuated for 10 min and the material replaced at 4°C. This procedure is repeated for at least 1 week, or until a pressure of 0.05 mm Hg can be obtained on evacuation. When drying is complete, the tubes are placed inside 150 × 12-mm test tubes, which have a cotton-wool plug in the bottom covering a layer of silica gel and $CoCl_2$ dryness indicator. A second plug is placed over the inner tube, and a narrow neck is drawn above this by means of a gas/air blowtorch. When cool, each tube is placed on the manifold and evacuated to a pressure of 0.05 mm Hg, and the neck is sealed. The ampoules are then tested for retention of vacuum, and stored at 4°C. Provided the vacuum is maintained, lyophilised cultures should retain their viability for many years.

Brucella cultures may also be preserved by freeze-drying. The suspending medium and post-drying procedure are as described for vacuum drying. Preservation by storage at liquid nitrogen temperature is also very effective. Further details are given by Corbel *et al.* (1979*a*).

Tests for Dissociation

Brucella cultures may undergo changes in colonial morphology which are accompanied by alterations in antigenic structure, phage-susceptibility, and virulence. This process is termed dissociation and is particularly likely to occur when smooth strains are grown in static liquid culture. The variants produced can range from extremely mucoid or rough forms to others which show only a mild degree of dissociation, intermediate or smooth-intermediate forms. Cultures in the non-smooth phase may also be recovered on primary isolation. This is invariably the case with *B. canis* and *B. ovis* isolates.

As the correct interpretation of serological and phage-typing tests is dependent on accurate knowledge of the colonial phase of the culture, this should always be determined. A variety of methods is available for detecting dissociation in brucella cultures, and details are given by Corbel *et al.* (1979*a*). For routine purposes the following three methods suffice.

Direct observation using reflected light

Plates of GDA are streaked with the culture to produce areas of dense growth as well as isolated colonies. After 4 days incubation, the growth is examined under a stereoscopic microscope with a clear glass stage. This is illuminated by a concave mirror placed to reflect light from a tungsten lamp up through the stage. The plate is moved around the stage so that individual colonies are examined under oblique illumination (Henry 1933).

Under these conditions smooth colonies of brucellas are convex, about 2 mm in diameter, with an entire edge and a blueish, glistening homogenous appearance. They are easily emulsified when touched with a wire. Rough colonies are yellow or brownish, opaque, granular and tend to break up when touched with a wire. Mucoid colonies are glistening, greyish in colour and have a slimy or sticky consistency. Intermediate colonies are the most difficult to classify. They resemble smooth colonies in colour and consistency but are rather more opaque and granular.

Agglutination by acriflavine

The test is done by placing a drop of neutral 0.1% w/v acriflavine solution on a clean slide. A colony is then touched with a straight wire and emulsified in the acriflavine. Smooth colonies emulisfy easily to produce a homogeneous suspension of cells undergoing Brownian movement. Rough colonies are difficult to emulsify and/or are agglutinated immediately, whereas mucoid colonies form threads. Intermediate colonies behave like smooth colonies or may produce a fine, granular agglutination.

Agglutination by anti-serum to rough brucellas

Anti-serum prepared in rabbits against the R antigen of *B. ovis*, as already described, is used; normal rabbit serum known to be free of brucella antibodies should be used as a control. The reconstituted anti-serum is diluted 1:5 in normal rabbit serum and a drop placed on a clean microscope slide. A colony is picked off with a straight wire and emulsified in the serum. Agglutination with the anti-serum but not with the control serum indicates that the organism is probably a rough strain. Agglutination with acriflavine but not with anti-R serum indicates that it probably is not brucella.

General Procedure for the Identification of Cultures

Preliminary examination

Before proceeding with the methods for identification at the species and biovar levels, it must first be established that an isolate belongs to the genus *Brucella*. In most instances this can be done by simply checking that its morphological, cultural and antigenic properties are consistent with *Brucella*. The following procedures are recommended.

(1) In heat-fixed smears of representative colonies, stained by Gram's method, brucellas will appear as Gram negative cocci, cocco-bacilli or short rods. If the organisms are vibrios, spirillae, large rods, occur in long chains or are Gram positive, brucellas can be excluded. If required, further examinations of the culture for bipolar staining, spores, capsules, flagella and acid-fastness may be made. These should give negative results with brucellas. Experience is required in performing and interpreting these staining reactions.

(2) A loopful of a suspension of culture in distilled water with a loopful of 0.1% w/v acriflavine on a clean glass slide is examined for agglutination. If no auto-agglutination occurs, a loopful of culture suspension is added with a loopful of unabsorbed anti-serum to smooth brucellas on another slide and agitated gently for 30 s. If no agglutination occurs, the culture is unlikely to be a brucella. If auto-agglutination occurs in the presence of acriflavine, a loopful of culture suspension is mixed with a loopful of rough brucella anti-serum and examined after shaking gently for 30 s. The absence of agglutination should exclude brucellas. A negative control should be set up using brucella-negative serum.

(3) If the isolate gives a positive reaction with the appropriate brucella anti-serum, it is sub-cultured into two tubes of Albimi brucella broth using a heavy inoculum. One of these is incubated at 37°C in air plus 10% v/v CO_2 and the other at 22°C in a similar atmosphere. (A candle jar or anaerobic jar filled with the appropriate gas mixture may be used if a CO_2 incubator is not available.) After 24–48 h incubation both cultures are examined for motile organisms by the hanging drop method. Motile organisms are not brucellas.

(4) Cultures are also made on SDA, sheep blood agar and MacConkey agar plates. These are incubated at 37°C in air plus 10% v/v CO_2. Replicate SDA plates should also be incubated at 37°C in an anaerobic jar in an atmosphere of CO_2 plus H_2 or H_2 alone, and aerobically at room temperature (18–22°C). The plates are examined daily.

Brucella strains will not grow under strictly anaerobic conditions, nor will they produce visible colonies in 24–48 h when incubated aerobically at room temperature. After 48–72 h incubation at 37°C in air plus 10% v/v CO_2, smooth brucella strains will grow on SDA to produce circular, convex colonies, 1–3 mm in diameter, with a smooth glistening surface. The colonies are a transparent honey colour in transmitted light and have a blueish-white translucent appearance in reflected light. Rough brucella strains produce colonies of a similar size and shape but of a more opaque off-white colour and often with a rather granular surface. On blood agar growth is slower than on SDA, with the production of non-haemolytic greyish-white glistening colonies 0.5–1 mm in diameter after 48–72 h incubation. On MacConkey agar, little or no growth may occur or small lactose-negative colonies may develop after 5 days incubation. Rapidly growing, haemolytic or lactose-fermenting colonies on the appropriate medium are not brucellas.

(5) Cultures grown on blood-free medium are tested for catalase production by transferring about 0.5 ml of a heavy suspension of the bacteria into a test tube plugged with cotton wool and adding about 1 ml of 3% w/v hydrogen peroxide with a Pasteur pipette inserted through the plug. Vigorous gas production indicates a positive reaction. A negative reaction will exclude brucellas.

(6) The cytochrome *C* oxidase test may also be done at this stage by smearing a loopful of culture grown on blood-free medium accross a piece of filter paper impregnated with 1% w/v tetramethyl *p*-phenylenediamine solution. Most brucella strains will give a positive reaction, but a few, including *B. ovis* and *B. neotomae*, give negative reactions.

(7) The following media are inoculated and the appropriate tests carried out: (a) glucose–peptone medium (Abd-el-Malek & Gibson 1948) for the methyl red and V–P test; (b) Koser's citrate medium for growth; (c) nutrient broth for indole production; (d) *o*-nitrophenyl β-D-galactoside broth (Lowe 1962) for the ONPG test; (e) nitrate broth (0.1% w/v KNO_3 in nutrient broth) for nitratase production; (f) glucose–peptone water with an indicator and a Durham tube for gas production; and (g) gelatin stabs for liquefaction.

Cultures (a) and (b) are incubated for 48 h, (f) for up to 14 days, and the others for 5 days at 37°C. After incubation each is examined for the methyl red and Voges–Proskauer reactions. Positive results in these will exclude brucella.

Growth in Koser's citrate is atypical of brucellas, although occasional strains give positive results.

A positive indole reaction in nutrient broth will exclude brucellas.

The majority of brucella strains are negative for β-galactosidase. The development of a slight yellow colour does not, however, exclude brucellas.

Nearly all brucella cultures, except those of *B. ovis* and a few of *B. canis*, will give positive nitrate reduction tests. A negative reaction should be checked by addition of zinc powder, as some strains will also reduce nitrites. The production of acid or gas in glucose–peptone water and liquifaction of gelatin excludes brucellas.

These tests will permit an organism to be identified or dismissed as a brucella strain in the majority of cases. If inconclusive results are obtained, then supplementary tests involving examination of DNA composition, protein migration patterns, antigenic specificity or fatty acid composition will need to be applied. Details of these are given by Corbel *et al.* (1979*a*).

Brucella typing procedure

Once an isolate has been identified as a brucella it may then be typed to identify its species and biovar. In some cases it may also be necessary to determine if the isolate has the characteristics of one of the live vaccine strains of *B. abortus* or *B. melitensis*.

It is essential to include in the tests at least the brucella reference strains *B. abortus* 544, *B. melitensis* 16M and *B. suis* 1330 as a check on media and methods. If the strains under test are non-smooth and/or thought to belong to one of the other *Brucella* species, then at least one strain typical of these should be included in the tests.

(1) The culture is grown on GDA to give isolated colonies and confluent growth on the same plate, and incubated for 47–72 h at 37°C in air plus 10% v/v CO_2. The plate is then examined in obliquely transmitted light as described in the section *Tests for Dissociation*.

(2) If all colonies are smooth, or there is a mixture of smooth and non-smooth colonial forms, a smooth colony is picked and sub-cultured on three SDA slopes. If only non-smooth colonies are present then a typical colony is picked and sub-cultured on three SDA slopes. In either case, one of the slopes is incubated at 37°C in air and the other two in air plus 10% v/v CO_2. A strip of lead acetate paper is placed in the mouth of one of the latter two tubes but without allowing it to come into contact with the agar.

(3) The lead acetate papers are examined daily for the next few days. The presence or absence of any blackening is recorded and the strips are renewed daily.
(4) After 72 h incubation the slope incubated in air is examined to record any growth as not requiring CO_2. Absence of growth on this slope but growth on the slopes incubated in the CO_2-supplemented atmosphere indicates a positive CO_2 requirement.
(5) After 48 h incubation, the growth from one of the slopes incubated in air plus 10% v/v CO_2 is suspended in about 0.5 ml of sterile saline to give an opacity equivalent to approximately 10^{10} organisms/ml. This suspension is used for the procedures described below.
(6) A Christensen's urea agar slope is inoculated with one loopful, incubated at 37°C, and examined after 15 min, and 1, 2, and 18 h for urease production.
(7) Dye plates that have been incubated overnight as a sterility check are quartered to permit four samples to be tested on one plate. One loopful of the suspension is used to inoculate each quarter with five consecutive streaks without recharging the loop. The loop is sterilised between samples. After drying, the plates are incubated for 48 h in air plus 10% v/v CO_2. Normally incubation for 48 h will be sufficient, but for slowly growing strains it may be necessary to incubate for up to 4 days.
(8) Drops of A, M and R antisera and 0.1% w/v acriflavine are placed on a clean microscope slide. To each drop a similar drop of the bacterial suspension prepared is added. The slide is rocked gently for about 0.5–1 min and examined for agglutination with each reagent.
(9) An SDA plate is inoculated with loopfuls of suspension to produce an area of confluent growth. Discrete drops of phage suspension are placed carefully on the inoculated area to form an isometric pattern. With this arrangement each strain can be examined for sensitivity to the phages Tb, Wb, Fi, and Bk_2 on one plate. The plate should be kept undisturbed on a flat surface for 1 h to allow the drops of phage suspension to absorb into the agar. It is incubated at 37°C in air plus 10% v/v CO_2 for 48 h and then examined for lysis. It may be re-incubated for a further 24 h before final examination is made.
(10) The suspension may also used at this stage to inoculate Roux flasks of SDA medium to prepare suspensions for oxidative metabolism tests if required.

The pattern of results obtained in these tests may be compared with those shown in Table 1. If it does not conform with any of those in the table, the procedure described in the section *Preliminary examination*, together with the supplementary methods recommended by Corbel & Morgan (1975), should be followed carefully to check that the organism is a member of the genus *Brucella*.

Notes on the interpretation of results

Production of hydrogen sulphide

The papers should be changed each day whether blackening has occurred or not, to exclude the possibility of false positive or negative results. In general, blackening of the paper on day 1 only is not regarded as significant at this laboratory. It should be noted that strains may lose the ability to produce hydrogen sulphide on prolonged storage without sub-culture.

Agglutination reactions

If the strain under test is found to be non-smooth (i.e. agglutinates in acriflavine and/or in anti-R serum), then any positive results obtained with A and M anti-sera should be ignored, as they may be due to auto-agglutination. The identity of such strains should be checked by other means, as agglutination in acriflavine is not specific for brucellas, and on rare occasions agglutination of organisms other than brucellas has been observed in anti-R serum.

Dye plates

Care should be taken that the suspension used for the inoculation of the dye plates contains not more than about 10^{10} organisms/ml, or a false positive results may be obtained. In general, the appearance of growth on the first two streaks only is not regarded as significant at this laboratory. If growth appears on three or more streaks, the organism is regarded as resistant to inhibition by the dye.

Urease test

In general, *B. canis*, *B. neotomae* and *B. suis* will give a strongly positive urease reaction, turning Christensen's medium a magenta colour in less than 15 min. Most *B. abortus* cultures will give a positive reaction in 1–2 h, but some, including *B. abortus* 544 biovar 1 reference strain, give a negative urease reaction. Most *B. melitensis* strains will hydrolyse urea in 1–2 h, but some may give negative reactions. *Brucella ovis* strains do not hydrolyse urea at all under these conditions.

Identification of vaccine strains

It will sometimes be necessary to examine isolates of brucella to determine whether they result from vaccination with the attenuated strains *B. abortus* strain 19 and *B. melitensis* Rev. 1. Neither of these strains is differentiated from typical biovar 1 strains of its respective species by the routine typing tests. Details of methods which can be used for the identification of the vaccine strains are given by Corbel *et al.* (1979*a*).

Phage Typing

The brucella phages used routinely for typing cultures differ in their ability to replicate in, and form plaques on, the various *Brucella* species. Thus Tb phage will replicate efficiently only in smooth or smooth-intermediate cultures of *B. abortus*. It will undergo limited replication, with formation of very small plaques, in smooth *B. neotomae* cells but not in those of other *Brucella* spp. At high concentrations, equivalent to 10^4 RTD, Tb phage will produce lysis-from-without on smooth cultures of *B. abortus*, *B. neotomae* and *B. suis* but not *B. melitensis*. It does not affect non-smooth cultures of any species.

Phage Wb replicates with high efficiency in, and forms plaques on, smooth cultures of *B. abortus*, *B. neotomae* and *B. suis*. It does not lyse non-smooth cultures of these and is not active on *B. melitensis*, *B. canis* or *B. ovis* isolates.

Phage Fi replicates with high efficiency in, and forms plaques on, smooth *B. abortus* and *B. neotomae* cultures. It replicates with much lower efficiency in *B. suis* cultures, forming relatively few plaques. It does not produce lysis of *B. melitensis*, *B. ovis* or *B. canis*, nor of non-smooth brucella strains generally.

Phage Bk_2 replicates in, and forms plaques on, smooth cultures of *B. abortus*, *B. neotomae*, *B. suis* and *B. melitensis*. Its efficiency of plating on *B. melitensis* is rather lower than on the other *Brucella* species, and the plaques formed tend to be fewer, smaller and slower to develop. Phage Bk_2 is very sensitive to dissociation in the host culture and does not produce plaques on non-smooth brucella strains generally, including *B. canis* and *B. ovis*.

Phage R replicates in, and forms plaques on, non-smooth cultures of *B. abortus*. It does not lyse smooth cultures of any brucella strain nor non-smooth cultures of the other species. During replication in the rough culture, a proportion of phage variants are produced which have a similar host range to phage Wb. The effect of this is minimal if phage R is used at the RTD for rough *B. abortus* cultures.

Phage R/O replicates in, and forms plaques on, smooth and non-smooth cultures of *B. abortus* and on *B. ovis*. Its relative plating efficiency on various strains varies between batches and each should be checked to ensure effective lysis of *B. ovis*.

Phage R/C replicates in, and forms plaques on, non-smooth *B. abortus*, *B. ovis* and *B. canis*. Lysis of the latter is most effective on young cultures. Gross lysis of smooth brucella cultures is not produced and plaque formation on these is rare and difficult to detect.

Procedure

Phage typing is most conveniently performed as an integral part of the routine identification procedure. For all of the phages, preparations standardised at the RTD on the respective propagating strain are used routinely. For cultures which present problems in identification, particularly as a result of partial dissociation, phage preparations standardized at 10^2 and 10^4 RTD may give useful results if the RTD fails to produce lysis. If only smooth cultures are to be examined, the Tb phage may be used alone at RTD and 10^4 RTD. If non-smooth cultures are to be typed, then phage strains R, R/O and R/C should be used in parallel.

The culture is inoculated on SDA medium to give an area of confluent growth by streaking a suspension containing about 10^{10} organisms/ml across the surface of the plate. The agar surface is then allowed to dry for a few minutes and the phage suspensions placed on the surface to form discrete well separated drops. Their position and identity should be marked on the base of the plate. The cultures are then incubated in the normal way. They should be examined for lysis after 24 h and again after 48 and 72 h. Some of the phages, particularly Bk_2 and the R series, may take several days to produce fully developed plaques on certain cultures.

References

Abd-el-Malek, Y. & Gibson, T. 1948 Studies in the bacteriology of milk II. The staphylococci and micrococci of milk. *Journal of Dairy Research* **15**, 249–260.

Brodie, J. & Sinton, G. P. 1975 Fluid and solid media for isolation of *Brucella abortus*. *Journal of Hygiene* **74**, 359–367.

Brown, G. M., Ranger, C. R. & Kelley, D. J. 1971 Selective media for the isolation of *Brucella ovis*. *Cornell Veterinarian* **61**, 265–280.

Buddle, M. B. 1956 Studies on *Brucella ovis* (n.sp.), a cause of genital disease of sheep in New Zealand and Australia. *Journal of Hygiene* **54**, 351–364.

Carmichael, L. E. & Bruner, D. W. 1968 Characteristics of a newly-recognized species of *Brucella* responsible for infectious canine abortions. *Cornell Veterinarian* **48**, 579–592.

CASTAÑEDA, M. R. 1947 A practical method for routine blood cultures in brucellosis. *Proceedings of the Society of Experimental Biology and Medicine* **64**, 114–115.

COLLINS, C. H. 1983 *Laboratory-acquired Infections*. London: Butterworths.

CORBEL, M. J. & MORGAN, W. J. B. 1975 Proposal for minimal standards for descriptions of new species and biotypes of the genus *Brucella*. *International Journal of Systemic Bacteriology* **25**, 83–89.

CORBEL, M. J. & THOMAS, E. L. 1980 *The Brucella-phages: Their Properties, Characterisation and Applications*. London: Ministry of Agriculture, Fisheries and Food.

CORBEL, M. J., BRACEWELL, C. D., THOMAS, E. L. & GILL, K. P. W. 1979*a* Techniques in the identification and classification of *Brucella*. In *Identification Methods for Microbiologists* 2nd edn., ed. Skinner, F. A. & Lovelock, D. W., pp. 71–122. Society for Applied Bacteriology Technical Series No. 14. London & New York: Academic Press.

CORBEL, M. J., GILL, K. P. W. & REDWOOD, D. W. 1979*b* *Diagnostic Procedures for Non-smooth* Brucella *Strains*. London: Ministry of Agriculture, Fisheries and Food.

FARRELL, I. D. 1974 The development of a new selective medium for the isolation of *Brucella abortus* from contaminated sources. *Research in Veterinary Science* **16**, 280–286.

HENRY, B. S. 1933 Dissociation in the genus *Brucella*. *Journal of Infectious Diseases* **52**, 374–402.

HUDDLESON, I. F. 1929 The differentiation of the species of the genus *Brucella*. *Michigan, Agricultural Experiment Station, Technical Bulletin* No. 100.

HUGHES, L. 1893 Sur une forme de fièvre frequente sur les Côtes de la Méditerranée. *Annales de l'Institut Pasteur, Paris* **7**, 628–639.

LOWE, G. H. 1962 The rapid detection of lactose fermentation in para-colon organisms by the demonstration of β-D-galactosidase. *Journal of Medical Laboratory Technology* **19**, 21–25.

MEYER, K. F. & SHAW, E. B. 1920 A comparison of the morphologic, cultural and biochemical characteristics of *Br. abortus*, and *Br. melitensis*. *Journal of Infectious Diseases* **27**, 173–184.

SCHMIDT, J. & WEIS, F. 1901 *Bakterierne Naturhistorik Grundlag for det bakteriologiske Studium*. Copenhagen: Morten Potsild.

STOENNER, H. G. & LACKMAN, D. B. 1957 A new species of *Brucella* isolated from the desert wood rat, *Neotoma lepida* Thomas. *American Journal of Veterinary Research* **18**, 942–951.

Yersinia enterocolitica and Related Species

R. A. BREWER AND M. J. CORBEL

Ministry of Agriculture, Fisheries and Food, Central Veterinary Laboratory, Weybridge, Surrey, UK

The Genus *Yersinia*

This genus contains genetically related bacteria, all species of which are agents of disease in man and other animals. They are Gram negative coccobacilli or short rods, usually with rounded ends and straight or convex sides, ranging in size from 0.5–1 μm by 1–2 μm. They do not form endospores, but capsules may be produced. Motility depends on incubation temperature, being absent at 37°C but occurring at 22°C except in *Yersinia pestis*. Flagella are peritrichous. Bipolar staining and pleomorphism are common, particularly when growing *in vivo*. They will grow on simple nutrient media and will tolerate bile salts, but the growth of pathogenic strains is improved by blood or tissue extracts. Growth may occur at temperatures down to 0°C. They are facultative anaerobes and are catalase positive and oxidase negative. Metabolism may be oxidative or fermentative. Acid is produced by fermentation of various carbohydrate substrates. Amino acids may be decarboxylated. Proteolytic activity is variable, as is the production of urease. Nitrates are reduced. The yersinias possess a large number of internal antigens in common, some of which are shared with other members of the Enterobacteriaceae. The lipopolysaccharide somatic antigens vary between serotypes. Pathogenic strains synthesise characteristic (V and W) antigens and show a capacity for intracellular growth. The DNA base composition is within the range 46–49% GC.

At present the genus contains three accepted species, *Y. pseudotuberculosis* [(Pfeiffer) van Loghem 1944], *Y. pestis* [(Lehmann and Neumann) van Loghem 1944] and *Y. enterocolitica* [(Schliefstein and Coleman) Frederiksen 1964].

Isolation and Identification of
Micro-organisms of Medical and
Veterinary Importance

Copyright © 1985 by the Society for Applied Bacteriology
All rights of reproduction in any form reserved
0-12-181460-2

In addition to these, the species *Y. ruckeri* has been proposed for the enteric redmouth bacterium (ERMB), a pathogen of salmonid fish (Ross *et al.* 1966; Ewing *et al.* 1978) and the species *Y. philomiragia* for an organism isolated from muskrats (Ritter & Gerloff 1966; Jensen *et al.* 1969). The status of *Y. ruckeri* is still undecided, but current evidence suggests that *Y. philomiragia* should not be included in the genus *Yersinia* (Ohara *et al.* 1974; Ursing *et al.* 1980; Corbel *et al.* 1982). Recently, the biochemical heterogeneity of the organisms classified as *Y. enterocolitica* has prompted proposals for the sub-division of this into the species *Y. enterocolitica sensu strictu*, *Y. frederiksenii*, *Y. intermedia* and *Y. kristensenii* (Brenner 1979; Bercovier *et al.* 1980*a*,*b*). It has also been suggested that *Y. pseudotuberculosis* and *Y. pestis* should be re-classified as *Y. pseudotuberculosis* ssp. *pseudotuberculosis* and *Y. pseudotuberculosis* ssp. *pestis* (Bercovier *et al.* 1980*c*).

Clinical Importance and Distribution

The three currently recognized *Yersinia* species are zoonotic agents, infecting wild or domestic animals with occasional transmission to man. *Yersinia pestis* is the causative agent of plague and is maintained in wild rodent populations in many parts of Asia, Africa, and North and South America. Man is an accidental host, infection usually being transmitted from the natural host via arthropod vectors, frequently the rat flea *Xenopsylla cheopsis*. Under suitable conditions, man to man transmission can occur by the dispersion of infected aerosols from pneumonic cases of the disease. Aerosol infection may also occur in the laboratory, from cultures or infected tissues.

Yersinia pseudotuberculosis has a worldwide distribution and has been isolated from a wide range of wild and domestic animals as well as from environmental sources, particularly soil and water. Infection is common in wild rodent and bird populations, which probably act as the main reservoirs of infection.

Transmission to man and domestic animals tends to be sporadic, but large outbreaks may occur, particularly in animals kept under intensive conditions. The organism is of some significance as a minor cause of abortion in sheep and has been implicated in severe outbreaks of haemorrhagic enteritis in farmed deer in New Zealand. Animals in zoological parks are also at risk, and the infection can be a major cause of loss in these establishments. Infection probably occurs by the oral route in most cases. Insect transmission is a possibility, although probably of little practical importance.

In man, the infection is usually associated with mesenteric adenitis or ileitis which often simulates acute appendicitis. Septicaemic and typhoidal forms of the infection have been described but are rare. A scarlatinal form has also been reported from the USSR.

Yersinia enterocolitica and the related species *Y. intermedia*, *Y. frederiksenii* and *Y. kristensenii* also have a wide distribution, having been isolated in practically every country in which they have been sought. *Yersinia enterocolitica* has been implicated in a wide variety of disease presentations. Most frequently, however, it is associated with enteric disease, particularly gastroenteritis, terminal ileitis, pseudo-appendicitis and mesenteric adenitis. Septicaemic infection also occasionally occurs, and a severe reactive arthritis may follow infection at other sites (Bottone 1977). The related species *Y. frederiksenii* and *Y. intermedia* are rarely implicated in intestinal disease but have been associated with wound and skin infection. *Yersinia kristensenii* has been infrequently associated with human disease and is probably environmental in origin.

Yersinia ruckeri is the cause of enteric redmouth disease of salmonid fish. It may have a wider host range than this implies, as organisms reported as "*Y. enterocolitica*-like" have been described in other fish diseases. *Yersinia philomiragia* has only been implicated as a cause of death in muskrats in one locality. Its general significance is unknown.

Isolation

Enrichment media

Phosphate-buffered saline (PBS; pH 7.6) is used for cold enrichment of yersinias. It can also be used as a transport medium. It contains (g/litre in distilled water): NaCl, 0.085; $NaH_2PO_4 \cdot 2H_2O$, 2.03; $Na_2HPO_4 \cdot 12H_2O$, 31.16. The solution is sterilized by autoclaving at 121°C for 15 min. The pH should be checked before use. The PBS is conveniently dispensed in 10-ml volumes in 25-ml screw-capped (Universal) bottles. These can be used for transport of specimens, either on swabs or simply added to the medium. Enrichment is carried out by incubating the sample in PBS at 4°C for the appropriate time. This medium can be supplemented with sorbitol, 10 g/litre, and bile salts (Oxoid no. 3), 1.5 g/litre, to enhance recovery from pork and reduce the incubation period (Mehlman *et al.* 1978). The further addition of haemoglobin, 1.0 g/litre and potassium oxalate, 2 g/litre, may facilitate isolation from vegetable products (Mehlman & Aulisio 1978).

Enhanced recovery from human faecal samples has been reported with PBS (pH 7.3) containing 0.1% w/v mannitol (Pai *et al.* 1979; Weissfeld & Sonnenwirth 1980).

Other modifications of the PBS enrichment medium include supplementation of PBS (pH 7.5) with malachite green (20 μg/ml), carbenicillin (10 μg/ml) and sodium selenite (2 mg/ml). This is incubated at 22°C for 2–3 days and is effective for isolation from beef and pork samples (Lee *et al.* 1980).

Alkali treatment of cultures enriched by incubation in PBS can improve the isolation rate (Aulisio *et al.* 1980). This is done by treating the PBS suspension, after incubation at 26°C for 1–2 days, with 0.5% w/v KOH in 0.5% w/v NaCl. This is suitable for lightly contaminated food samples, but for heavily contaminated material, cold enrichment for 2–3 weeks at 4°C is recommended, before the alkali treatment.

A recently described enrichment medium consists of trypticase soy broth supplemented with novobiocin (30 μg/ml), polymyxin B sulphate (20 units/ml) and crystal violet (2 μg/ml). Incubation in this medium at 18°C for 4 days is followed by alkali treatment (Park *et al.*, 1981). It is unsuitable for isolation of *Y. enterocolitica* serogroup 0:12,25 and possibly some other strains (Swaminathan *et al.* 1982).

Enrichment of food samples in PBS at 4°C for 14 days, followed by selective enrichment in bile–oxalate–sorbose broth (BOS) at 22°C for 5 days, is more likely to be successful than alkali treatment where the number of *Yersinia* organisms present is small (Schiemann 1982, 1983). BOS medium is prepared by dissolving 17.25 g $Na_2HPO_4 \cdot 7H_2O$, 5.0 g sodium oxalate, and 2.0 g bile salts (Difco) in 619 ml distilled water together with 10 ml 10% w/v NaCl and 10 ml 0.1% w/v $MgSO_4 \cdot 7H_2O$. The pH is adjusted to 7.6 with 5 *N* HCl and the solution autoclaved at 121°C for 15 min. The following filter-sterilised solutions are then added: 100 ml 10% w/v sorbose, 100 ml 1.0% w/v asparagine, 100 ml 1% w/v methionine, 10 ml 0.25% w/v metanil yellow, 10 ml 0.25% yeast extract, 10 ml 0.5% w/v pyruvic acid (sodium salt), 1 ml 0.4% Irgasan DP 300 (2,4,4′–trichloro–2′–hydroxy diphenyl ether; (iba–Geigy) in 95% ethanol. The pH is adjusted to 7.6 with 5 *N* NaOH or HCl and 10 ml 0.1% w/v sodium furadantin is added on the day of use.

Unmodified trypticase soy or brain–heart infusion broth are suitable for enrichment of *Y. pseudotuberculosis* in uncontaminated material.

Solid media

Selective plating media

Blood agar, trypticase soy agar (TSA) or other nutritionally adequate medium is suitable for plating of uncontaminated samples containing *Y. pseudotuberculosis*. For contaminated samples, MacConkey agar, Endo agar, or eosin methylene blue (EMB) agar may give positive isolations. *Yersinia pseudotuberculosis* will also grow on many of the selective media recommended for *Y. enterocolitica*, although the recovery rates from small inocula tend to be poor.

A large number of selective plating media have been described for *Y. enterocolitica*. The apparent ease of use of these fails to take into account the biochemical heterogeneity of the organisms covered by this species. Readily available enteric media include *Salmonella–Shigella* agar (SSA), which tends to be inhibitory for some strains and gives poor recovery with small inocula, and MacConkey agar.

Bismuth sulphite agar (Hanna *et al.* 1977) has been recommended for isolation of *Y. enterocolitica* from canine faeces, but some strains are inhibited.

The two media recommended at present are "Y" medium (Soltesz *et al.* 1980) and Cefsulodin–Irgasan–Novobiocin agar (CIN agar; Schiemann 1979).

"Y" medium

This is modified from Soltesz *et al.* (1980) and contains (g/litre): peptone, 12; casein hydrolysate (Oxoid L41), 8; yeast extract (Oxoid L21), 2; lactose, 10; Tween 80, 10; NaCl, 5; $CaCl_2$, 0.02; sodium oxalate, 16; bile salts (Oxoid), 5; neutral red, 0.008; Bacto agar (Difco), 20.

All the ingredients except the $CaCl_2$ are suspended in water and boiled to dissolve, then autoclaved at 121°C for 15 min. The $CaCl_2$ solution (2 ml of a 1% w/v solution) is then added and the plates are poured. They may be stored at room temperature for up to 10 days. Novobiocin, to 50 mg/litre, may be added.

"Y" medium has been modified and extensively tested in this laboratory. It will support the growth of all the known serogroups of *Y. enterocolitica*, although the growth of one strain of serogroup 0:2a,2b,3 was restricted in one series of tests (Brewer & Corbel 1983).

CIN medium

This is available in dehydrated form from Oxoid and Difco. It contains (g/litre): peptone, 20; yeast extract, 2; mannitol, 20; sodium pyruvate, 2; NaCl, 1; $MgSO_4{\cdot}7H_2O$, 0.01; sodium deoxycholate, 0.5; agar, 12.

The ingredients are suspended in 949 ml of distilled water and boiled to dissolve. The solution is cooled to about 80°C and 1 ml of 0.4% w/v Irgasan in 95% ethanol is added. After mixing and cooling to 50–55°C, the following solutions are added: 5 *N* NaOH, 1 ml; neutral red (3 mg/ml), 10 ml; crystal violet (0.1 mg/ml), 10 ml; cefsulodin (1.5 mg/ml), 10 ml; novobiocin (0.25 mg/ml), 10 ml. Ten millilitres of a filter-sterilized solution of 10% w/v strontium chloride solution is then added slowly with stirring, the pH is adjusted to 7.4, and plates are poured.

CIN medium is reported to give higher recovery rates of *Y. enterocolitica* from human faeces than "Y" medium (Head *et al.* 1982). In our tests, however, the media was inhibitory for one strain of serogroup 0:6,30 although others of this serogroup grow well on it. In trials with bovine and porcine faeces samples, CIN and Y medium gave comparable results.

For the isolation of *Y. ruckeri*, trypticase soy agar (TSA), alone or supplemented with yeast extract (5 g/litre), has been used. Little information is available about selective media for this organism.

Yersinia philomiragia has been isolated on only one occasion in one area. Brain–heart infusion agar and cystine–heart infusion agar supplemented with 1% w/v haemoglobin were used. We have obtained luxuriant growth of the organism on a medium composed of brain–heart infusion agar supplemented with 20% v/v heated sheep blood, 1% w/v Panmede (Paines & Byrne Ltd. Greenford, UK), 0.4% w/v glucose and 0.1% w/v cystine. This is incubated for 4 days at 37°C in air supplemented with 10% w/v CO_2.

Isolation procedures

The methods described relate mainly to the isolation and subsequent identification of *Y. enterocolitica* and *Y. pseudotuberculosis*. Where appropriate, reference to the other *Yersinia* species will be made. Methods for the isolation of *Y. pestis* tend to be specialized, requiring strict safety precautions and the use of special techniques, including the examination of ectoparasites and the use of animal inoculation. Work with this organism is hazardous, and suspected isolates should be submitted to a reference laboratory for identification. Details of methods for the isolation and identification of *Y. pestis* are given by Quan and others (1981).

Selection of samples

For the isolation of *Y. pseudotuberculosis* from human cases, blood, faeces or solid tissues may be examined. The latter most frequently include mesenteric lymph nodes and appendix, but any tissue removed post-mortem or for biopsy may be examined.

For animals, a similar range of materials may be examined, but those most frequently submitted include blood, faeces and samples of internal organs removed post-mortem, particularly liver, spleen, lymph nodes and bone marrow.

For *Y. enterocolitica*, most attempts at isolation from human cases are made on faeces, but blood, sputum, cerebrospinal fluid, urine and solid tissues may also give positive cultures.

In pigs, the samples most likely to give positive isolations include tongue, tonsil, mesenteric lymph nodes and caecal contents.

In cattle and sheep, isolations have been made from lymph nodes, abortion material, faeces and occasionally local lesions in specific organs. Raw bovine milk has also given frequent positive cultures in some studies.

Isolations of *Y. enterocolitica* have also been made from water and a variety of processed and unprocessed foods.

For the isolation of *Y. ruckeri*, the kidney tissue of diseased fish is recommended as the most suitable material. *Yersinia pestis* is most likely to be recovered from the blood, lymph node (bubo) aspirates and sputum of plague patients and from the blood and viscera of infected rodents. The ectoparasites of the latter are also likely to be infected and should be examined.

Too few isolations of *Y. philomiragia* have been made to permit recommendations on the most suitable material for culture. Those strains so far isolated were recovered from dead muskrats and from surface water in the same locality.

Transport of samples

Where desiccation is likely to occur, samples should be transported immersed in PBS. Samples should be packed in ice if they are likely to encounter high ambient temperatures in transit.

Cultural procedures

Yersinia pseudotuberculosis

Uncontaminated samples, such as blood or tissues removed under aseptic conditions, are plated directly on blood or trypticase–soy agar. For enrichment the material is incubated aerobically and anaerobically at both 22 and 37°C in trypticase–soy or brain–heart infusion broth for up to 10 days before plating on solid medium.

The isolation of *Y. pseudotuberculosis* from contaminated samples, including faeces, can be difficult. It is recommended that a 10% v/v suspension of material in PBS be enriched by incubation at 4°C for up to 6 weeks, with subculture onto solid medium at weekly intervals. MacConkey agar and eosin methylene blue agar are suitable for sub-culture. "Y" medium or CIN agar may also be used, although *Y. pseudotuberculosis* does not grow well on these from small inocula. Cultures should be incubated at 22 and 37°C aerobically and anaerobically.

Colonies of *Y. pseudotuberculosis* are about 1 mm in diameter, translucent, dome-shaped, with an entire edge and non-haemolytic after 24 h incubation on blood agar. On MacConkey agar, they are slightly smaller and do not ferment lactose. Both smooth and rough forms exist. The latter appear more granular and irregular than the smooth colonial type.

Yersinia enterocolitica

For isolation from uncontaminated material, direct plating on ordinary nutrient media will usually give satisfactory results. Where the numbers of organisms in the samples are likely to be small, it is best to enrich the culture by incubation in trypticase–soy broth at 22°C before plating onto solid medium.

For recovery from contaminated material, direct plating in combination with enrichment and subsequent plating are recommended to recover the full range of serotypes. For direct plating, the sample is inoculated across one-third of the area of the plate and then streaked out across the remaining two-thirds.

Both "Y" medium and CIN agar are recommended for direct plating. The best isolation rates are obtained when both media are used.

Colonies of *Y. enterocolitica* on CIN agar are raised and 2–5 mm in diameter after 48 h incubation at 22°C. They vary somewhat in general appearance and may have a smooth regular edge or be very irregular. The colour ranges from pink, with a darker centre and more transparent edge, to dark red. The red colonies tend to be smaller than the more lightly coloured ones and have virtually no transparent border.

Colonies of *Y. enterocolitica* on "Y" medium after 48 h incubation at 22°C are about 1 mm in diameter, raised, translucent, and pink or colourless.

Enrichment is most simply achieved by incubating a 10% v/v suspension of sample in PBS at 4°C for 3 weeks. This should be subcultured onto solid medium at weekly intervals. The solid medium may be non-selective, but best results are obtained using enrichment followed by plating onto CIN and "Y" medium. The modified PBS cold enrichment media listed in the materials section may also be used for specific types of sample.

Where gross contamination, particularly with *Pseudomonas* or spreading bacteria, is a problem, alkali treatment can be used on the enriched samples. Not all *Y. enterocolitica* strains will withstand this treatment, however.

Identification Methods

Identification can be done with bacteriological identification kits of the Enterotube or API 20E types, but these kits do not take account of the biochemical heterogeneity of *Y. enterocolitica* and its related species. Conventional cultural and biochemical tests performed as described by Cowan & Steel (1970) give the best results.

Preliminary examination

The yersinias are Gram negative coccobacilli or bacilli, often tending to be more pleomorphic in older cultures. Bi-polar staining is characteristic of fresh isolates of *Y. pseudotuberculosis* and *Y. pestis*.

Initial tests for identification at genus level should include determination of Gram reaction and morphology, catalase and oxidase reactions and examination for motility (Table 1). Strains should be grown at both 22 and at 37°C for the motility test, and it is also recommended that the flagella staining method developed by Kodaka and others (1982) be used (see Fig. 1). Ryu stain is prepared by mixing 10 volumes of a mordant solution (5% phenol, 10 ml; tannic acid, 2 g; saturated aluminium potassium sulphate·$12H_2O$, 10 ml) with 1 volume of a saturated solution of crystal violet in ethanol (12 g/100ml) and is stable indefinitely at ambient temperature. A single colony is picked up with an inoculating needle, taking care not to pick up agar, and is lightly touched onto the centre of a drop of PBS on a clean slide. The drop preparation is dried in air at room temperature (*not* heated) and flooded with stain. After 5 min the slide is thoroughly washed in running tap water, dried, and examined under oil immersion, starting at the periphery and moving towards the centre. Positive and negative controls should be used, and it should be borne in mind that unusually high ambient temperatures will affect the results unless incubators are equipped with cooling equipment.

A second series of tests is used to identify the species. Rapid presumptive identification of *Y. pestis* is achieved by assaying for isocitrate lyase activity (Quan *et al.* 1982). Secondary tests for differentiating *Y. enterocolitica* and *Y. pseudotuberculosis* are shown in Table 1, and *Y. ruckeri* is distinguished by its ability to decarboxylate lysine. A further set of tests (Table 1) can be used to characterise isolates further.

All tests are carried out at 22°C. Uncharacteristic reactions and loss of plasmids may occur if these organisms are grown at higher temperatures.

The secondary tests should be examined daily for 1 week, with the exception of the decarboxylase media, which can be discarded after 3 days. Sugar media should be examined for 2 weeks, and tests for indole production and nitrate reduction should be completed after 2 days incubation. Those for phenylalanine deamination are completed after 1 day and KCN broths are discarded after 2 days.

Biochemical typing

Various sub-divisions of the species *Y. enterocolitica* have been proposed on the basis of biochemical reactions. Typing schemes have been put forward, however, on the basis of results obtained under differing conditions, notably

TABLE 1
Tests for preliminary identification of Yersinia enterocolitica *and* Yersinia pseudotuberculosis

Test	*Y. enterocolitica*	*Y. pseudotuberculosis*
Morphology	Coccobacilli	Short, ovoid, bi-polar staining
Catalase	+	+
Oxidase	−	−
Motility		
37°C	−	−
22°C	+	+
Urease production	+†	+
H_2S production on TSI	−	−
O/F (glucose)	F	F
Glucose fermentation		
acid production	+	+
gas production	−/+*	−
Aesculin hydrolysis	−/+	+
Citrate utilisation	−/+*	−
Gelatin hydrolysis	−	−
ONPG reaction	+†‡	+
Arginine dihydrolase	−	−
Lysine decarboxylation	−	−
Ornithine decarboxylation	+†	−§
Control decarboxylation	−	−
Indole production	+/−‖	−
Growth in KCN broth	−	−
V–P test	+/−‖	−
Nitrate reduction	+†	+
Phenylalanine deamination	−	−
Cellobiose	+	+
Melibiose	Variable	−
Rhamnose	−/+*	+
Salicin	+/−‖	+
Sorbitol	+†	−
Sucrose	+†	−
Xylose	+†	−

* Most strains are negative.
† Rarely negative.
‡ May be slow.
§ Rarely positive.
‖ Most strains are positive.

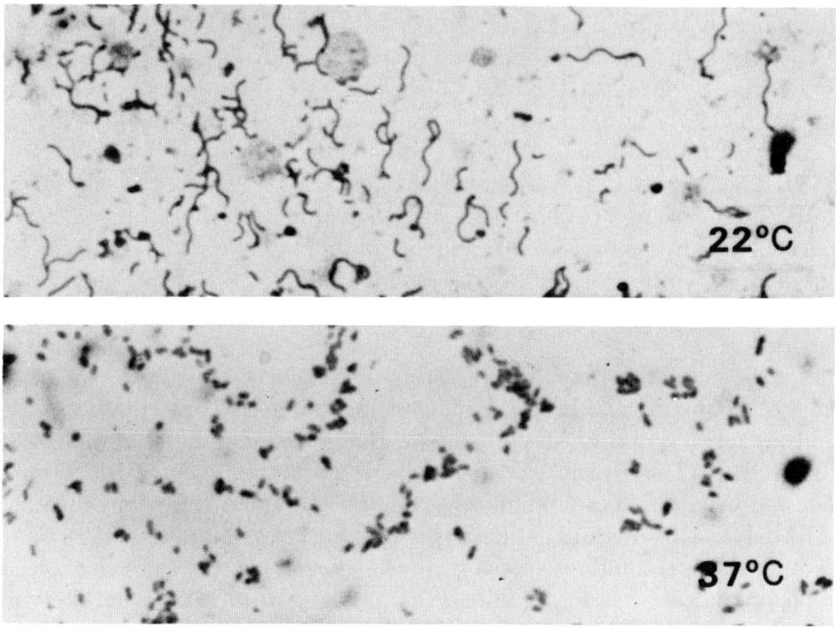

FIG. 1. Determination of motility of *Yersinia enterocolitica*, grown at 22 and 37°C, with Ryu stain. Many flagella have become detached but are clearly visible at 22°C whilst they are absent at 37°C.

temperature, so the position is somewhat confused. A scheme adapted from those proposed by Niléhn (1969) and by Wauters (1970) is shown in Table 2. It should be noted that most incubations were done at 37°C.

More recently, a series of papers by Brenner (1979) and Bercovier and others (1980*a,b*) have proposed sub-division of *Y. enterocolitica* strains into several species. This proposal has not been generally accepted *in toto* although it has helped to clarify biogroups within the species. The reactions used to differentiate these species are listed in Table 3. In addition to the three newly proposed species, the five biotypes of Niléhn (1969) and of Wauters (1970) are retained as the species *Y. enterocolitica sensu strictu*.

Disc electrophoresis

This technique is a useful supplement to biochemical typing. Strains of equivocal status can usually be assigned to a particular species or alternatively rejected as a member of the genus (Corbel *et al.* 1982). The recommended procedure is outlined in the following sections.

Preparation of extracts

Bacterial suspensions are prepared by washing off slope cultures into ice-cold 0.15 mole/litre NaCl. The organisms are then deposited by centrifugation (15,000 **g**, 15 min) and washed three times in the same diluent. Finally, packed cell deposits should be resuspended in phenol:acetic acid:water (4 : 2 : 1, w/v/v) at a concentration of 100 μg cells/100 μl solvent. After storing suspensions at room temperature overnight, cellular debris is removed by centrifugation (40,000 **g** for 30 min). The supernatant fluid can be stored at 4°C until required.

Preparation of gels

A stock solution is prepared consisting of 7.5 g acrylamide, 30 g urea and 0.2 g *N*,*N*^{1}-methylene-bis-acrylamide made up to 100 ml with 35% v/v aqueous acetic acid. For gel preparation, 10 mg of ammonium persulphate is dissolved in 12 ml of the stock solution which is then degassed for 5 min followed by addition of 0.06 ml of *N*,*N*,*N*^{1},*N*^{1}-tetramethylethylene diamine (TEMED). This mixture is then dispensed into glass running tubes (~7 × 85 mm). The gel solutions are overlaid with a few drops of water and allowed to polymerise at 37°C for 90 min. (Any problems with gelling are likely to arise from the TEMED reagent, which tends to be unstable.)

TABLE 2
Classical biotyping of Yersinia enterocolitica*

Reaction	Biotype 1	2	3	4	5
Aesculin	+	−	−	− M	−
Salicin	+	−	−	−	−
Lecithinase	+	−	−	−	−
Indole	+	(+)	−	−	−
Lactose (O/F)	(+0)	+ 0	+ 0	−	−
Xylose	+	+	+	−	−
Nitrate reduction	+	+	+	+ M	−
Trehalose	+	+	+	+	−
Ornithine decarboxylase	+	+	+	+	− M
ONPG (25°C)	+	+	+ M	+ M	− M
Sorbitol	+	+	+ M	+	− M
V–P (25°C)	+ M	+ M	+ M	+ M	− M
Sucrose	+	+ M	+ M	+	+ M
Sorbose	+	+	+	+	Variable

* All tests done at 37°C unless otherwise stated; (+) indicates slow reaction; M indicates that a minority of strains differ.

TABLE 3
Differentiation of proposed new species of Yersinia enterocolitica*

Test	*Y. intermedia*	*Y. frederiksenii*	*Y. kristensenii*
Lipase (Tween 80)	+	+	v
DNA-ase	–	–	v
Citrate (Simmons)	+	v	–
V–P	+	+	–
Indole	+	+	v
Sucrose	+	+	–
Rhamnose	+	+	–
Melibiose	+	–	–
Raffinose	+	–	–
α-Methylglucoside	+	–	–
Ornithine decarboxylase	+	+	+
Xylose	+	+	+
Trehalose	+	+	+
Sorbose	+	+	+
Inositol	+	+	+

* All tests done at 28°C; v indicates variable reaction.

Electrophoresis

Gel tubes are positioned in the electrode tanks with 10% v/v aqueous acetic acid as the reservoir fluid. Tracer stain such as methyl green is added to the top (anode) reservoir to permit visualisation of the protein migration front. A 50-μl volume of bacterial extract is then carefully layered onto the top of each gel column. A constant current of 3 mA per gel should be applied for 15 min, after which time it can be increased to 5 mA per gel. Electrophoretic separation takes about 1.5–2 h.

When the migration front approaches the lower end of the gel column, the current is switched off and the running tubes are removed. The gel columns are extracted by carefully prising them from their tubes with a syringe needle whilst keeping the columns lubricated with distilled water from the syringe.

Staining

Gel columns are immersed in 1% w/v amido black in 7% v/v aqueous acetic acid for 30 min before removing background stain in successive changes of 30% v/v aqueous acetic acid for approximately 48 h. Alternatively, electrophoretic destaining can be used.

Interpretation

Controls should be included in all runs, and only gels run simultaneously should be compared. Differentiation of strains into any of the classical species of *Yersinia* is possible on the basis of their protein migration patterns. *Yersinia ruckeri* appears to be a *bona fide* member of the genus on this evidence, although *Y. philomiragia* patterns show little similarity to those of other members of the genus (Fig. 2). Further division of the *Y. enterocolitica* group is not possible on the basis of the gel electrophoretic patterns.

Serotyping

Methods for the preparation of specific anti-sera for serotyping *Y. enterocolitica* and *Y. pseudotuberculosis* have been described in detail by Winblad (1978) and Thal (1978).

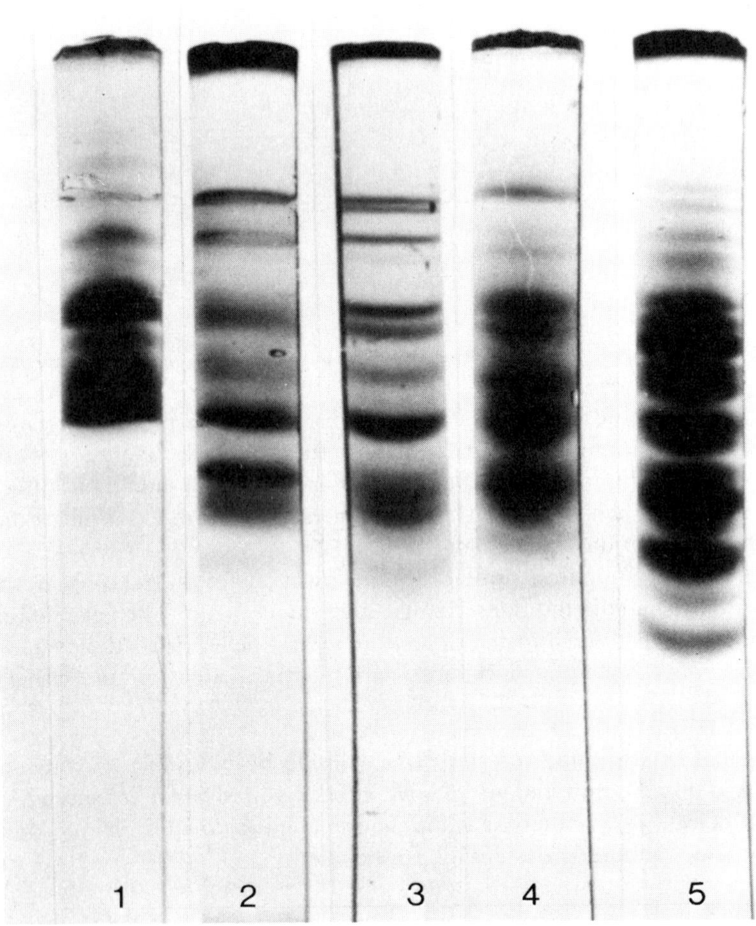

FIG. 2. Disc electrophoresis of phenol:acetic acid:water (4:2:1, w/v/v) extracts of (1) *Y. pestis* TS, (2) *Y. pseudotuberculosis* O-group V, (3) *Y. enterocolitica* 296/68, (4) *Y. ruckeri* 2396–61, (5) *Y. philomiragia* 067.

TABLE 4

Serotypes of Yersinia pseudotuberculosis

O-Group	Sub-group	Antigens: Rough	Antigens: O (thermostable)	Antigens: H (thermolabile)	Epidemiology	Cross reactions
I	A	1	2, 3	a, c	Europe, North America, Asia	*Salmonella* B
	B	1	2, 4	a, c		
II	A	1	5, 6	a, d		
	B	1	5, 7	a, d		
III		1	8	a		
IV	A	1	9, 11	b; a, b	Europe, Japan	*Salmonella* D
	B	1	9, 12	a, b, d		
V	A	1	10, 14	a; a, e (b)		
	B	1	10, 15	a		
VI		1	13	a	Japan	*Escherichia coli* O:55

TABLE 5
Serotypes of Yersinia enterocolitica

O Factor	Bio-group	Pathogenicity* Man	Animal	Cross reactions
O : 1, 2a, 3	Biotype 3 (B3)		+	
O : 2a, 2b, 3	B5		+	
O : 3	B4	+	+	
O : 4, 32	B1			
O : 4, 33	B1			
O : 5	B1	(+)	+	
O : 5, 27	B2	+	+	
O : 6, 30	B1	+		
O : 6, 31	B1	+		
O : 7, 8	B1			
O : 8	B1	+	+	
O : 9	B2	+		*Brucella*/*Salmonella* O : 30/*Vibrio*/ *Morganella morganii* O : 43
O : 10	B1			
O : 11, 23	*Y. kristensenii*	(+)		
O : 11, 24	*Y. kristensenii*	(+)		
O : 12, 25	*Y. kristensenii*	(+)		*S. bergen* O : 47
O : 12, 26	*Y. kristensenii*	(+)		
O : 13, 7	B1			
O : 14	B1			
O : 15	B4			
O : 16	*Y. kristensenii*			
O : 16, 29	*Y. frederiksenii*			
O : 17	*Y. intermedia*	(+)		*M. morganii* O : 44
O : 18	B1	(+)		
O : 19, 8	B1			
O : 20	B1	(+)		
O : 21	B1			
O : 22	B1	(+)		
O : 28	*Y. kristensenii*			
O : 34	Identical to O : 10			
O : 35	*Y. frederiksenii*			
O : 36	*Y. intermedia*			
O : 37	*Y. intermedia*			
O : 38	*Y. frederiksenii*			
O : 39	*Y. frederiksenii*			
O : 40	B1			
O : 41, 42	B1			
O : 41, 43	B1			
O : 44	*Y. frederiksenii*			
O : 44, 45	*Y. frederiksenii*			

TABLE 5 (*Continued*)

O Factor	Bio-group	Pathogenicity* Man	Animal	Cross reactions
O : 46	*Y. kristensenii*			*Y. pseudotuberculosis* IV
O : 47	B3			
O : 48	*Y. intermedia*			
O : 49, 51	B1			
O : 50, 51	*Y. kristensenii*			
O : 52	*Y. kristensenii*			
O : 52, 53	*Y. intermedia*			
O : 52, (54)	*Y. intermedia*			
O : 55	*Y. frederiksenii*			
O : 56	B3			
O : 57	*Y. intermedia*			

* (+), Rare.

A common enterobacterial antigen of the Kunin type is present in *Yersinia* species. Grouping anti-sera therefore need to be absorbed with other Gram negative organisms such as *Escherichia coli, Salmonella* spp., *Proteus* spp. or *Pseudomonas aeruginosa*.

Strains of both *Y. pseudotuberculosis* and *Y. enterocolitica* can be serogrouped according to the serological specificity of their somatic antigens. This is particularly useful for epidemiological studies, and certain serogroups of *Y. enterocolitica* have been linked consistently with disease whereas others appear to be relatively harmless environmental or commensal strains.

Both *Y. pseudotuberculosis* and *Y. enterocolitica* also possess flagella antigens, but these are of secondary importance in typing schemes.

Yersinia pseudotuberculosis has been divided into six groups on the basis of O antigens. Of these, O-groups I, II, IV and V are further divided into subgroups A and B in each case (Table 4). In Europe, O-group I is most common, although O-groups II and III are not unusual. All three O-groups have been isolated in North America and Asia. O-Groups IV and V are found rarely in Europe and Japan. To date, O-group VI has only been found in Japan. Antigenic cross reactions occur between O-group II and *Salmonella* group B, between O-group IV and *Salmonella* group D, and between O-group VI and *Escherichia coli* O-group 55. Cross agglutination also may occur with some *Y. enterocolitica* serogroups, particularly among rhamnose-positive strains. *Yersinia pseudotuberculosis* strains possess the rough antigen (1) which is also shared with *Y. pestis*. Although smooth strains of *Y. pestis* and *Y. pseudotuberculosis* do not cross agglutinate, it can be very difficult to distinguish rough strains serologically.

At least 57 O-groups have been identified in *Y. enterocolitica* (Table 5).

Some of these are common to *Y. enterocolitica sensu strictu* and the newly designated species *Y. kristensenii*, *Y. frederiksenii*, and *Y. intermedia*. Only a few serogroups are consistently associated with disease in man or domesticated animals. These tend to show a particular geographical distribution. Thus human infections caused by O:3 strains have been identified in most European countries, the USSR, South Africa, Zaire, Australia and Canada. Group O:5B infections are fairly common in Canada and the UK but rare elsewhere, whereas group O:8 is most frequent in the USA. Group O:9 has been isolated in Finland, Belgium, Sweden and the Netherlands and infrequently elsewhere. All of these human pathogenic strains have also been isolated from pigs, dogs, monkeys, and occasionally other species. Other O-groups implicated as human pathogens include 6, 11, 12, 17, 18, 20 and 22. A number of surveys have indicated that many O-groups can be isolated from both wild and domesticated animals but most do not cause recognisable disease. Those O-groups which have been implicated as animal pathogens of veterinary importance include 1, 2, 3, 5b and 8.

Some strains of *Y. enterocolitica* possess a fimbrial or K antigen, and this has to be taken into account in serotyping schemes as some K-bearing strains are not agglutinated by O anti-sera.

In general, serotyping of yersinia isolates is best performed by reference laboratories. Details of the methods used are given by Thal (1978) and Winblad (1978).

Serological cross reactions between *Y. enterocolitica* serogroups and other organisms have been documented. That between O-group 9 and *Brucella* species is best known. Others are listed in Table 5.

Phage Typing

Methods for the isolation, propagation and use of phages are given by Bergan (1978) and Baker & Farmer (1982).

Two schemes have been quite widely studied. The Swedish scheme (Niléhn & Ericson 1969) recognizes 12 lysotypes, including the major groups I–VIII and some sub-groups. The French scheme (Mollaret & Nicolle 1965; Nicolle *et al.* 1976) has probably received the most extensive application. This recognizes 10 lysotypes, designated phage groups I–X. Strains that do not conform to the lytic pattern for these groups are included in group XI. In addition, lysotypes IX and X are sub-divided into sub-types.

Neither of these schemes has been developed sufficiently to produce a large number of distinct epidemiological types. Many isolates tend to fall within a single lysotype. Thus in the French system one phage pattern accounted for 57% of the isolates (Mollaret & Nicolle 1965), and in the

Swedish system, a single phage pattern accounted for 87% of the isolates (Niléhn & Ericson 1969).

Recently, an improved system has been described by Baker & Farmer (1982). This is claimed to permit strain identification in *Y. enterocolitica sensu strictu*, *Y. frederiksenii*, *Y. intermedia* and *Y. kristensenii*. The most common phage type accounted for 22% of the strains examined.

Phages have been used for the differentiation of *Y. pseudotuberculosis* and *Y. pestis* (Gunnison *et al.* 1951).

Other Tests

Pathogenicity tests are useful for the differentiation of *Yersinia* species where other methods give inconclusive results. Thus *Y. pestis* is lethal for white rats and guinea pigs, *Y. pseudotuberculosis* is lethal only for guinea pigs (except for rare toxigenic strains of O-group III which will also kill rats), and *Y. enterocolitica* does not kill either species following sub-cutaneous injection. The pathogenicity of many *Y. enterocolitica* strains is often lost rapidly following cultivation *in vitro*. Reference to methods of determining pathogenicity of *Y. enterocolitica* isolates is made by Brewer & Corbel (1983).

DNA hybridisation methods have been used for examination of *Yersinia* strains by Brenner and others (1980). They are not suited to the routine laboratory. Antibiotic resistance patterns can be determined by standard methods. They have been found of limited epidemiological value in the *Y. enterocolitica* group (Baker & Farmer, 1982).

Reference Laboratories for Yersinias

Prof. H. H. Mollaret, Centre Nationale des Yersinia, Institut Pasteur, 28 Rue du Dr Roux, 75724 Paris Cedex 15, France.

Prof. S. Winblad, Department of Medical Microbiology, University of Lund, Malmo, Sweden.

Dr S. Toma, Ontario Ministry of Health, Public Health Laboratories, Toronto, Canada.

The Director, Public Health Laboratory, Leicester, United Kingdom.

References

Aulisio, C. C. G., Mehlman, I. J. & Sanders, A. C. 1980 Alkali method for rapid recovery of *Yersinia enterocolitica* and *Yersinia pseudotuberculosis* from foods. *Applied and Environmental Microbiology* **39**, 135–140.

BAKER, P. M. & FARMER, J. J. 1982 New bacteriophage typing system for *Yersinia enterocolitica*, *Yersinia kristensenii*, *Yersinia frederiksenii*, and *Yersinia intermedia:* Correlation with serotyping, biotyping and antibiotic susceptibility. *Journal of Clinical Microbiology* **15**, 491–502.

BERCOVIER, H., BRENNER, D. J., URSING, J., STEIGERWALT, A. G., FANNING, G. R., ALONSO, J. M., CARTER, G. P. & MOLLARET, H. H. 1980*a* Characterisation of *Yersinia enterocolitica sensu strictu*. *Current Microbiology* **4**, 201–206.

BERCOVIER, H., URSING, J. BRENNER, D. J., STEIGERWALT, A. G., FANNING, G. R., CARTER, G. P. & MOLLARET, H. H. 1980*b Yersinia kristensenii:* A new species of Enterobacteriaceae composed of sucrose-negative strains (formerly called atypical *Yersinia enterocolitica* or *Yersinia enterocolitica*-like). *Current Microbiology* **4**, 219–224.

BERCOVIER, H., MOLLARET, H. H., ALONSO, J. M., BRAULT, J., FANNING, G. R., STEIGERWALT, A. G. & BRENNER, D. J. 1980*c* Intra- and interspecies relatedness of *Yersinia pestis* by DNA hybridisation and its relationship to *Yersinia pseudotuberculosis*. *Current Microbiology* **4**, 225–229.

BERGAN, T. 1978 Bacteriophage typing of *Yersinia enterocolitica*. In *Methods in Microbiology*, ed. Bergan, T. & Norris, J. R., Vol. 12, pp. 25–36. New York: Academic Press.

BOTTONE, E. 1977 *Yersinia enterocolitica:* a panoramic view of a charismatic microorganism. *CRC Critical Reviews in Microbiology* **5**, 211.

BRENNER, D. J. 1979 Speciation in *Yersinia*. *Contributions to Microbiology and Immunology* **5**, 33–43.

BRENNER, D. J., URSING, J., BERCOVIER, H., STEIGERWALT, A. G., FANNING, G. R., ALONSO, J. M. & MOLLARET, H. H. 1980 Deoxyribonucleic acid relatedness in *Yersinia enterocolitica* and *Yersinia enterocolitica*-like organisms. *Current Microbiology* **4**, 195–200.

BREWER, R. A. & CORBEL, M. J. 1983 Characterisation of *Yersinia enterocolitica* strains isolated from cattle, sheep and pigs in the United Kingdom. *Journal of Hygiene* **90**, 425–533.

CORBEL, M. J., BREWER, R. A. & HENDRY, D. MCL. F. D. 1982 Disc electrophoresis of phenol: acetic acid: water-soluble proteins for the identification of *Yersinia* species. *Research in Veterinary Science* **33**, 43–46.

COWAN, S. T. & STEEL, K. J. 1970 *Manual for the Identification of Medical Bacteria*. London: Cambridge University Press.

EWING, W. H., ROSS, A. J., BRENNER, D. J. & FANNING, G. R. 1978 *Yersinia ruckeri*, the Redmouth (RM) bacterium. *International Journal of Systematic Bacteriology* **28**, 37–44.

FREDERIKSEN, W. 1964 A study of some *Yersinia tuberculosis*-like bacteria ("Bacterium enterocoliticum" and "*Pasteurella X*"). In *Proceedings of the XIVth Scandinavian Congress of Pathology and Microbiology* pp. 103–104. Oslo: Universitetsforlaget.

GUNNISON, J. B., LARSON, A. & LASARUS, A. S. 1951 Rapid differentiation between *Pasteurella pestis* and *Pasteurella pseudotuberculosis* by action of bacteriophage. *Journal of Infectious Diseases* **88**, 254.

HANNA, M. O., STEWART, J. C., CARPENTER, Z. L. & VANDERZANT, C. 1977 Development of *Yersinia enterocolitica*-like organisms in pure and mixed cultures on different bismuth sulphite agars. *Journal of Food Protection* **40**, 676–677.

HEAD, C. B., WHITTY, D. A. & RATNAM, S. 1982 Comparative study of selective media for recovery of *Yersinia enterocolitica*. *Journal of Clinical Microbiology* **16**, 615–621.

JENSEN, W. I., OWEN, C. R. & JELLISON, W. L. 1969 *Yersinia philomiragia*, a new member of the *Pasteurella* group of bacteria, naturally pathogenic for the muskrat (*Ondatra zibethica*). *Journal of Bacteriology* **100**, 1237–1241.

KODAKA, H., ARMFIELD, A. Y., LOMBARD, G. L. & DOWELL, V. R. 1982 Practical procedure for demonstrating bacterial flagella. *Journal of Clinical Microbiology* **16**, 948–952.

LEE, W. H., HARRIS, M. E., MCCLAIN, D., SMITH, R. E. & JOHNSTON, R. M. 1980 Two modified selenite media fo; the recovery of *Yersinia enterocolitica* from meats. *Applied and Environmental Microbiology* **39**, 205–209.

MEHLMAN, I. J. & AULISIO, C. C. G. 1978 *Yersinia enterocolitica, Yersinia pseudotuberculosis* and related bacteria. *Bacteriological Analytical Manual.* Washington DC: Food and Drug Administration.

MEHLMAN, I. J., AULISIO, C. C. G. & SANDERS, A. C. 1978 Problems in the recovery and identification of *Yersinia* from food. *Journal of the Association of Official Analytical Chemists* **61**, 761–771.

MOLLARET, H. & NICOLLE, P. 1965 Sur la fréquences de la lysogénic dans l'espèce nouvelle *Yersinia enterocolitica. Comptes Rendus Hebdomadaires des Séances de l'Académie des Sciences* **260**, 1027–1029.

NICOLLE, P., MOLLARET, H. H. & BRAULT, J. 1976 La lysotypie de *Yersinia enterocolitica.* Arguments géographiques, zoologiques, antigéniques et biochimiques, plaidant en sa faveur. *Bulletin de l'Académie Nationale de Médecine (Paris)* **160**, 404–408.

NILÉHN, B. 1969 Studies on *Yersinia enterocolitica* with special reference to bacterial diagnoses and occurrence in human acute enteric disease. *Acta Pathologica et Microbiologica Scandinavica, Supplement* No. 206, 1–48.

NILÉHN, B. & ERICSON, C. 1969 Studies on *Yersinia enterocolitica.* Bacteriophage liberated from chloroform treated cultures. *Acta Pathologica et Microbiologica Scandinavica* **75**, 177–187.

OHARA, S., SATO, T. & HOMMA, M. 1974 Serological studies on *Francisella tularensis, Francisella novicida, Yersinia philomiragia* and *Brucella abortus. International Journal of Systematic Bacteriology* **24**, 191–196.

PAI, C. H., SORGER, S., LAFLEUR, L., LACKMAN, L. & MARKS, M. 1979 Efficacy of cold enrichment techniques for recovery of *Yersinia enterocolitica* from human stools. *Journal of Clinical Microbiology* **9**, 712–715.

PARK, C. E., LANDGRAF, M. & STANKIEWICZ, Z. K. 1981 A new sensitive and rapid procedure for the isolation of *Yersinia enterocolitica* from food, particularly from low calory food such as vegetables. In *Psychrotrophic Microorganisms in Spoilage and Pathogenicity*, ed. Roberts, T. A., Hobbs, G., Christian, J. H. B. & Skovgaard, N., pp. 425–429. London & New York: Academic Press.

QUAN, T. J., BARNES, A. M. & POLAND, J. D. 1981 Yersinioses. In *Diagnostic Procedures for Bacterial, Mycotic and Parasitic Infections* 6th edn., ed. Balows, A. & Hausler, W. J., pp. 723–745. Chicago, Illinois: American Public Health Association.

QUAN, T. J., VANDERLINDEN, J. J. & TSUCHIYA, K. R. 1982 Evaluation of a qualitative isocitrate lyase assay for rapid presumptive identification of *Yersinia pestis* cultures. *Journal of Clinical Microbiology* **15**, 1178–1179.

RITTER, D. B. & GERLOFF, R. K. 1966 Deoxyribonucleic acid hybridisation among some species of the genus *Pasteurella. Journal of Bacteriology* **92**, 1838–1839.

ROSS, A., RUCKER, R. & EWING, W. 1966 Description of a bacterium associated with Redmouth disease of rainbow trout (*Salmo gairdneri*). *Canadian Journal of Microbiology* **12**, 763–770.

SCHIEMANN, D. A. 1979 Synthesis of a selective agar medium for *Yersinia enterocolitica. Canadian Journal of Microbiology* **25**, 1298–1304.

SCHIEMANN, D. A. 1982 Development of a two-step enrichment procedure for recovery of *Yersinia enterocolitica* from food. *Applied and Environmental Microbiology* **43**, 14–27.

SCHIEMANN, D. A. 1983 Alkalotolerance of *Yersinia enterocolitica* as a basis for selective isolation from food enrichments. *Applied and Environmental Microbiology* **46**, 22–27.

SOLTESZ, L. V., SCHALEN, C. & MÅRDH, P. A. 1980 An effective medium for *Yersinia enterocolitica* containing sodium oxalate. *Acta Pathologica et Microbiologica Scandinavica, Section B: Microbiology* **88**, 11–16.

SWAMINATHAN, B., HARMON, M. C. & MEHLMAN, I. J. 1982 A review: *Yersinia enterocolitica. Journal of Applied Bacteriology* **52**, 151–183.

THAL, E. 1978 The identification of *Yersinia pseudotuberculosis.* In *Methods in Microbiology*, ed. Bergan, T. & Norris, J. R., Vol. 12, pp. 15–24. New York: Academic Press.

URSING, J., STEIGERWALT, A. G. & BRENNER, D. J. 1980 Lack of genetic relatedness between *Yersinia philomiragia* (the "Philomiragia" bacterium) and *Yersinia* species. *Current Microbiology* **4**, 231–233.

VAN LOGHEM, J. J. 1944 The classification of plague bacillus. *Antonie van Leeuwenhoek* **10**, 15–16.

WAUTERS, G. 1970 Contribution à l'étude de *Yersinia enterocolitica*. Thèse de l'obtention du grade d'agrégé de l'enseignement superieur, Université catholique de Louvain, Belgium: Vander.

WEISSFELD, A. S. & SONNENWIRTH, A. C. 1980 *Yersinia enterocolitica* in adults with gastrointestinal disturbances: need for cold enrichment. *Journal of Clinical Microbiology* **11**, 196–197.

WINBLAD, S. 1978 *Yersinia enterocolitica*. In *Methods in Microbiology*, ed. Bergan, T. & Norris, J. R., Vol. 12, pp. 37–50. New York: Academic Press.

Legionellas

A. E. WRIGHT AND P. J. DENNIS

Environmental Microbiology and Safety Reference Laboratory, Public Health Laboratory Service, Centre for Applied Microbiology and Research, Porton Down, Salisbury, UK

The outbreak of pneumonia which took place in 1976 in Philadelphia, now known as Legionnaire's disease, caused 34 deaths in the 221 cases reported. Such a mortality (15%) occurring in one incident was sufficient to initiate what has been described as the largest epidemic investigation in the history of the Center for Disease Control in Atlanta, Georgia, USA. The investigation has been well documented elsewhere (Fraser *et al.* 1977) and the details will not be repeated here.

The discovery of the organism now known to be the cause not only of Legionnaire's disease but also of Pontiac fever was the result of 6 months' endeavour which is well recognised and acknowledged by microbiologists everywhere. Those 6 months of work before the description of the organism was published (McDade *et al.* 1977) have been followed by 6 years of steady development in this field. During this period a number of outbreaks of Legionnaire's disease have been recorded, an intense study of the natural habitat of the organism has been launched, and clinical and experimental facets of the organism and the disease have been published.

This chapter briefly reviews these advances but concentrates on cultural methods of isolation of the organism from environmental specimens.

Species of *Legionella*

No fewer than 23 species of legionella have been described (Macfarlane 1983) and others are being investigated. The type species is *Legionella pneumophila*, which is divided into eight serogroups with two further possible

Isolation and Identification of
Micro-organisms of Medical and
Veterinary Importance

Copyright © 1985 by the Society for Applied Bacteriology
All rights of reproduction in any form reserved
0-12-181460-2

serogroups. Serogroup 1, the cause of the original outbreak, remains the commonest cause of pneumonia in man and is the most frequently isolated organism from environmental specimens.

Other species include *L. bozmannii*, which has been isolated from lung tissue from a case of pneumonia but has not yet been isolated from the environment. It differs from the type species both by serology and by DNA homology. *Legionella dumoffii* causes pneumonia in man and has been isolated from lung tissue and water. *Legionella gormanii* has not been implicated in human disease directly but has been recognised as a cause of serological conversion. It has been isolated from soil. These three organisms, *L. bozmanii*, *L. dumoffii* and *L. gormanii*, all show fluorescence under ultraviolet (u.v.) light, thus differing from other species of legionellas. For this reason it has been suggested that they should be split off to form a separate genus *Fluoribacter*.

Legionella micdadei has been isolated from human blood and lung and also from nebulisers used in respiratory therapy and from waters. It is the causal agent of Pittsburgh pneumonia and hence referred to as the Pittsburgh pneumonia agent. It is also known as *Tatlockia micdadei*. Other species of importance include *L. longbeachae* (serogroups 1 and 2), *L. jordanis*, *L. oakridgensis* and *L. wadsworthii*.

All these organisms are Gram negative non-sporing small bacilli with rather pointed ends. They have the ability to survive in nature, usually in water, over a wide range of temperature. Described as thermophilic, they have a growth range of 20–45°C and can survive for some time at temperatures in excess of 50°C. They are thus well adapted to life in naturally warm and thermally polluted waters. They have also some association with amoebae, although their precise relationship to these protozoa has yet to be established (Rowbotham 1980).

Bearing in mind the wide temperature tolerance of these organisms, it is not surprising that legionellas are found in water systems such as cooling towers and calorifiers and that they are able to colonise hot water systems in hotels and private houses. It is from these situations that the organism is able to infect man, causing either a mild allergic type reaction known as Pontiac fever or a pneumonia which may often be severe.

Legionellosis

Legionnaire's disease may present with fever, malaise and chills. Cough, although a feature in more than half the cases, is not usually severe and is rarely productive. Although this picture is not unique, many clinicians (Miller 1979) consider that a recognisable pattern of signs and symptoms can be discerned due to the multiplicity of systems affect and to the neurological abnormalities observed (Weir *et al.* 1982). The presenting picture may be mild but the patient's condition may deteriorate rapidly within a few days. Liver, kidneys and other systems may be involved, and diarrhoea and neurological signs may become a feature. The chest radiograph may progress from a patchy alveolar infiltration to consolidation.

Pontiac fever is very different. The symptoms are usually confined to fever, malaise, chills and headache occurring within a few hours of exposure. There may be muscular pains, dizziness and photophobia, but penumonia does not occur. Recovery is often complete in 48 h.

Epidemiology and mode of transmission

Although legionellosis is thought of as an epidemic disease, many cases are sporadic. It is not yet clear what proportion of all cases of pneumonia these sporadic cases of legionnaires disease constitute. Workers in Nottingham (Macfarlane *et al.* 1982) found that 15% of 127 consecutive cases of pneumonia were due to legionellas, which was the commonest group after *Streptococcus pneumoniae.* This finding was the result of a long term follow-up of patients for seroconversion, which may often occur late in this disease (Macfarlane *et al.* 1982). Surveys with shorter periods of follow-up yield figures of a lower order (White *et al.* 1981). In the USA it is estimated that 1% of all cases of pneumonia are caused by legionellas representing some 25,000 cases each year (Foy *et al.* 1979). A survey at the Ruchill Hospital in Scotland (Kennedy & Borland 1983) incriminated *L. pneumophila* as the commonest infecting organism in 96 cases.

A survey of the 281 cases which occurred in 1981–1982 in the UK (Anon. 1983) shows that 34 patients died, giving a case fatality rate of 12%. This does not differ appreciably from that observed in the original outbreak in Philadelphia, although the UK cases included 12 patients already suffering from neoplastic, renal, chronic respiratory and cardiovascular disease, who became infected in hospitals.

In the normal population, antibody studies revealed that only 0.4% of 740 sera had an antibody titre of >256 to *L. pneumophila* serogroup 1 (Fallon 1982). These results from Scotland are similar to those recorded in England by Macrae *et al.* (1979) who detected 31 positive sera among 2023 examined, an incidence of 1.5%. These findings are in contrast with those of Yonke *et al.* (1982), who showed that between 11.8 and 21.8% of sera collected from 600 healthy residents in Michigan USA had antibody to *L. pneumophila* sergroup 1.

The mode of transmission remains to be elucidated, but strong circumstantial and experimental evidence suggests an airborne route. In 1978 (Dondero *et al.* 1980) an outbreak occurred in which 39 cases of Legionnaire's disease were associated with the use of an auxillary cooling tower.

This belief in the airborne route received support when Tobin *et al.* (1980) isolated the organism from shower heads in a hospital in Oxford where infections had occurred. In 1979–1980 a cooling tower was thought to have been the source of infection in an outbreak in a district general hospital. The water system was subsequently shown to be the real culprit (Fisher-Hoch *et al.* 1981). There is now adequate evidence that many water systems are colonised and that when this occurs organisms may multiply in the system and the rubber washers of taps (Colbourne *et al.* 1983). It is not clear how colonisation of water systems takes place, but isolation of legionellas from soil, streams and ponds suggest that it would not be difficult to contaminate air conditioning systems, condensers, cooling towers and hot water systems. Units which are unused for a period, especially in summer months, may be a suitable site for multiplication of the organism. Lack of maintenance in the form of descaling and disinfection may play a part in encouraging growth. It is not yet known if the chemical composition of the local water is important nor if the method of construction of the water system and the materials used have any influence.

These factors are being studied in the UK by a Working Party of the Public Health Laboratory Service with financial support from the government. A survey of this type has only become possible because of the recent development of culture methods which enable small numbers of legionellas in environmental samples to be detected. The significance of small numbers of organisms in a water system is difficult to determine, but it appears that the organism is ubiquitous and that small numbers are unlikely to cause disease. Such evidence is of course only available from animal experiments, and these have shown that counts in the air of the order of 10^4/ml (Fitzgeorge *et al.* 1983) are required to infect animals. Person to person transmission, although suspected (Love *et al.* 1978), has not been confirmed (Yu *et al.* 1983).

Isolation

Diagnosis of clinical cases is usually made retrospectively by the indirect fluorescent antibody (IFA) test on serum or by the demonstration of the organism using the direct fluorescent method (DFA). Sputum, tracheal washings or aspirate, pleural fluid or lung tissue are all suitable, but sputum in pneumonia due to legionellas is often scanty.

The examination of a single serum specimen may be unhelpful, and the collection of a second specimen is best delayed for 4 weeks. This retrospective diagnosis has little clinical relevance, but now that efficient culture media is available, attempts to isolate the organism from patients should be made. Sputum, as already indicated, may be scanty, and tracheal aspirates are less commonly attempted in the UK than in the USA. Isolation by blood culture has been made perhaps fortuitously (Chester *et al.* 1983).

Environmental investigations and experimental work with animals has now become possible because improved culture methods permit enumeration of the organism. The original isolation in egg yolk (McDade *et al.* 1977) was followed by isolation through the guinea pig. The step to solid media was made by Feeley *et al.* (1978) when a Mueller–Hinton agar supplemented by 1% haemoglobin and 1% isoVitalex was found to be successful. Incubation was at 35°C in 2.5% CO_2. Analysis of the medium showed the haemoglobin supplied iron which could also be provided by ferric salts. The essential ingredient if isoVitalex is L-cysteine HCl, as demonstrated by Feeley *et al.* (1978) after a complicated series of experiments in which all 12 chemicals contained in isoVitalex were omitted or added to the Mueller–Hinton agar. Feeley and co-workers also substituted yeast extract for casein hydrolysate, added charcoal and omitted the starch and beef extracts. The new medium, now known as charcoal yeast extract or CYE medium, was further modified and improved by the introduction of ACES buffer (Pascule *et al.* 1980). Antibiotics (Edelstein & Finegold 1979) are essential if one is to examine heavily contamined environmental specimens such as those from cooling towers. In the USA the antibiotics employed were cefamandole, polymixin, and anisomycin (Edelstein 1981), the latter being included to exclude yeasts. In the UK yeasts do not seem to be a problem and as in any event anisomycin is difficult to obtain and extremely expensive, cefamandole, polymyxin and cyclohexamide may be used. In 1981 Edelstein also added α-ketoglutarate which enhanced the growth of the organism, and in the same year Wadowsky & Yee (1981) suggested the addition of glycine, vancomycin and polymixin to the buffered CYE medium.

The medium which was finally shown to be the most effective (Edelstein 1982) consists of a buffered charcoal yeast extract (BCYE) containing L-cysteine, ferric ions, ACES buffer, α-ketoglutarate, glycine and antibiotics (see below). Although this medium is extremely effective, contamination still occurred. This is understandable as samples not only on cooling towers and evaporative condensers, but from lakes, streams, and even mud were examined. Bopp *et al.* (1981) attempted a pretreatment method in which they exposed the specimens to a pH of 2.2 using a HCl–KCl buffer for a period of 5 min. This reduced the total microbial flora by about 100-fold but in their hands the guinea-pig/egg inoculation method remained superior to cultural methods, although there was often a delay of up to 1 year between the isolation procedures. As the authors themselves point out, "Legionella may not have been still viable in some of the water samples". We have been able to confirm the usefulness of Bopp's acid treatment but we also considered simpler and more rapid methods of eliminating contaminants before culture. A project set up to study the temperature sensitivity of *Legionella* spp. (Dennis *et al.* 1984) soon showed that a temperature of 50°C for a period of 30 min had little effect on many species of legionellas but reduced the count of contaminating organisms dramatically. *Legionella micdadei* was an exception, and for that reason reliance should not be placed entirely on pretreatment with heat. Other species of legionellas yet to be described may exhibit similar heat sensitivity. It is interesting that Edelstein *et al.* (1982) experimented with heat to reduce contamination in clinical specimens. These workers used 60°C for 1–2 min, a period of time difficult to control. They concluded that "this technique should not be generally used". In our hands, however, the use of 50°C for 30 min for environmental specimens is extremely effective Dennis *et al.* (1983).

Preparation of samples

Five litres of water collected in a plastic container are filtered through a polycarbonate or nylon membrane filter, pore size 0.22 or 0.45 μm. The filter membrane is placed in a sterile honey jar and shaken vigorously in 50 ml of filtrate.

Isolation procedures

Samples may be cultured without further treatment, or, to reduce the numbers of other organisms, they may be heated or exposed to a low pH.

Heat treatment

Ten millilitres of the concentrated sample is place in a water bath at 50°C for 30 min. After mixing, 0.1 ml is cultured on BCYE medium, using a glass spreader. The plates are incubated aerobically for 7 days. Colonies of legionellas begin to appear after 3 days, but growth may not be apparent on this medium for up to 7 days.

Low pH treatment

An HCl–KCl buffer is used. It consists of 3.9 ml of 0.2 mole/litre HCl mixed with 25 ml of 0.2 mole/litre KCl. The pH must be adjusted to 2.2 with 1 mole/litre KOH, as there appears to be a discrepancy in the original formula (Bopp *et al.* 1981) which results in a lower pH.

Ten millilitres of the concentrated sample is centrifuged at 2500 rev/min for 20 min, using sealed centrifuge buckets to avoid disperal of aerosols. The supernatant fluid is removed, leaving about 1 ml in the tube. Nine millilitres of the buffer are added, and after shaking the mixture is allowed to stand for 5 min, when 0.1–ml amounts are cultured on BCYE medium, using a glass spreader. The plates are incubated aerobically for 3–7 days (see above).

Culture media

Buffered charcoal yeast extract medium plus α-ketoglutarate

Buffered charcoal yeast, extract (BCYE) agar is the same as charcoal yeast extract (CYE) agar except for a difference in the preparation procedure and the addition of α-ketoglutarate monopotassium and ACES buffer. It contains (g/litre): yeast extract (Oxoid), 10.0; agar (Oxoid no. 2), 12.0; activated charcoal (Norit A), 1.5; α-ketoglutarate monopotassium, 1.0; ACES buffer (Sigma), 10.0; KOH pellets, 2.8; L-cysteine (Sigma), 0.4; ferric pyrophosphate, 0.25 (obtained from ICN Pharmaceuticals, ICN Life Science Group, Kodak Ltd., Kirby, Liverpool).

It is essential that the ACES buffer be hydrated in the following way to avoid possible denaturation of the yeast extract. The buffer, 10 g, is dissolved in 500 ml of distilled water in a water bath at 45–50°C. It is then mixed with 440 ml of distilled water to which ~40 ml of 1 mole/litre KOH is then added (or with 48 ml of distilled water in which 2.8 g of KOH pellets have been dissolved). This mixture is used to hydrate the Norit charcoal, yeast extract, α-ketoglutarate and agar. After autoclaving and cooling to 50°C, the cysteine and the ferric pyrophosphate each dissolved in 10 ml of distilled water are added. The final pH should be 6.9.

Selective medium

This is BCYE agar containing polymixin, vancomycin, cycloheximide and glycine.

Polymixin B. The stock solution contains 200 mg/100 ml of distilled water. It is sterilized by filtration (0.2 μm membrane) and bottled in 5.5-ml amounts.
Vancomycin. Vancomycin sulphate, 100 mg, is dissolved in 20 ml of distilled water, filtered as for polymixin, and bottled in 1-ml amounts.
Cycloheximide. This is a 2% w/v solution in distilled water, filter-sterilized as for polymixin. It is bottled in 4-ml amounts. All the antibiotic solutions are stored at −20°C.

BCYE medium is melted and cooled to 54°C. One bottle each of the antibiotics and 3 g of glycine (ammonia-free) is added to each litre before pouring plates. The final concentrations of antibiotics are: polymixin, 79 200 IU/litre; vancomycin, 5 mg/litre; cycloheximide, 80 mg/litre. (Note: Buffered charcoal yeast extract agar is now available from commercial sources. A reliable medium is produced by the Oxoid Company. Selective cocktails are also provided which can be added to the rehydrated autoclaved base.)

Identification

Cultures are incubated at 35°C in air or 2.5% CO_2 although CO_2 is not essential. The incubator should be humidified by placing a tray of water in the base, and the plates are incubated for 2 weeks and read at intervals of 3, 5, 7 and 14 days. Colonies with a grey-blue or purple appearance, an entire edge and a ground glass texture are subcultured to BCYE and BCYE without L-cysteine and ferric pyrophosphate.

Gram negative organisms which grow on BCYE but fail to grow on BCYE without cysteine and ferric pyrophosphate are tested by an indirect immunofluorescence method with antilegionella sera.

Indirect immunofluorescence method for confirmation of legionellas

(1) A colony is picked from the suspected growth and suspended in 1.0 ml of 2% v/v neutralised formal saline.
(2) Antigen suspension, 5 μl, is placed in each of the 15 wells of a Multitest slide (Flow) and air-dried at 37°C. It is fixed in acetone and allowed to dry in air.
(3) Doubling dilutions, as required, of rabbit antisera are made in phosphate-buffered saline (PBS), and 10 μl of each is added to the appropriate spots on the slide.
(4) The slide is incubated in a moist chamber at 37°C for 30 min, then rinsed and washed with PBS for 15 min on an orbital shaker.
(5) It is then rinsed in distilled water and dried at 37°C, and 5 μl of the fluorescein-conjugated anti-rabbit immunoglobulin serum at working dilution is added. After 30 min the slide is washed again with PBS on the orbital shaker, dried and the mounted in buffered glycerol (see below) at pH 9.0 for examination by epi-illumination.

The buffered glycerol mounting medium contains glycerol (neutral), 9 parts, and buffered saline, pH 8.5, 1 part.

The buffered saline consists of 10 parts of M/15 KH_2PO_4 in 0.5% NaCl (K_2HPO_4, 1.161 g, dissolved in 0.85% saline) and 1 part of M/15 K_2HPO_4 in saline (KH_2PO_4, 0.907 g dissolved in 100 ml of 0.85% saline).

Legionella bozmanii, *L. dumoffii* and *L. gormanii* all show a blue-white fluorescence on CYE medium under u.v. light (36 μm).

Other tests

Biochemical tests are not rewarding as a means of identification of the different *Legionella* species.

Confirmation by testing for specific fatty acids by gas–liquid chromatography should be done.

For further information, see Harrison & Taylor (1982) and the Association of Clinical Pathologists Broadsheet (Anon. 1981).

References

ANON. 1981 *Laboratory Diagnosis of Legionnaires' Disease* ISSN 0308 2555. London: British Medical Association.

ANON. 1983 *Epidemic Legionellosis in England and Wales. 1979–82* CDR 83–25.

BOPP, C. A., SUMNER, J. W., MORRIS, G. K. & WELLS, J. G. 1981 Isolation of *Legionella* spp. from environmental water samples by low pH treatment and use of a selective medium. *Journal of Clinical Microbiology* **13**, 714–719.

CHESTER, B., POULOS, E. G., DEMRCY, M. J., ALBIN, E. & PRILUCIK, T. 1983 Isolation of *Legionella pneumophila* serogroup 1 from blood with nonsupplemented blood culture bottles. *Journal of Clinical Microbiology* **17**, 195–197.

COLBOURNE, J. S., SMITH, M. G., HARPER, D. & FISHER-HOCH, S. P. 1983 Source of *Legionella pneumophila* in a hospital hot water system: materials used in water fittings capable of supporting the growth of *Legionella pneumophila*. In *Proceedings of the 2nd International Symposium on Legionella*. Washington DC: American Society for Microbiology.

DENNIS, P. J., TAYLOR, J. A., FITZGEORGE, R. D., BARTLETT, C. L. R. & BARROW, G. I. 1982 *Legionella pneumophila* in water plumbing systems. *Lancet* **1**, 949–951.

DENNIS, P. J., BARTLETT, C. L. R. & WRIGHT, A. E. 1983 A comparison of isolation methods for Legionellaceae. In *Proceedings of the 2nd International Symposium on Legionella*. Washington DC: American Society for Microbiology.

DENNIS, P. J., GREEN, D. & JONES, B. P. C. 1984 A note on the temperature tolerance of *Legionella*. *Journal of Applied Microbiology* **54**, 349–350.

DONDERO, T. J., RENDTORFF, R. C., MALLISON, G. F., WEEKS, R. M., LEVY, J. S., WONG, E. S. & SCHAFFNER, W. 1980 An outbreak of Legionnaire's disease, associated with a contaminated air-conditioning cooling tower. *New England Journal of Medicine* **302**, 365–370.

EDELSTEIN, P. H. 1981 Improved semiselective medium for isolation of *L. pneumophila* from contaminated clinical and environmental specimens. *Journal of Clinical Microbiology* **14**, 298–303.

EDELSTEIN, P. H. 1982 Comparative study of selective media for isolation of *Legionella pneumophila* from potable water. *Journal of Clinical Microbiology* **16**, 697–699.

EDELSTEIN, P. H. & FINEGOLD, S. J. 1979 Use of a semiselective medium to culture *Legionella pneumophila* from contaminated lung specimens. *Journal of Clinical Microbiology* **10**, 141–143.

EDELSTEIN, P. H. SNITZER, J. B. & BRIDGE, J. A. 1982 Enhancement of recovery of *Legionella pneumophila* from contaminated respiratory tract specimens by heat. *Journal of Clinical Microbiology* **16**, 1061–1065.

FALLON, R. J. 1982 Legionella infections in Scotland. *Journal of Hygiene* **89**, 439–48.

FEELEY, J. C., GORMAN, G. W., WEAVER, R. E., MACHEL, D. C. & SMITH, H. W. 1978 Primary isolation media for the Legionnaire's disease bacterium. *Journal of Clinical Microbiology* **8**, 320–325.

FISHER-HOCK, S. P., BARTLETT, C. L. R., TOBIN, J. O'H. *ET AL.* 1981 Investigation and control of an outbreak of Legionnaire's disease in a district general hospital. *Lancet* **1**, 932–936.

FITZGEORGE, R. B., BASKERVILLE, A., BROSTER, M., HAMBLETON, P. & DENNIS, P. J. 1983 Aerosol infection of animals with strains of *Legionella pneumophila* of different virulence: comparison with intraperitoneal and intranasal routes of infection. *Journal of Hygiene* **90**, 81–89.

FOY, H. M., BROOME, C. V., HAYES, P. S., ALLEN, I., COONEY, M. K. & TROBE, R. 1979 Legionnaire's disease in a prepaid medical-care group in Seattle 1963–75. *Lancet* **1**, 767–770.

FRASER, D. W., TSAI, T. F., ORENSTEIN, W., PARKIN, W. E., BEECHAM, H. J., SHARRAR, R. G., HARRIS, J., MALLISON, G. F., MARTIN, S. M., MCDADE, J. E., SHEPARD, C. C. & BRACKMAN, P. S. 1977 Field investigation team: Legionnaire's disease: description of an epidemic of pneumonia. *New England Journal of Medicine* **297**, 1189–1197.

HARRISON, T. G. & TAYLOR, A. G. 1982 Diagnosis of *Legionella pneumophila* infection by means of formalised yolk sac antigens. *Journal of Clinical Pathology* **35**, 211–214.

KENNEDY, D. H. & BORLAND, W. 1983 How common is Legionnaire's disease. Letter. *Lancet* **1**, 360–361.

LOVE, W. C., CHAUDHURI, A. K. R., CHIN, K. C. & FALLON, R. J. 1978 Possible case-to-case transmission of Legionnaire's disease. Letter. *Lancet* **ii**, 1249.

MCDADE, J. E., SHEPARD, C. C., FRASER, D. W., TONI, T. F., REDUS, M. A. & DOWDLE, W. R. 1977 Laboratory Investigation Team: Legionnaire's Disease: Isolation of a bacterium and demonstration of its role in other respiratory disease. *New England Journal of Medicine* **297**, 1197–1203.

MACFARLANE, J. T. 1983 Leader—Legionnaire's Disease: Update. *British Medical Journal* **287**, 443–444.

MACFARLANE, J. T., FINCH, R. G., WARD, M. J. & MACRAE, A. D. 1982 Hospital study of adult community-acquired pneumonia. *Lancet* **ii**, 255–258.

MACRAE, A. D., APPLETON, P. N. & LAVERICK, A. 1979 Legionnaire's disease in Nottingham England. *Annals of Internal Medicine* **90**, 580–583.

MILLER, A. C. 1979 Early clinical differentiation between Legionnaire's disease and other sporadic pneumonias. *Annals of Internal Medicine* **90**, 526–528.

PASCULE, A. W., FEELEY, J. C., GIBSON, R. J., CORDES, L. G., MYEROWITZ, R. L. PATTON, C. M. GORMAN, G. W., CARMACH, C. L., EZZEL, J. W. & DOWLING, J. N. 1980 Pittsburg pneumonia agent: direct isolation from human lung tissue. *Journal of Infectious Diseases* **141**, 727–732.

ROWBOTHAM, T. C. 1980 Preliminary report on the pathogenicity of *Legionella pneumophila* for freshwater and soil amoebae. *Journal of Clinical Pathology* **33**, 1179–1183.

TOBIN, J. O'H., BEARE, J., DUNNILL, M. S., FISHER-HOCH, S., FRENCH, M., MITCHELL, R. G., MORRIS, P. J. & MUIRS, M. F. 1980 Legionnaire's disease in a transplant unit: Isolation of the causative agent from shower baths. *Lancet* **ii**, 118–121.

WADOWSKY, R. M. & YEE, R. B. 1981 Glycine containing selective medium for the isolation of Legionellacae from environmental water supplies. *Journal of Applied and Environmental Microbiology* **14**, 768–772.

WEIR, A., BONE, I. & KENNEDY, D. H. 1982 Neurological involvement in legionellosis. *Journal of Neurology, Neurosurgery and Psychiatry* **45**, 603–608.

WHITE, R. J., BLAINEY, A. D., HARRISON, K. J. & CLARKE, S. K. R. 1981 Causes of pneumonia presenting to a District General Hospital. *Thorax* **36**, 566–570.

YONKE, C. A., STIEFEL, H. E., WENTWORTH, B. B. & WILSON, D. L. 1982 Prevalence of antibody to serogroups 1–4 of *Legionella pneumophila*. *American Journal of Epidemiology* **115**, 633–639.

YU, V. L., ZURAVLEFF, J. J., GAVLIK, L. & MAGNUSSEN, M. H. 1983 Lack of evidence for person-to-person transmission of Legionnaire's disease. *Journal of Infectious Diseases* **147**, 362.

Gardnerella vaginalis

C. S. F. Easmon and C. A. Ison

Department of Medical Microbiology, Wright-Fleming Institute, St. Mary's Hospital Medical School, London, UK

In recent years there has been renewed interest in the role of *Gardnerella vaginalis* in non-specific vaginitis. Controversy has surrounded the microbiology and clinical significance of this organism ever since the work of Gardner & Dukes (1955). The taxonomic debate has been settled with the creation of a new genus, *Gardnerella*. There is still uncertainty, however, about the aetiology and pathogenesis of non-specific vaginitis. In this chapter, in addition to describing methods for the isolation and identification of *G. vaginalis*, we consider the need for cultural techniques in the diagnostic laboratory.

Taxonomy

Gardner & Dukes (1955) described the organism now known as *Gardnerella vaginalis* as a Gram variable to Gram negative bacillus. Initially isolated on blood-containing media, it was placed in the genus *Haemophilus* as *H. vaginalis*. This proved unsatisfactory for a number of reasons:

(1) Neither blood nor X and V factors are necessary for its growth.
(2) It had been described as being Gram positive and its cell wall character is still not fully resolved.
(3) There is no antigenic relationship or DNA homology with other members of the genus *Haemophilus*.

An alternative proposal that *G. vaginalis* should be placed classified in the genus *Corynebacterium*, as *C. vaginale*, was made on the grounds that it was really Gram positive (Zinnerman & Turner, 1963). This too was unsatisfactory for the following reasons:

Isolation and Identification of
Micro-organisms of Medical and
Veterinary Importance

Copyright © 1985 by the Society for Applied Bacteriology
All rights of reproduction in any form reserved
0-12-181460-2

(1) There is controversy about the cell wall and the Gram reaction.
(2) It is catalase negative.
(3) Typical corynebacterial cell wall components, arabinose and teichoic acid, are lacking.
(4) There is no clear antigenic relationship or DNA homology with other members of the genus *Corynebacterium*.

This taxonomic debate was finally resolved by Greenwood & Pickett (1980) who proposed the creation of a new genus *Gardnerella* (in honour of H. L. Gardner) with to date a single species, *G. vaginalis*. This was supported by a detailed taxonomic study of *G. vaginalis* by Piot *et al.* (1980).

A summary of the main features of *G. vaginalis* as presented by Greenwood & Pickett (1980) is given next. It is a Gram variable to Gram negative bacillus with laminated cell walls. Acetic acid is the major end product of fermentation. Acid is produced from from dextrin, fructose, galactose, glucose, maltose, mannose, ribose, and starch. Variable results are given with lactose, sucrose, and xylose. Catalase, oxidase, indole and urease tests are negative. It is non-capsulated, non-motile, and facultatively anaerobic.

The DNA base composition is ~43 mole % GC. The optimum temperature is 35–37°C, but growth occurs at 25 and 40°C.

The colonies are β-haemolytic on human and rabbit blood agar but non-haemolytic on sheep blood agar. The optimum pH is between 6.0 and 6.5; growth is poor at pH 4.5 and there is no growth at pH 4.0.

The type strain is 594 of Gardner and Dukes: ATCC 14018; NCTC 10287.

Isolation of *Gardnerella vaginalis*

Gardnerella vaginalis can be isolated on a variety of blood-based media. Casman's blood agar and chocolate agar have been used extensively, particularly in earlier studies. The main problem with these media is that they give little help in distinguishing *G. vaginalis* from the rest of the vaginal flora, and if they are used relatively more time must be spent on identifying *G. vaginalis*. Over the past decade two approaches have been used in the isolation of *G. vaginalis*, both of which have also offered the basis of an identification system. The first, pioneered by Dunkleberg *et al.* (1970), uses peptone–starch–dextrose (PSD) medium. The second uses human blood agar. Both media can be incubated either anaerobically or in 5% CO_2 for 48 h.

Peptone–starch–dextrose agar (PSD)

This medium (Dunkleberg *et al.* 1970) contains (g/litre): proteose peptone no. 3, 20; soluble starch, 10; glucose, 2; Na_2HPO_4, 1; NaH_2PO_4, 1; agar 15.

To make 1 litre, the starch is added to 100 ml cold distilled water. This is then added to 400 ml boiling water and the remaining constituents are added. The pH is adjusted to 6.8 and the medium is autoclaved at 121°C for 15 min.

We have no experience with PSD agar for the primary isolation of *G. vaginalis*. With pure cultures we have found considerable interstrain differences in growth, some clinical isolates growing far less well than stock laboratory cultures.

PSD agar was designed to be used for clinical purposes as part of a comprehensive identification system (outlined in Fig. 1). This system is time-consuming and requires considerable operator skill. If this scheme is not followed, the use of PSD agar for clinical purposes cannot be recommended. It is, however, very useful for the growth of *G. vaginalis* without the use of serum or blood.

Human blood agar

Columbia agar is used as the base, as this provides the necessary growth factors. The components (polypeptone, biosate peptone, myosate peptone, corn starch and NaCl) are critical.

Two-layer plates are best, with 7 ml of the Columbia agar as base and an upper layer of Columbia agar containing 5% time-expired human bank blood.

On human blood agar, *G. vaginalis* produces a characteristic diffuse β-haemolysis (also seen with rabbit blood). When horse or sheep blood is used,

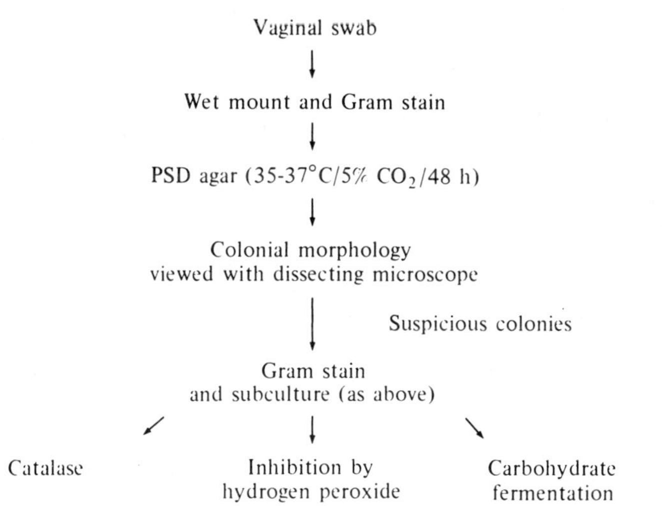

FIG. 1. Identification of *Gardnerella vaginalis* based on peptone–starch–dextrose agar (Dunkleberg *et al.* 1970).

haemolysis is weak or absent and we have noticed that growth is much poorer.

If single-layer plates are used, some *G. vaginalis* isolates may show α- rather than β-haemolysis. Totten *et al.* (1982) recommended the use of Tween 80 to improve the β-haemolysis. By lysing blood in the stored plates, this will reduce their shelf life.

Antibiotic supplements

With vaginal swabs the use of an antibiotic supplement will help to suppress growth of other bacteria. Either of two mixtures may be used: (1) colistin, 10 mg/l; nalidixic acid, 15 mg/l; and amphotericin, 2 mg/l and (2) gentamicin sulphate, 4 mg/l; nalidixic acid, 30 mg/l; and amphotericin, 2 mg/l. These should be added to both layers.

Identification

A scheme for the identification of *G. vaginalis* using PSD was shown in Fig. 1. Another, using human blood agar is given in Fig. 2.

Microscopy

The Gram stain, which will normally form a basis for the identification of most bacterial species, is only of limited value with *G. vaginalis*. This organism is truly Gram variable. Stock laboratory strains and old cultures tend to be Gram negative; fresh clinical isolates and younger cultures are Gram positive. This is irrespective of the decolourising agent and whether neutral red or carbol fuchsin are used as counter stains.

In smears made from vaginal samples, large numbers of small Gram variable bacilli are suggestive of *G. vaginalis*. These are often adherent to epithelial cells forming the clue cell characteristic of non-specific vaginitis.

Specific immunofluorescence staining has been used to identify *G. vaginalis* (Redmond & Kotcher 1963; Vice & Smaron 1973; Svarva & Moeland 1982). We have found this to be a useful technique, and antisera may be applied in other ways (e.g. coagglutination). Unfortunately such antisera are not commercially available.

Culture

Peptone–starch–dextrose agar system

This depends on the recognition of gardnerella colonial morphology on PSD agar under a dissecting microscope. Colonies with the correct morphology after 48 h incubation (dull, white, convex colonies 0.5–2.0 mm in

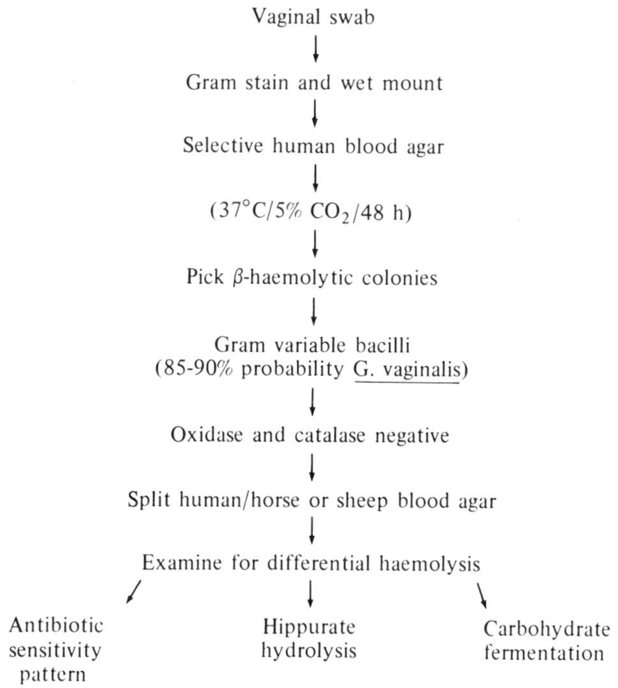

FIG. 2. Identification of *Gardnerella vaginalis* based on human blood agar. For routine purposes, differential haemolysis and Gram staining are probably sufficient. Carbohydrate fermentation tests are not recommended for diagnostic use.

diameter with few if any irregularities) are subcultured. After subculture, *G. vaginalis* is identified by (a) negative catalase reaction, (b) inhibition of growth by hydrogen peroxide, and (c) carbohydrate fermentation.

Human blood agar system

Split bilayer plates, one-half containing human blood, the other containing horse or sheep blood, are used to demonstrate differential β-haemolysis. Colonies exhibiting this pattern can then be further identified by hippurate hydrolysis, antibiotic sensitivity patterns and carbohydrate fermentation reactions (Greenwood *et al.* 1977; Ison *et al.* 1982).

Other tests

Catalase and oxidase tests

These are done using standard methods. With human blood agar a positive catalase test must be treated with caution. Both tests should be negative with G. vaginalis.

Hydrogen peroxide inhibition

Gardnerella vaginalis is inhibited by hydrogen peroxide. A 3% solution can be used to test this, 10 μl of which are placed on a blank disc.

Hippurate hydrolysis

The substrate solution contains: sodium hippurate, 1g; KH_2PO_4 (0.067 *M*), 73.2 ml; Na_2HPO_4 (0.067 *M*), 26.8 ml. The final pH is 6.4.

This solution is sterilised by filtration and dispensed in 0.5-ml amounts. It can be stored at −20°C.

Tubes are inoculated with a loopful of overnight culture from a solid medium. After incubation for 2 h at 37°C in a waterbath, 0.2 ml of 3.5% ninhydrin solution (in equal parts of acetone and butanol) is added. The reaction, which should be negative, can be read either by eye or at 580 nm after 5 min at 37°C.

Carbohydrate fermentation tests

Two media have been described for *G. vaginalis* fermentation tests.

The first is that of Dunkleberg *et al.* (1970). The basal medium contains (g/litre): proteose peptone no. 3, 20; agar, 5. Bromocresol purple (1.6% w/v in ethanol), 10 ml, is added.

The second medium is a modification of that of Bailey *et al.* (1979). The basal medium is that same as above, but Tween 80 (10% v/v), 2 ml, and phenol red (1.6% v/v stock solution), 2 ml, are added.

Both media are prepared by mixing reagents (apart from carbohydrate) in one-tenth the final volume of distilled water and steaming at 100°C to dissolve. After cooling, the pH is adjusted to 7.4 before autoclaving at 115°C for 10 min.

The carbohydrates (glucose, maltose, raffinose and starch) are made up to a 10% w/v concentration and sterilised by membrane filtration and then added to the basal medium. Starch requires the addition of sodium hydroxide to help dissolution.

Fermentation reactions can be carried out in small tubes or bottles, or in Microtitre plates. With the addition of extra agar to a final concentration of 1.1% they can also be done in petri dishes.

Cultures are incubated at 37°C in 5% CO_2 for up to 48 h.

Even in experienced hands these carbohydrate reactions can be difficult to reproduce. They are not recommended for routine use.

Antimicrobial susceptibility tests

Most strains of *G. vaginalis* are sensitive to metronidazole (50-μg disc) and trimethoprim (5-μg disc), and bile (10%) but are resistant to sulphonamides (1- to 2-mg disc).

The metronidazole and sulphonamide discs can be obtained from Oxoid, Difco or BBL and trimethoprim from a variety of sources. Even with the high concentration of sulphonamide used, there may still be problems with blood-containing media. For bile sensitivity testing, 10 μl of a 10% aqueous solution is placed on a blank disc.

Growth and identification of Gardnerella vaginalis *in the diagnostic laboratory. Is it necessary?*

Although the isolation and identification of *G. vaginalis* is straightforward, albeit time-consuming, the question that remains is whether the diagnostic laboratory, as opposed to the reference or research laboratory, needs to do it at all.

Gardnerella vaginalis is associated with the common mild condition of non-specific vaginitis (NSV) or vaginosis. There is no conclusive proof that it is the cause of this disease. Other organisms, in particular *Bacteroides* spp. and anaerobic cocci and vibrios, are also associated with NSV (Spiegel *et al.* 1980; Sprott *et al.* 1983). *Gardnerella vaginalis* can be isolated from a significant minority of women who have no signs or symptoms of NSV (Amsel *et al.* 1983; Ison *et al.* 1983).

Even if *G. vaginalis* is not the cause of NSV, its presence is a reasonable marker for the condition. If culture is relied upon to make the diagnosis, however, some normal women could be diagnosed as having NSV.

Recent work suggests that the diagnosis of NSV should be based on the following:

(1) A homogeneous, often foul smelling vaginal discharge.
(2) The presence of clue cells and/or Gram variable bacilli.
(3) A vaginal pH of 4.5 or more.
(4) A positive amine test (release of a fishy odour) on the addition of 10% potassium hydroxide to vaginal discharge.

In many sexually transmitted disease clinics, experienced staff will be able to carry out all these tests without the assistance of the laboratory. Where this is not the case the main role of the laboratory will be in examining Gram stained films of vaginal samples.

At present, with the inconvenience of obtaining human blood and the length of time taken to culture and identify *G. vaginalis*, we cannot recommend culture as a routine procedure for the diagnosis of NSV. If it is undertaken, carbohydrate fermentation tests should be avoided. In most cases differential β-haemolysis and Gram staining will be satisfactory. Culture may be indicated in difficult cases (e.g. intractable vaginitis) or when *G. vaginalis* is suspected as a cause of more serious infection.

Our work on *G. vaginalis* is supported by a grant from the Medical Research Council.

References

AMSEL, R., TOTTEN, P. A., SPIEGEL, C. A., CHEN, K. C. S., ESCHENBACH, D. & HOLMES, K. K. 1983 Non-specific vaginitis. Diagnostic criteria and microbial and epidemiologic associations. *American Journal of Medicine* **74**, 14–22.

BAILEY, R. K., VOSS, J. L. & SMITH, R. F. 1979 Factors affecting isolation and identification of *Haemophilus vaginalis* (*Corynebacterium vaginale*). *Journal of Clinical Microbiology* **9**, 65–71.

DUNKLEBERG, W. E., JR. & MCVEIGH, I. 1969 Growth requirements of *Haemophilus vaginalis*. *Journal of Microbiology and Serology* **35**, 129–145.

DUNKLEBERG, W. E., JR., KELLOGG, D. S., JR. & SKAGGS, R. 1970 Method for isolation and identification of *Corynebacterium vaginale* (*Haemophilus vaginalis*). *Applied Microbiology* **19**, 47–52.

GARDNER, H. L. & DUKES, C. D. 1955 *Haemophilus vaginalis* vaginitis. *American Journal of Obstetrics and Gynecology* **69**, 962–976.

GREENWOOD, J. R. & PICKETT, M. J. 1980 Transfer of *Haemophilus vaginalis* to a new genus *Gardnerella: G. vaginalis* (Gardner and Dukes) camb nov. *International Journal of Systematic Bacteriology*.

GREENWOOD, J. R., PICKETT, M. J., MARTIN, W. J. & MACK, E. G. 1977 *Haemophilus vaginalis* (*Corynebacterium vaginale*): method for isolation and rapid biochemical identification. *Health Laboratory Science* **14**, 102–106.

ISON, C. A., DAWSON, S. G., HILTON, J., CSONKA, G. W. & EASMON, C. S. F. 1982 Comparison of culture and microscopy in the diagnosis of *Gardnerella vaginalis* infection. *Journal of Clinical Pathology* **35**, 550–554.

ISON, C. A., EASMON, C. S. F., DAWSON, S. G., SOUTHERTON, G. & HARRIS, J. R. W. 1983 Non-volatile fatty acids in the diagnosis of non-specific vaginitis. *Journal of Clinical Pathology* **36**, 1367–1379.

PIOT, P., VAN DYCK, E., GOODFELLOW, M. & FALKOW, S. 1980 A taxonomic study of *Gardnerella vaginalis* (*Haemophilus vaginalis*. *Journal of General Microbiology* **119**, 373–396.

REDMOND, D. L. & KOTCHER, E. 1963 Comparison of cultural and immunofluorescent procedures in the identification of *Haemophilus vaginalis*. *Journal of General Microbiology* **33**, 89–94.

SPIEGEL, C. A., AMSEL, R., ESCHENBACH, D., SCHOENKNECHT, F. & HOLMES, K. K. 1980 Anaerobic bacteria in non-specific vaginitis. *New England Journal of Medicine* **303**, 601–606.

SPROTT, M. S., INGHAM, H. R., PATTMAN, R. S., EISENSTADT, R. L., SHORT, G. R., NARANG, H. K., SISSON, P. K. & SELKON, J. B. 1983 Characteristics of motile curved rods in vaginal secretions. *Journal of Medical Microbiology* **16**, 175–182.

SVARVA, P. L. & MOELAND, J. A. 1982 Identification of *Gardnerella vaginalis* by a fluorescent antibody test. *Acta Pathologica et Microbiologica Scandinavica, Section B: Microbiology* **90B**, 453–455.

TOTTEN, P. A., AMSEL, R., HALE, J., PIOT, P. & HOLMES, K. K. 1982 Selective differential human blood bilayer medium for isolation of *Gardnerella (Haemophilus) vaginalis*. *Journal of Clinical Pathology* **15**, 141–147.

VICE, J. L. & SMARON, M. F. 1973 Indirect fluorescent-antibody method for the identification of *Corynebacterium vaginale*. *Applied Microbiology* **25**, 908–916.

ZINNERMAN, K. & TURNER, G. C. 1963 The taxonomic position of *Haemophilus vaginalis* (*Corynebacterium vaginale*). *Journal of Pathology and Bacteriology* **85**, 213–219.

Campylobacters

K. P. LANDER AND K. P. W. GILL

Ministry of Agriculture, Fisheries and Food, Central Veterinary Laboratory, New Haw, Weybridge, Surrey, UK

Campylobacters have recently attracted much attention. It is in fact some 75 years since they were first described, but progress in understanding the genus was slow for several reasons: their study requires the use of techniques which, although not difficult in themselves, were not in routine use in most laboratories; they frequently occur in contaminated sites from which they can be difficult to isolate; and the diseases associated with them were often insidious and, with a few exceptions, not obviously important socially or economically. Furthermore, much of the earlier literature was confusing because of the diversity of names used, the inadequate identification of strains and some false claims regarding their association with disease. The recent attention has added to our knowledge of the genus, but much still remains to be learned.

The aims of this chapter are to state the current taxonomic position of *Campylobacter* species; to clarify, where possible, their role in disease production; and to give broad guidelines on laboratory methods for their detection, isolation and identification. For more detailed accounts of particular techniques the reader will need to consult the relevant literature.

The Genus *Campylobacter* and Its Species

Bacteria in the genus *Campylobacter* are slender, spirally curved, Gram negative rods, 0.2–0.8 μm in diameter and 0.5–8.0μm in length. Rods may have one or more spirals or may appear S-shaped or as "gull wings". In older cultures coccoid forms may be present. They do not form spores. They are motile by single polar flagella at one or both ends of the cell and have a

Isolation and Identification of
Micro-organisms of Medical and
Veterinary Importance

Copyright © 1985 by the Society for Applied Bacteriology
All rights of reproduction in any form reserved
0-12-181460-2

characteristic darting, corkscrew motility. They are microaerophilic, requiring an O_2 concentration of 3–15%. Some will grow anaerobically. Their metabolism is respiratory. Acid is not produced from carbohydrates. They are oxidase positive and reduce nitrates. The guanine and cytosine content of their DNA is 29–36 mole %. They may be found in the alimentary and reproductive systems of man, animals and birds. Some are pathogenic in man, animals and birds (Smibert 1974).

Nomenclature

These organisms were included in the genus *Vibrio* until Sebald & Véron (1963) demonstrated several fundamental differences between the true vibrios and the microaerophilic vibrios and proposed the new genus *Campylobacter* for the latter. The genus currently contains the following species.
Catalase-positive group: Campylobacter fetus ssp. *fetus; C. fetus* ssp. *venerealis; C. coli; C. jejuni; *C. laridis; *C. faecalis; *C. hyointestinalis;* *"aerotolerant campylobacters".
Catalase-negative group: Campylobacter sputorum ssp. *sputorum; C. sputorum* ssp. *bubulus; *C. sputorum* ssp. *mucosalis; *C. concisus*. Those marked with an asterisk are not included in the approved lists of Skerman *et al.* (1980).

Habitat

Campylobacters are not thought to be free-living but are obligate parasites of man, animals and birds. Most are found in the alimentary system of animals and birds, the exceptions being *C. fetus venerealis*, which has been found only in the reproductive tract of cattle, and *C. sputorum bubulus*, which has been found only in the reproductive tract of cattle and sheep.

Pathogenicity

Campylobacters have been associated with a wide range of diseases in a variety of warm-blooded vertebrates. Their role as causative agents in many of these diseases has not been clear. For this reason each of the following sections is divided into one or more subsections: (1) *definite*, where the role of *Campylobacter* as a cause of the disease is beyond reasonable doubt; (2) *probable*, where the association with disease seems very close but the causal role is not yet proven; and (3) *possible*, where little evidence currently exists for a significant association with disease.

Disease in man

Definitely pathogenic

The *C. jejuni* group (*C. jejuni, C. coli* and *C. laridis*) is one of the commonest recognised causes of enteritis in man. The pattern of disease in developed countries differs from that in developing countries. In the latter, infection is often acquired early in life and disease is thus commonest in infants. There is a relatively high proportion of asymptomatic carriers. In the former, infection generally occurs rather later. The highest incidence of disease occurs in children and young adults, and there are few asymptomatic carriers (Butzler & Skirrow, 1979).

The disease is characterised particularly by abdominal pain and diarrhoea.

Campylobacter fetus fetus has, on rare occasions, been isolated from diseased organs in man. Isolation sites have included blood, joints, heart, liver, brain and meninges, and aborted fetuses. The patients involved are often compromised by some intercurrent disease such as immunodeficiency, cancer or diabetes.

Possibly pathogenic

Campylobacter sputorum sputorum and *C. concisus* have been isolated from the gingival crevice and dental plaque and may play some part in the production of dental caries.

Disease in cattle

Definitely pathogenic

Campylobacter fetus venerealis causes infertility and abortion, the syndrome known as "bovine venereal vibriosis". It is a true venereal infection, carried in the prepuce of the bull and the cervix and vagina of the cow, and spread at coitus. These organisms have been isolated only from the reproductive tract or aborted fetuses of cattle.

Campylobacter fetus fetus causes sporadic abortion in cattle, probably as a result of haematogenous spread from the intestine to the gravid uterus.

Probably pathogenic

Campylobecter fetus fetus has also been isolated from the reproductive tract of bulls and cows, and there is some evidence that it may cause infertility, as in the classical "bovine venereal vibriosis" syndrome. In fact, the dividing line between *C. fetus fetus* and *C. fetus venerealis*, both clinically and biochemically, appears to be imprecise.

Campylobacter jejuni has been isolated sporadically from aborted bovine fetuses.

Possibly pathogenic

Both *C. jejuni* and *C. fetus fetus* have been isolated from cattle with diarrhoea. They are, however, also frequently present in clinically normal cattle. While it is possible that, alone or in combination with other factors, they may be involved in the causation of diarrhoea, their isolation in such cases should be interpreted with care.

Campylobacter jejuni has been shown experimentally to be a potential cause of mastitis, but there is as yet no clear proof of its association with naturally occurring cases of the disease.

The aerotolerant campylobacters have been isolated from aborted and healthy bovine fetuses, and from cases of mastitis.

Disease in sheep

Definitely pathogenic

Campylobacter fetus fetus and *C. jejuni* are causes of outbreaks of abortion in sheep. In Britain they are the third most common cause of abortion.

Possibly pathogenic

Campylobacter fetus fetus and *C. jejuni* may occasionally play some part in causing enteritis in sheep, but isolation of these organisms is not proof of causation. Aerotolerant campylobacters have been isolated from normal and aborted fetuses.

Disease in pigs

Probably pathogenic

Campylobacter sputorum mucosalis is frequently found in association with a group of enteric diseases in pigs called the "porcine intestinal adenomatosis" complex. While observations of gross and microscopic pathology are the most common means of diagnosis of the condition, isolation of the organisms is considered strong supportive evidence. Recently *C. hyointestinalis* has also been found in the porcine intestinal adenomatosis complex. The aetiology of this disease is not yet clear.

Possibly pathogenic

The aerotolerant campylobacters have been isolated from normal and aborted fetuses of pigs.

Disease in other animals

Possibly pathogenic

Campylobacter jejuni is said to be the cause of "vibrionic hepatitis" in poultry. Organisms of the *C. jejuni* group have been isolated from a wide range of animals and birds with and without diarrhoea. It is possible that in some cases of diarrhoea, particularly in dogs and cats, the campylobacters are wholly or partially responsible for the disease. Recently, *C. jejuni* has been reported as the cause of a large outbreak of abortion in mink and two outbreaks of enteritis in young mink (Hunter & Prescott 1983; Hunter *et al.* 1983).

Collection of Samples

Clinical samples for the detection of campylobacters are, in most instances, collected by standard methods. Two types of sample, however, used in the diagnosis of *C. fetus* infection in cattle, deserve special mention. Both present particular problems to the veterinary clinician, and in both instances the accuracy of diagnosis can be greatly affected by the quality of sample sent to the laboratory. The two samples are preputial washings from bulls and vaginal mucus from cows.

Preputial washings from bulls

Campylobacter fetus resides deep in the folds of mucosa covering the penis and inner surface of the prepuce. They are not easily dislodged from these sites. Consequently, if a lavage technique is used, it is important to massage vigorously the fluid instilled into the preputial sac in order to ensure that the sample contains *C. fetus* in adequate numbers for the diagnosis of infection.

Vaginal mucus from cows

The vaginal mucus is frequently scanty in volume and tightly adherent to the walls of the vagina. As a result the current method of collection, which entails sucking the mucus into a tube, is often unsuccessful. A newly devised technique consists of using a spray to dislodge the mucus from the wall of the vagina and then withdrawing the mucus suspension with a syringe (Lander 1983).

Transport and preservation

All campylobacters die rapidly on exposure to air, heat and drying, and in the presence of post-mortem invaders. Therefore carcasses such as animal foetuses should be fresh, and clinical samples should be kept moist and cool, and, if possible, air should be excluded from the sample. They should reach the laboratory as soon as possible after collection. Once again *C. fetus* presents particular problems, especially in preputial washings from bulls where it remains viable for no longer than 6–8 h.

Commercially available transport media (e.g. Stuart's, Amies) have been used successfully for preserving campylobacters in suitable clinical samples. Other media have been devised purely for campylobacters. Details are given below.

Detection

Light microscopy

Because of their characteristic shape and motility, it is sometimes possible to make a presumptive diagnosis of a campylobacter infection from dried stained smears or wet preparations of clinical material (e.g. human faeces, ovine and bovine fetal stomach contents). Because of the low refractive index of the organisms it may be preferable to examine wet preparations by dark-ground or phase-contrast microscopy. Such presumptive diagnoses should always be confirmed by isolation or other methods of identification of the organism.

Immunofluorescence

Staining organisms with a specific antibody conjugated to a fluorescent dye can be a useful technique for demonstrating campylobacters. Its advantages are that it is rapid, does not depend on the organisms being viable and is not affected by the presence of other contaminating organisms. With the multiplicity of serotypes present among campylobacters (see *Identification, Serotyping*), however, the technique has not been widely used except for the demonstration of *C. fetus*, chiefly in preputial washings from bulls and, occasionally, vaginal mucus in cows and fetal stomach contents. Even here the test is not specific for *C. fetus venerealis*, because many strains of *C. fetus fetus* have antigens common to *C. fetus venerealis*. Consequently immunofluorescence for *C. fetus* gives only a presumptive diagnosis which should be supported by other evidence of infection.

Isolation

General principles

Campylobacters are fairly fastidious organisms, particularly with regard to oxygen tension but also in their temperature range and nutrient requirements.

Gaseous requirement

Campylobacters are microaerophilic, requiring an oxygen concentration of 3–15% (optimal 4–8%). The growth of most species is improved by raising the carbon dioxide level to 5–10%. Some species (e.g. *C. sputorum*) have an obligate requirement for hydrogen, while in others (e.g. *C. hyointestinalis*) growth is improved by the presence of hydrogen; in addition the "H_2S in FBP" test (see *Identification, Biotyping*) is more dependable if the organisms are cultured in an atmosphere containing hydrogen.

Methods of producing a suitable microaerobic atmosphere include the following:

A gassed incubator. The incubator is connected to cylinders containing nitrogen and carbon dioxide. An oxygen electrode measures the oxygen concentration and automatically allows the nitrogen to enter the cabinet until the oxygen concentration reaches the required (preset) level. A carbon dioxide electrode similarly controls the concentration in the cabinet. In theory it should be possible to replace the nitrogen with hydrogen but the dangers associated with the use of the latter render this alternative unacceptable.

These incubators are expensive but have the great advantage that the contents are easily accessible. It is worth noting that their use is not restricted to microaerophilic organisms—the atmosphere should support the growth of most aerobic and facultative organisms, as well as carbon dioxide-dependent bacteria such as *Brucella* species.

Gas-generating kits. These commercially available (e.g. BBL, Oxoid) kits use an anaerobic jar into which a gas-generating envelope is placed. When activated the envelope contents generate hydrogen and carbon dioxide. If a catalyst is placed inside the jar, the hydrogen reacts rapidly with oxygen in the jar and anaerobic conditions result. Originally these systems were designed to produce only anaerobiosis. Microaerobic conditions can be produced by using the system without a catalyst, so that the reaction between hydrogen and oxygen takes place much more slowly, thus allowing suitable conditions for cultivation of campylobacters. Now, however, there are envelopes made especially for *Campylobacter* culture (e.g. Campy-Pak, BBL). These are more efficient, accurate and safer than anaerobic systems without a catalyst.

These kits are convenient to use, do not require vacuum pumps and do not create aerosols. It should be remembered, however, that gas mixtures of hydrogen and carbon dioxide and of hydrogen and oxygen are flammable, and care should be taken to keep the jars away from naked flames, sparks, etc.

Evacuation and replacement. For this method an anaerobic jar without a catalyst is used. The jar is evacuated with a vacuum pump to a pressure of about 250 mm Hg. The air is replaced with a mixture of 10–15% carbon dioxide in nitrogen or hydrogen. This should give an atmosphere containing about 6%

oxygen and 8–10% carbon dioxide. This method is fairly accurate but requires a vacuum pump, a pressure gauge and gas cylinders. A further disadvantage is that when the gas mixture is being fed into the evacuated jar quite marked air turbulence occurs, aerosols are created, and plate contamination can be a problem. In addition if hydrogen is used in the replacement gas a flammable mixture results.

Candle jar. As the candle burns in the jar it uses oxygen and forms carbon dioxide, which, being heavier than air, displaces the air from the bottom of the jar. Thus a microaerobic, carbon dioxide-rich atmosphere is created. The flame is eventually extinguished by lack of oxygen. The technique has been shown to be effective in the isolation of *Campylobacter* spp. Oxygen and carbon dioxide concentrations depend on the number of plates in the jar, the height of the candle, and the turbulence of the atmosphere, and is therefore difficult to standardise. Nevertheless, the technique is inexpensive and has obvious advantages in remote areas and developing countries.

Temperature range

This is fairly narrow and varies with different species. All of the currently recognised species will grow at 37°C. The *C. jejuni* group has an optimal temperature of about 42°C, and many medical microbiologists prefer to use this temperature for the isolation of *C. jejuni*. It should be remembered, however, that the *C. fetus* group either will not grow at 42°C or will grow only poorly.

Their rate of growth is slower than many of the common pathogens. The *C. jejuni* group is clearly visible on solid media after 24 h at 42°C, but at 37°C may take 48 h for growth to appear. The *C. fetus* group usually requires 72 h at 37°C.

Selective techniques

The isolation of campylobacters from clinical and other samples usually entails contending with large numbers and varieties of contaminating organisms. Therefore selective methods are vital to success. Two types of selective technique are commonly used, filtration and selective media.

Filtration

The slender, strongly motile campylobacters can pass through a membrane filter of average pore size 0.65 μm which does not allow the passage of many of the larger contaminants. Filters are held in a Swinney adaptor fitted to a syringe (or disposable filters and holders are available commercially). The sample is forced gently through the filter onto a culture plate or other medium. Alternatively, a sterile membrane filter (pore size 0.65 or 0.45 μm) is placed on a blood (or selective) agar plate. A drop of sample is put on the filter and left for 15–30 min. The filter is removed and the inoculum under the filter is spread over the plate.

Selective media

Several media have been developed for the isolation of campylobacters. Most of them are available commercially. For details, see *Culture media*.

Solid media include, for the *C. jejuni* group, Skirrow, Preston, Butzler, Blaser, Campy-BAP; for the *C. fetus* group, Skirrow, Blaser, Campy-BAP; for *C. hyointestinalis*, Skirrow, Blaser, Campy-BAP; for *C. fecalis*, Skirrow; for the aerotolerant strains, Neill, Skirrow, Blaser; and for *C. sputorum*, Lawson & Rowland.

Most of these can be adapted to make liquid media for the enrichment of campylobacters. We have achieved greatest success with Preston enrichment broth for the *C. jejuni* group and Weybridge transport and enrichment broth for the *C. fetus* group.

For samples which are likely to harbour fungal contaminants (e.g. from herbivorous or omnivorous animals, vegetarian populations or environmental samples), a fungistat or fungicide should be added to those media which do not already contain one.

Culture

Non-selective media

While campylobacters do not necessarily require enriched media, most of them grow better on media containing blood or serum. Almost any of the commercially available enriched media may be used. The following are recommended.

Sheep blood agar (SBA)

Blood agar base (Oxoid no. 2) is added to distilled water according to the maker's instructions and heated to dissolve. The clear base is autoclaved at 121°C for 15 min, cooled to 50°C and sheep blood added to give 5%. The medium is mixed well and dispensed into petri dishes.

Blood broth

Veal infusion broth (Difco) is dispensed in 15-ml volumes in 25-ml screw-capped (Universal) bottles and autoclaved at 121°C for 15 min. Before use, 2–3 ml of fresh defibrinated sheep blood is added.

Mehlman & Romero (1982) medium

This contains (g/litre) in distilled water: proteose peptone (Difco no. 3), 15; yeast extract, 7.5; casamino acids, 5; K_2HPO_4, 5; $(NH_4)_2SO_4$, 1.5; agar, 1.5; soluble starch, 1. The ingredients are dissolved in the water and the pH is adjusted to 7.3 with 0.1 *M* potassium hydroxide. The medium is dispensed in 20-ml volumes in 25-ml screw-capped bottles (Universal containers) and autoclaved at 121°C for 15 min. A 2.5% solution of sodium sulphite is prepared in sterile distilled water and filtered through a 0.2 μm membrane filter. Volumes of 0.5 ml of this solution are added aseptically to 20 ml of the completed medium.

Selective media

Stock solutions of antibiotics and other inhibitors are prepared in sterile distilled water at 500-fold or 1000-fold their final concentration. Non-sterile solutions may be sterilised by filtration through a 0.2 μm membrane filter. They are stored at 4°C until required. Most are readily soluble in water except for the following:

1. Trimethoprim. The lactate salt is readily soluble. If this is unavailable a few drops of lactic acid will aid solubility. After dissolving, the pH of the solution should be readjusted to neutral using 0.1 *M* NaOH.
2. 5-Fluorouracil. 5-Fluorouracil, 1 g, is dissolved in approximately 50 ml of distilled water containing 1–2 ml of 2 *M* NaOH by heating in a waterbath at 56°C with frequent agitation. This may take up to 2 h. The pH is then adjusted to 7.4–7.6 using 1 *M* HCl and the volume made up to 100 ml with distilled water. The sterile solution is stored in 15- to 20-ml volumes. It is usually necessary to redissolve the 5-fluorouracil before use. This should again be done in a waterbath at 56°C. The solution should never be heated above 56°C.
3. Rifampicin. Rifampicin, 1 g, is dissolved in 10 ml of dimethyl formamide. Distilled water is added to 100 ml. The sterile solution is stored frozen in the dark.

Where required, and if not already included, any of the selective media may be supplemented by the addition of a fungicide or fungistat, for example, actidione (cycloheximide) at 100 μg/ml, or amphotericin B at 2 μg/ml.

Skirrow's medium

This is recommended (Skirrow 1977) for the isolation of all catalase-positive campylobacters from contaminated sources. The following bases have been used: blood agar base no. 2 (Oxoid), Columbia agar base, *Brucella* medium base, and Mueller–Hinton agar. Basal media containing nutritionally rich ingredients are likely to contain sufficiently high levels of thymidine to antagonise the inhibitory effect of trimethoprim. Thymidine phosphorylase, which is present in lysed horse blood, can break down thymidine to thymine which is less antagonistic. The antagonistic effect of thymidine levels greater than 10 μg/ml cannot be reversed by lysed horse blood. Thymidine does not inhibit the synergistic effect shown by trimethoprim and polymyxin B. The agar base is melted and cooled and the following are added: lysed horse blood, 5–7%; vancomycin, 10 μg/ml; trimethoprim, 5 μg/ml; and polymyxin B, 2.5 IU/ml.

The antibiotic mixture is available from Oxoid, Gibco Europe, or Mast Laboratories. The concentration of the antibiotics may be increased to as much as three times normal strength for the isolation of *C. fetus* from heavily contaminated sources.

Preston medium (solid)

Good isolation rates for *C. coli/jejuni* and *C. fetus fetus* strains have been achieved with this medium (Bolton & Robertson 1982). It is nutrient broth no. 2 (Oxoid) containing the following: New Zealand agar, 1.2%; lysed horse blood, 5%; polymyxin B, 5 IU/ml; rifampicin, 10 μg/ml; and trimethoprim, 10 μg/ml.

The blood and antibiotic solutions are added to sterile, melted and cooled agar base. The antibiotic supplement is available commercially.

Preston liquid enrichment medium

This medium contains the same base and antibiotics as Preston solid medium. It is widely used and well tried for the isolation of *C. coli/jejuni* and *C. fetus fetus*. It does not contain agar but it does have an added aerotolerant supplement (FBP) containing (g/litre): sodium pyruvate, 0.25; sodium metabisulphite, 0.25; and ferrous sulphate (hydrated salt) 0.25. This supplement is available commercially.

Weybridge transport and enrichment medium (TEM)

This medium will support the growth of all *C. fetus* subspecies (Lander & Gill, unpublished observations). It is a Mueller–Hinton broth to which are added: bacteriological charcoal, 0.5%; sodium pyruvate, 0.025%; ferrous sulphate (hydrated salt), 0.025%; and sodium metabisulphite, 0.025%.

The base and supplement are prepared in 100-ml volumes and sterilised at 121°C for 15 min. For use, lysed horse blood (7%) and antibiotic supplement are added to the following concentrations: vancomycin, 40 μg/ml; trimethoprim, 20 μg/ml; polymyxin B, 10 IU/ml; actidione, 100 μg/ml; 5-fluorouracil, 500 μg/ml. The complete medium is mixed well to disperse the charcoal evenly and dispensed in 10-ml volumes in small screw-capped bottles.

Butzler's medium

This is blood agar base no. 2 (Oxoid) containing whole horse blood, 5%; and bacitracin, 25 IU/ml; cephazolin, 15 μg/ml; colistin, 10 IU/ml; novobiocin, 5 μg/ml; and cycloheximide, 50 μg/ml (Lauwers *et al.* 1978).

The horse blood and antibiotics are added to the sterile melted and cooled base. The antibiotics are available commercially as a supplement.

Blaser's medium

This medium (Blaser *et al.* 1978) is blood agar base no. 2 (Oxoid) containing the following: lysed horse blood, 5%; and cephalothin, 15 μg/ml; polymyxin B, 2.5 IU/ml; trimethoprim, 5 μg/ml; vancomycin, 10 μg/ml; and amphotericin B, 2 μg/ml. The blood and antibiotics are added to the sterile melted and cooled base. The antibiotics are available commercially as a supplement.

Campy-BAP medium

Brucella agar base (BBL) is used with sheep blood at 10% and the following antibiotics: cephalothin, 15 μg/ml; polymyxin B, 2.5 IU/ml; trimethoprim, 5 μg/ml; vancomycin, 10 μg/ml; and amphotericin B, 2 μg/ml (Blaser *et al.* 1979). The antibiotics are used at the same concentration as in Blaser's medium. (Campy BAP is available from BBL as ready-poured plates.)

Campylobacter sputorum mucosalis *medium*

Lawson & Rowland's (1974) isolation medium is blood agar base no. 2 (Oxoid) containing: horse blood, 5%; novobiocin, 5 μg/ml; and brilliant green, 1/60,000. The blood, novobiocin and dye are added to the melted and cooled base.

Maintenance medium for pure cultures

This is a Columbia agar base with 7% horse blood or Hartley's digest agar with 5% horse blood.

Campylobacter hyointestinalis *isolation medium*

Mueller-Hinton agar base is melted, cooled to 56°C, and to this sterile medium are added: sheep blood, 5%; trimethoprim, 2.5 μg/ml; and sulphamethoxazole, 47.5 μg/ml. Pure cultures are maintained on the same medium without the antibiotics.

Media for the isolation of aerotolerant strains

The liquid selective enrichment medium (Neill *et al.* 1980) is Leptospira EMJH medium (Difco) with the addition of: agar, 0.15%; EMJH Leptospira enrichment supplement (Difco), 10%; 5-fluorouracil, 100 μg/ml; and rabbit serum, 1%. The sloppy agar base is melted, sterilised and cooled according to the manufacturer's instructions and the supplement, 5-fluorouracil, and serum added. The medium is bottled aseptically in 20-ml volumes in 1-oz Universal containers.

The solid selective medium is prepared from blood agar base (Oxoid no. 2) by adding to the sterile melted and cooled base: lysed horse blood, 7% and carbenicillin, 125 μg/ml.

Preservation (FBP) medium

Cultures remain viable in this medium for a few weeks at room temperature or a few months at −20°C. This medium is also used for the "H_2S in FBP" test. It is nutrient broth (Oxoid no. 2) containing (g/litre): agar (Difco), 12; yeast extract, 1.0; sodium pyruvate, 0.5; sodium metabisulphite, 0.5; ferrous sulphate (hydrated), 0.5; glycerol, 150 ml.

The ingredients are dissolved in water using heat to melt the agar. The medium is dispensed into 4-ml screw-capped bottles, which are filled to the top to exclude air before autoclaving at 121°C for 15 min.

Liquid nitrogen storage medium

Solution A contains NaH_2PO_4, 8.0 g/litre; solution B contains Na_2HPO_4, 9.5 g/litre. These are mixed in the proportions 39 volumes of A and 61 volumes of B. Dimethyl sulphoxide, 15% v/v, is added and the medium is sterilised at 121°C for 15 min.

Low-temperature (−70°C) storage medium

This is sterile Mueller–Hinton broth with fresh defibrinated ovine blood added to approximately 16%.

A dense suspension of bacteria is mixed into the medium, which can then be stored at −70°C. To recover organisms the vial is removed from storage, a small amount of medium is scraped off the top and plated out, and the remaining suspension is replaced in the −70°C storage. Thus it is not necessary to thaw and refreeze the stored bacteria (J. H. Bryner, personal communication).

Identification

Colonial morphology

Unless otherwise stated, the following descriptions refer to cultures on blood agar incubated at 37°C for 48 h.

Campylobacter fetus (72 h). Diameter 1–3 mm, convex, entire-edged, translucent and buff-coloured. On moist agar, swarming may occur.

Campylobacter jejuni, C. coli, C. laridis. Diameter 2–4 mm, low convex. Edges are usually entire but the colony may extend along the line of inoculation. Sometimes two colony types occur in one culture, one the low convex type and the other smaller, domed and entire-edged; subculture of either of these colony types will also produce two types of colony. Colonies are translucent and buff-coloured to tan-coloured. Older cultures frequently develop a metallic sheen. Swarming may occur on moist agar.

Campylobacter fecalis. Variable in size but most commonly up to 3.5 mm diameter, low convex or pyramidal, entire-edged. They are translucent and flesh-coloured to dirty grey.

Campylobacter hyointestinalis. Diameter, 1.5–2.0 mm, low convex, entire-edged, translucent and buff-coloured.

"Aerotolerant" campylobacters. Diameter 1–3 mm, flat, transparent colonies; when incubated in air at 30°C the colonies are 1–2 mm diameter, convex, entire-edged, translucent and cream-coloured.

Campylobacter sputorum ssp. *sputorum.* Diameter 1–2 mm, raised flat colonies, round with a thin irregular spreading edge, translucent, greyish, occasionally produce α-haemolysis.

Campylobacter sputorum ssp. *bubulus.* Diameter 1–3 mm, low convex to flat, entire-edged, translucent to opaque and buff-coloured with a greenish tinge. Frequently α-haemolytic causing greening of the medium.

Campylobacter sputorum ssp. *mucosalis.* Diameter 1.5 mm, raised with flat surface or umbonate, entire-edged but colony may extend along the line of inoculation, translucent and yellowish-grey. Sometimes α-haemolysis occurs, most obviously under the heaviest part of the inoculum. Swarming may occur on moist agar.

Campylobacter concisus. Diameter 1 mm, convex with entire edge, translucent, greyish. Some strains may produce swarming growth on some media.

Cellular morphology

Campylobacters are Gram negative, non-acid-fast and non-sporing. Because they tend to stain weakly it may be necessary either to increase the concentration of stain or to prolong staining time.

All campylobacters are curved rods, but species and strain variations occur in size, curvature and the tendency to produce coccoid forms or long spiral filaments.

Most young cultures contain a mixture of comma-shapes, "seagull wings", S-shapes and short spirals. The coccoid or filamentous forms become more prominent in older cultures. Some strains tend to lose their curvature after frequent subculturing in the laboratory. Cells of *C. fetus* are generally larger than those of the *C. jejuni* group, with a longer wavelength to the curve and with fairly blunt ends. In older cultures longer spiral filaments become more frequent and coccoid forms are rare.

Cells of the *C. jejuni* group usually are smaller with a tighter curve and tapering ends. In older cultures (more than 48 h) coccoid forms become very common, even to the near-total exclusion of curved and spiral cells. The cocci of *C. jejuni* and *C. coli* tend to be small while those of *C. laridis* tend to be large. Predominantly coccoid cultures may be difficult to sub-culture.

Cells of *C. hyointestinalis* show some variation in morphology. Most cells are smaller and more tightly curved than *C. fetus* but less so than *C. jejuni*.

Aerotolerant campylobacters vary in morphology with conditions of culture. If grown at 37°C the cells are fairly straight slender rods with a proportion showing some curvature. At 30°C most cells are curved with a long wavelength and, particularly in liquid media, long spiral forms occur. These may reach 20 μm in length.

The *C. sputorum* group generally shows less regular morphology than young cultures of *C. fetus* and *C. jejuni*. The organisms are less curved and cells with more than two spirals are uncommon, although in older cultures coccoid and filamentous forms may occur.

Campylobacter concisus has curved rods with occasional straight or helical forms, with rounded ends.

Flagella stains show that most campylobacters have single polar flagella at one or both ends of the cell. However, cells of the *C. jejuni* group and *C. fetus venerealis* may have one to four flagella per pole. Occasionally non-flagellated variants occur.

Campylobacters are generally very actively motile with rapid, darting movements. Motility is best demonstrated by dark-ground or phase-contrast microscopy where it can be seen to be a corkscrew-like motion. *Campylobacter sputorum* is said to be sluggishly motile with only a small proportion of cells being active. All campylobacters tend to become more sluggish in older cultures.

Biotyping

Table 1 shows a biotyping scheme for *Campylobacter* species.

The methods and the media listed next (see also *Culture*) are those used at the Ministry of Agriculture, Fisheries and Food, Central Veterinary Laboratory, Weybridge. Other laboratories may use minor variations to these techniques.

The inoculum for biotyping media is prepared by inoculating a 5-ml vol-

TABLE 1

The identification of Campylobacter *species*[*],[†]

	C. fetus species			*C. coli/jejuni* group					*C. sputorum* species					
	C. fetus fetus	*C. fetus venerealis*	*C. fetus venerealis* biotype *intermedius*	*C. jejuni* biotype 1	*C. jejuni* biotype 2	*C. coli*	*C. laridis* (NARTC)	*C. fecalis*	*C. sputorum sputorum*	*C. sputorum bubulus*	*C. sputorum mucosalis*	*C. hyointestinalis*	Aerotolerant campylobacters	*C. concisus*
Catalase	+	+	+	+	+	+	+	+	−	−	−	+	+	−
Morphology														
Wavelength	Long	Long	Long	Short	Short	Short	Short	Long	Medium long	Medium long	Medium long	Long	Long	Medium
Coccal forms	Rare	Rare	Rare	Common	Common	Common	Very common	Rare	Rare	Rare	Rare	Rare	Rare	NK
Growth in air	−	−	−	−	−	−	−	−	−	−	−	−	+	−
Temperature growth, °C														
25	+	+	+	−	−	−	−	−	+	d	+	+	+	NK
30.5	+	+	+	−	−	+	+	+	+	+	+	+	+	NK
43	−	−	−	+	+	+	+	+	−	−	−	−	−	NK
Sensitivity to														
Nalidixic acid	R	R	R	S	S	S	R	NK	NK	NK	R(d)	R	S	R
Cephalothin	S	S	S	R	R	R	S	NK	NK	NK	S	S	NK	NK
1.5% NaCl	NK	NK	d	S	S	S	R	R	R	R	R	S	NK	NK
3.0% NaCl	S	S	S	S	S	S	S	R	S	R	S	S	NK	NK
1.0% glycine	R	S	S	R	R	R	R	R	R	R	S	R	R(d)	NK
TTC	S	S	S	S	S	R	S	NK	NK	NK	R(d)	S	S	NK
H_2S production														
Triple sugar iron agar	−	−	−	−	−	−	+	+	−d	+	+	+	−	NK
"Sensitive" medium	+	−	+	+	+	+	+	+	+	+	+	+	−	+
FBP	−	−	−	−	+	−	+	−	NK	NK	NK	NK	−	NK
Hippurate hydrolysis	−	−	−	+	+	−	−	−	−	−	−	−	NK	NK
Nitrite reduction	−	−	−	−	−	−	−	−	+	+	+	−	−	+

[*] R, resistant; S, sensitive; d, variable results; NK, not known; TTC, triphenyltetrazolium chloride.

[†] Prepared from the observations of Sebald & Véron (1963); Firehammer (1965); Loesche *et al.* (1965); Véron & Chatelain (1973); Smibert (1974, 1978); Neill *et al.* (1979); Skirrow & Benjamin (1980); Lawson *et al.* (1981); Tanner *et al.* (1981); Gebhart *et al.* (1983).

ume of Mueller–Hinton broth with a single colony of the isolate to be biotyped. The broth is incubated at 37°C on a shaker for 5 h. Each test medium is inoculated with two drops of this broth using a Pasteur pipette.

The basal medium referred to below is Mueller–Hinton broth made semi-solid by the addition of 0.1% agar. The required reagents are added to this as indicated and the media are dispensed in volumes of 7.5 ml in 6 × 0.5 in. test tubes with "Cap-O-Test" tops (Payne, Iveley Road, London SW4).

Inoculated media (unless otherwise stated) are incubated at 37°C in a microaerobic atmosphere. Tests may be read after 48, 72 and 120 h.

Tests for motility, catalase, oxidase, nitrate and nitrite reduction (in basal medium plus 0.1% KNO_3 or 0.001% $NaNO_2$, respectively) and H_2S in triple sugar iron agar (T.S.I.) are performed by standard methods (Cowan 1974).

Differential temperature tests

Basal medium is inoculated and incubated at 25 and 43°C.

"Sensitive" H_2S test

Basal medium and basal medium containing 0.02% cysteine hydrochloride are inoculated and lead acetate paper strips are suspended above the medium during incubation. H_2S production is indicated by blackening of the paper. Tests should be incubated for 120 h before a negative result is recorded.

Tolerance to NaCl and glycine

This is done in basal medium to which the reagent is added to give the required concentration. Tests are read visually.

Sensitivity to nalidixic acid, cephalothin and triphenyltetrazolium chloride (TTC)

One loopful of inoculum is streaked across a sheep blood agar plate. For nalidixic acid and cephalothin a disc containing 30 μg of the required reagent is placed in the centre of the streak. For TTC a filter-paper strip (8 × 1 cm), which has been soaked in TTC solution (20 mg/ml), dried and autoclaved, is placed at a right angle across the centre of the streak. For each of these tests up to six organisms can be tested on one plate.

Growth in air

An SBA plate is inoculated and incubated aerobically at 37°C.

H_2S in FBP

An SBA plate is inoculated and incubated in a microaerobic atmosphere containing hydrogen. A pea-sized pellet of growth is scraped from the plate and placed carefully in FBP broth taking care not to disrupt the pellet. The broth is incubated at 37°C in the dark for 4 h. H_2S production is indicated by blackening of the pellet.

Hippurate hydrolysis

The growth from a SBA plate is scraped off and suspended in 2 ml of distilled water. A volume of 0.5 ml of 5% sodium hippurate is added and the mixture is incubated at 37°C for 2 h. One millilitre of ninhydrin reagent (3.5 g of ninhydrin in 100 ml of a 1:1 mixture of butanol and acetone) is added. Hippurate hydrolysis is indicated by the rapid development of a deep purple colour.

Serotyping

The current system of classification of *Campylobacter* spp. is based on the biotyping scheme just given, or variations of it. Serotyping, particularly of the *C. jejuni* group, is receiving much attention and will be a valuable method of subdividing species for more accurate identification. The surface antigens of the major groups of campylobacters appear to be distinct with little or no cross reaction between groups.

Campylobacter fetus can be divided into two serogroups, A and B, on the basis of heat-stable antigens (Morgan 1959). Serogroup A includes all *C. fetus venerealis* and some *C. fetus fetus* strains; the remaining *C. fetus fetus* strains are serogroup B. The species contains a range of heat-labile antigens.

The *C. jejuni* group (*C. jejuni*, *C. coli* and *C. laridis*) has so far been divided into 60 serogroups based on heat-stable antigens and 54 serogroups based on heat-labile antigens. One serogroup may include strains of one, two or all three species. There appears to be no major antigen common to all organisms in this group. Knowledge of the antigenic structure of the group is developing rapidly, however, and progress in all aspects of serotyping can be expected.

The three subspecies of *C. sputorum* are antigenically distinct from each other and from other members of the genus. *C. sputorum mucosalis* has been divided into at least three serogroups. *C. concisus* is antigenically distinct from the type strains of *C. sputorum* and *C. fetus*.

Other methods

Other methods which have been used to study the genus include DNA homology studies, DNA hydrolysis, analysis of protein profiles, analysis of products of metabolism, enzyme studies and bacteriophage. All have yielded more or less valuable information.

References

BLASER, M., CRAVENS, J., POWERS, B. W., & WANG, W. L. 1978 Campylobacter enteritis associated with canine infection. *Lancet* **ii**, 979–981.

BLASER, M. J., BERKOWITZ, I. D., LA FORCE, F. M., CRAVENS, J., RELLER, L. B. & WANG, W. L. 1979 Campylobacter enteritis: clinical and epidemiological features. *Annals of Internal Medicine* **91**, 179–185.

BOLTON, F. J. & ROBERTSON, L. 1982 A selective medium for isolating *C. jejuni/coli*. *Journal of Clinical Pathology* **35**, 462–467.

BUTZLER, J. P. & SKIRROW, M. B. 1979 Campylobacter enteritis. *Clinics in Gastroenterology* **8**, 737–765.

COWAN, S. T. 1974 *Cowan and Steel's Manual for the Identification of Medical Bacteria* 2nd edn. Oxford: Cambridge University Press.

FIREHAMMER, B. D. 1965 The isolation of vibrios from ovine faeces. *Cornell Veterinarian* **55**, 482–494.

GEBHART, C. J., WARD, G. E., CHANG, K. & KURTZ, H. J. 1983 *Campylobacter hyointestinalis* (new species) isolated from swine with lesions of proliferative ileitis. *American Journal of Veterinary Research* **44**, 361–367.

HUNTER, B. D. & PRESCOTT, J. F. 1983 Natural and experimental campylobacter diarrhoea in mink. (Abstract.) In *Campylobacter II*, ed. Pearson, A. D., Skirrow, M. B., Rowe, B., Davies, J. R. & Jones, D. M., pp. 125–126. London: Public Health Laboratory Service.

HUNTER, B. D., PETTIT, J. R. & PRESCOTT, J. F. 1983 An outbreak of *Campylobacter jejuni* abortion in mink. (Abstract.) In *Campylobacter II*, ed. Pearson, A. D., Skirrow, M. B., Rowe, B., Davies, J. R. & Jones, D. M., p. 129. London: Public Health Laboratory Service.

LANDER, K. P. 1983 New technique for collection of vaginal mucus from cattle. *Veterinary Record* **112**, 570.

LAUWERS, S., DEBOECK, M. & BUTZLER, J. P. 1978 Campylobacter enteritis in Brussels. *Lancet* **i**, 604–605.

LAWSON, G. H. K. & ROWLAND, A. C. 1974 Intestinal adenomatosis in the pig: A bacteriological study. *Research in Veterinary Science* **17**, 331–336.

LAWSON, G. H. K., LEAVER, J. L., PETTIGREW, G. W. & ROWLAND, A. C. 1981 Some features of *Campylobacter sputorum* subsp. *mucosalis* subsp. nov., nom. rev. and their taxonomic significance. *International Journal of Systematic Bacteriology* **31**, 385–391.

LOESCHE, W. J., GIBBONS, R. J. & SOCRANSKY, S. S. 1965 Biochemical characteristics of *Vibrio sputorum* and relationships to *Vibrio bubulus* and *Vibrio fetus*. *Journal of Bacteriology* **89**, 1109–1116.

MEHLMAN, I. J. & ROMERO, A. 1982 Improved growth medium for *Campylobacter* species. *Applied and Environmental Microbiology* **43**, 615–618.

MORGAN, W. J. B. 1959 Studies on the antigenic structure of *Vibrio fetus*. Journal of Comparative Pathology and Therapeutics **69**, 125–140.

NEILL, S. D., ELLIS, W. A. & O'BRIEN, J. J. 1979 Designation of aerotolerant *Campylobacter*-like organisms from porcine and bovine abortions to the genus *Campylobacter*. *Research in Veterinary Science* **27**, 180–186.

NEILL, S. D., O'BRIEN, J. J., & ELLIS, W. A. 1980 The isolation of aerotolerant campylobacter. *Veterinary Record* **106**, 152–153.

SEBALD, M. & VÉRON, M. 1963 Teneur en bases de l'ADN et classification des vibrions. *Annales de l'Institut Pasteur, Paris* **105**, 897–910.

SKERMAN, V. B. D., MCGOWAN, V. & SNEATH, P. H. A. 1980 Approved lists of bacterial names. *International Journal of Systematic Bacteriology* **30**, 225–420.

SKIRROW, M. B. 1977 Campylobacter enteritis: a "new" disease. *British Medical Journal.* **2,** 9–11.

SKIRROW, M. B. & BENJAMIN, J. 1980 "1001" Campylobacters: cultural characteristics of intestinal campylobacters from man and animals. *Journal of Hygiene* **85,** 427–442.

SMIBERT, R. M. 1974 Campylobacter. In *Bergey's Manual of Determinative Bacteriology* 8th edn, ed. Buchanan, R. E. & Gibbons, N. E., pp. 207–211. Baltimore, Maryland: Williams & Wilkins.

SMIBERT, R. M. 1978 The genus *Campylobacter. Annual Review of Microbiology* **32,** 673–709.

TANNER, A. C. R., BADGER, S., LAI, C. H., LISTGARTEN, M. A., VISCONTI, R. A., & SOCRANSKY, S. S. 1981 *Wolinella* gen. nov., *Wolinella succinogenes* (*Vibrio succinogenes* Wolin *et al.*) comb. nov., and description of *Bacteroides gracilis* sp. nov., *Wolinella recta* sp. nov., *Campylobacter concisus* sp. nov., and *Eikenella corrodens* from humans with periodontal disease. *International Journal of Systematic Bacteriology* **31,** 432–445.

VÉRON, M. & CHATELAIN, R. 1973 Taxonomic study of the genus *Campylobacter* Sebald and Veron and designation of the neotype strain for the type species *Campylobacter fetus* (Smith and Taylor) Sebald and Veron. *International Journal of Systematic Bacteriology* **23,** 122–134.

Neisseria gonorrhoeae

A. E. JEPHCOTT AND S. I. EGGLESTONE

Gonococcus Reference Laboratory, Public Health Laboratory, Myrtle Road, Kingsdown, Bristol, UK

Until recently, the diagnosis of infections with *Neisseria gonorrhoeae* (gonococci) was regarded as a difficult, slow, and imprecise process. This is no longer true. Some fundamentally new techniques have been developed, but attention to detail and use of modern refinements of traditional methods will now produce rapid, accurate and reliable results.

Gonococcal infections usually involve the mucous membranes of the genital tract and rectum, and diagnostic techniques seek to detect the organisms at these sites. Infections of conjunctivae and pharynx may dictate use of similar techniques elsewhere. The occasional spread of infection to blood, joints and cerebro-spinal fluid (CSF) will require alternative techniques.

The widespread distribution of chromosomally mediated resistance to penicillin and other commonly used antibiotics makes it important that the susceptibility of cultured gonococci should, at least periodically, be monitored to confirm that treatment regimes employed are not allowing for increasing resistance, and, as different treatment may be required for penicillinase-producing gonococci (PPNG), some form of screening of all isolates for this ability must now be considered mandatory. Tests to confirm penicillinase production are desirable but need only be carried out at a distant reference laboratory.

Gonococcal strains can now be effectively typed. The techniques involved are most applicable to specialist laboratories, but they can yield much of interest both for local and international epidemiological studies.

Indirect evidence of infection was traditionally sought with the Gonococcal Complement Fixation Test. This test is insensitive and is liable to give false positive results after meningococcal meningitis and in cases of chronic bronchitis, but modern solid-phase enzyme-linked immunosorbent assay

Isolation and Identification of Micro-organisms of Medical and Veterinary Importance

Copyright © 1985 by the Society for Applied Bacteriology
All rights of reproduction in any form reserved
0-12-181460-2

(ELISA) techniques offer hope of greater sensitivity and specificity. These tests have a great advantage for population screening purposes in that they do not involve genital examination, but at present they should not be regarded as alternatives to cultural methods when investigating individual cases.

Sites for Examination

Sites that may profitably be examined in mucous membrane infections are: in the female, urethra, cervix (material taken with direct vision), high vagina (usually taken blindly), rectum (using a proctoscope); in the male, urethra, rectum (in homosexuals); and in both sexes, conjunctivae, pharynx. A sample of "first voided urine" may be examined instead of a urethral swab (Murray *et al.* 1979).

Sites for examination in disseminated gonococcal infections (DGI) are skin lesions, blood, cerebrospinal fluid and joint aspirates.

Direct Examination

Gonococci can be recognised as Gram negative bean-shaped diplococci in exudates from infected sites. Significant numbers will be seen to be inside pus cells early in an infection, but as time passes they become predominantly extracellular. They can also be seen in deposits from CSF and in smears of joint aspirates. Microscopical appearance can never be taken as unequivocal proof of a gonococcal infection, but almost 100% of male and 60% of female genital infections should be detected by this method.

Gram stain

Any good Gram stain technique will produce acceptable results. Care should be taken not to over-decolourise Gram positive cocci.

Fluorescent antibody (FA) stain

A fluorescein-conjugated anti-gonococcal serum can be used to reveal gonococci in smears of exudates previously fixed by gentle heat or methanol (Deacon *et al.* 1959). The technique is similar to that used for the FA identification of cultured gonococci but is not so reliable, and it is essential to use a naphthalene black counter stain (Sommerville 1968) and anti-staphylococcal serum (Lind 1964). Best results are obtained using "home-made" anti-sera which have been absorbed with tissue powders. This technique has the advantage that it will detect dead organisms after treatment has begun (Danielsson & Molin 1971) and it can be particularly useful in the examination of skin scrapings in DGI (Kahn & Danielsson 1969).

Solid-phase enzyme immunosassay

Gonococcal antigens in specimens from the urethra in men and the endocervix in women can be detected by a recently developed ELISA test (Gonozyme, Abbott Laboratories, Brighton Hill Parade, Basingstoke, Hants.). This method is not recommended, however, for the testing of rectal and pharyngeal material because of possible cross reactions with other neisserias. Aardoom *et al.* (1982) found that the sensitivity and specificity of this test was high in men with urethritis and could be accepted as a rapid screening method for the detection of gonorrhoea in women, as the sensitivity exceeded that of microscopy of specimens from the endocervix.

Culture

This technique is more sensitive and specific than direct microscopy. As gonococci are rapidly killed by dehydration and oxidation if left unprotected on cotton wool swabs, either rapid plating on culture media or protection in a transport medium is necessary once a specimen has been taken.

Swabs used to take specimens should be boiled in *M*15 phosphate buffer (pH 7.2) to neutralise any acids in wood or cotton wool and should be coated with charcoal if this is not incorporated in the transport medium.

Transport media

These provide non-nutrient semi-solid reducing conditions which hold organisms in suspended animation. Amies' (1967) medium is recommended. It has theoretical advantages over the Stuart's medium (Stuart *et al.* 1954) from which it evolved, and contains its own charcoal. Even this medium cannot be guaranteed to maintain viability of gonococci for more than 12 h and, as ambient temperatures rise, so its efficiency falls off (Korner 1962). The great majority of cultures will remain viable in it for several days, however, if they are stored in cool conditions. The original Stuart's medium works well also, particularly if made up from basic ingredients with scrupulous care. Cary–Blair medium (Cary & Blair 1964) is very similar. These media must be prepared with chlorine-free demineralised distilled water, as there are insufficient nitrogenous compounds in them to neutralise the traces of this bacteriocide which are often present in normal distilled water. Pure cultures will survive well on slopes of culture media, often better than they do in Amies' medium in hot weather, but contaminant overgrowth renders these unsuitable for clinical specimens unless they contain selective agents.

Culture media

These must be highly nutritious. Choice of basal medium is not critical, but the media should not be overheated during production, and additives are required for good growth. These may simply be 5% chocolated or 10% freeze/thaw lysed blood, but yeast dialysate, serum, ascitic fluid or Isovitalex (Difco) supplements or permutations thereof may be added. They may be made selective by the addition of vancomycin (3 μg/ml), colistin (7.5 μg/ml) and nystatin (12.5 μg/ml) with trimethoprim (3–8 μg/ml) if maximal selectivity is required, but lincomycin (1 μg/ml) and amphotericin B (1 μg/ml) can be used in place of the vancomycin and nystatin. Effective media have been described by Reyn (1965), Thayer & Martin (1966), Phillips *et al.* (1972) and Young (1978).

Selective culture media have the advantage that they will reveal gonococci which may be undetectable on non-selective media because of overgrowth by commensal bacteria picked up during the sampling procedure. Against this benefit must be weighed the disadvantage that a proportion of gonococci will prove susceptible to the antibiotics (particularly to the vancomycin) in the selective media and so will fail to grow on any but a non-selective medium (Reyn 1973). Culture on selective media must be considered essential for rectal and pharyngeal samples where the likelihood of overgrowth is high and a non-selective plate will have a poor chance of success. Using selective media will be counter-productive, however, on specimens from sites such as blood or joint aspirates with no commensal flora.

For the more usual situations, culture on both selective and non-selective media will yield best results, but if economic considerations dictate that only one be used, then a slightly less selective medium (e.g. one containing vancomycin and colistin only) will prove the best compromise.

Culture technique

Rapid processing of specimens is essential. Unprotected swabs must be inoculated on culture media at time of examination and the plates must be transported rapidly to the laboratory. Survival in transport medium provides only some respite, and best results are still obtained with short transit times. The inoculated plates should be incubated within 1 h at 36°C in a humid atmosphere containing 3–10% CO_2. Air-tight jars or fish tanks with sheet-glass lids can be used. Suitable atmospheres can be produced by lighted candles or by calculated quantities of sulphuric acid mixed with excess sodium bicarbonate, but the convenience of a purpose-built CO_2 incubator commends itself, and space requirements usually make these essential if large numbers of plates are to be processed.

Plates should be examined after incubation for 24 and 48 h for typical greyish-white 1–2 mm diameter colonies. These should be tested for a positive oxidase reaction with 0.3% aqueous tetramethyl-*p*-phenylenediamine (which does not kill the organisms immediately but will allow prompt sub-culture) and examined by Gram's method. If no suspicious colonies are seen at 48 h then the entire plate should be flooded with oxidase reagent and any positive colony investigated.

Transport-cum-culture systems

If plating and incubation of cultures can begin soon after a specimen has been taken, then a much more rapid diagnosis can be achieved than if these are delayed until after receipt in a distant laboratory. In the clinic it is possible to inoculate specimens on conventional culture plates, place these in candle jars and transport them to the laboratory. Survival times of gonococci on culture plates at room temperature or above compare well with the more traditional Stuart/Amies media (Peizer & Stefen 1943). Moreover, whilst the bugbear of contaminant overgrowth of nutrient transport media rendered these unsatisfactory in the past, skilled spreading of a whole plate will do much to overcome the problem. If skilled plating is not possible then a selective culture medium alone can be employed. Such a medium will fail to detect any vancomycin-sensitive gonococci, but this may well be an acceptable penalty. More recently the advantages of direct plating have been retained, whilst the logistical problems arising from transporting bulky candle jars have been avoided by the development of transport-cum-culture systems. These allow the clinician to inoculate a selective agar slope or mini-plate which is then provided with a CO_2-enriched atmosphere and enclosed in a gas-tight individual container. This is transported to the diagnostic laboratory. Survivals which compare favourably with non-nutrient transport systems are regularly achieved, and if pre-incubated before transit then the culture can be examined on receipt, considerably speeding the diagnostic process. The original system, named Transgrow (Difco) (Martin & Lester 1971), used a glass bottle pre-gassed with CO_2 and containing a slope of Thayer–Martin medium, but developments using microplates (Jembec) or slides (Till-u-Test, Tillomed Laboratories, Henlow, Bucks.) and which generate CO_2 from citric acid/sodium bicarbonate mixtures allow better visualisation and access to colonies for further manipulation and sensitivity testing than was afforded by the original system.

These systems find considerable favour in the USA, and their use should be seriously considered if delays before plating of specimens are likely to exceed 18 h. Transit times should be under 48–72 h and extremes of environmental temperatures should be avoided.

Identification of Gonococci

A "presumptive gonococci" identification (Schroeter & Pazin 1970), of oxidase-positive Gram negative cocci which grow on selective medium and have been isolated from a genital site, is relied on in many parts of the world and will rarely be erroneous. Confirmation by biochemical or antigenic characterisation is considered good practice, however, and is essential for isolates from non-genital sites.

Differentiation from other neisserias has traditionally depended on the "fermentation" of glucose but not maltose, sucrose or lactose. This, in fact, is an oxidative process and will not take place under anaerobic conditions. Usually slopes of rabbit serum enriched agar or cystine trypticase agar stab cultures were used. The serum-free medium of Flynn & Waitkins (1972), containing a sugar and an indicator and enriched with a defined supplement has now displaced these. Problems encountered are usually due to choice of an insufficiently nutritious basal medium, damage to the medium by overheating, or use of an inadequate inoculum. Ultra-rapid "fermentation" identification is possible by detecting pre-formed enzymes in suspensions of gonococci incubated in lightly buffered sugar solutions (Kellogg & Turner 1973). These tests work best if enzyme production has been induced by growth on carbohydrate-rich media beforehand. This technique has been devised to overcome the delays required for bacterial growth inherent in the "fermentation" methods, but at least 1 day incubation of a subculture is still required to obtain sufficient bacterial mass and to confirm purity.

Gonococcal antibodies

Use of anti-gonococcal antibodies can now provide alternative means of identification. Techniques employed have the advantages over sugar tests that they can be carried out after shorter incubation periods, do not require pure cultures, and take less than 1 h to perform, resulting in a vastly more rapid identification. Provided a sufficiently specific antibody is employed, results as good or even better than those obtained by fermentation are achievable. Fluorescent antibody and co-agglutination systems are now employed.

Fluorescent antibody (FAB) identification

Used with care this method will give reproducible results, but it is best with younger cultures. The technique hinges on the specific attachment to gonococci under examination of an anti-gonococcal antibody which has been chemically bonded to fluorescein molecules. When a gently fixed smear of gonococci has been overlaid with a solution of such an antibody and subsequently rinsed and examined under suitably filtered ultraviolet dark-ground or incident light illumination, then characteristic green ring forms will be seen. These can be accepted as evidence of positive identification provided that control slides of meningococci, a staphylococcus and a non-pathogenic neisseria, which have been prepared at the same time, fail to show any fluorescence. Before a batch of anti-serum is accepted its specificity should be checked against a wide range of bacterial species. Problems of non-specific staining of bacteria due either to free dye in the system or to lack of specificity of the anti-serum can be overcome by pre-absorbing the conjugated serum with tissue powders, but if this does not prove effective then the batch should be discarded. Non-specific fluorescence of objects on the slide is not usually a problem when examining films of bacterial cells taken from culture plates, but if the technique is being employed to detect gonococci in smears of exudate (the so-called direct method) then granules in pus cells which frequently auto-fluoresce can cause confusion. This phenomenon can be suppressed by mixing naphthalene black with the fluorescent stain. Staphylococci which contain protein A will cause all gamma globulin molecules to adhere to them by the FAB portions, and this will also result in a non-specific staining of any such staphylococci on the slide. This can be prevented by treating with an excess of unstained gamma globulin, either in the form of normal human serum or rabbit anti-staphylococcal serum, both being mixed with the stain at the time of the test.

Co-agglutination identification of gonococci

This technique capitalises on the ability of protein-A-containing staphylococci to adsorb antibody onto their surfaces. Anti-gonococcal antibodies are attached to suspensions of preserved staphylococci, and if these are exposed to gonococcal antigens then clumping of the staphylococci will occur (Danielsson & Kronvall 1974). This test has all the advantages of speed and the possibility of testing mixed cultures, but requires rather more material than the fluorescent method (this may necessitate reincubation or subcultures), and in practice it has also proved more successful with dead organisms.

The original techniques involved mixing whole gonococcal suspensions with staphylococci. These proved unsatisfactory if any degree of autolysis had occurred. Excellent consistent results can be obtained by employing as antigen the supernatant of a suspension of test organisms which have been heated in a boiling water bath for 10 min and then centrifuged. A commercially freeze-dried kit is now available (Phadebact: Pharmacia, UK) and may be relied on if it is reconstituted with scrupulous regard to the maker's instructions.

Whichever method of identification is employed, a small minority of isolates will give rise to concern. It was usual practice to confirm that organisms grow at 36 but not at 22°C, and whilst some strains will not grow on nutrient agar, many will do so. Some strains of *N. meningitidis* acidify maltose slowly. These can usually be recognised by their larger colony size but they should also be differentiated by their ability to grow on Dubos oleic acid medium (Diena *et al.* 1970). Some strains of gonococci "ferment" glucose very poorly and very late. These usually regain their ability if subcultured or if tested on a more nutrient medium, but it may be necessary to resort to antigenic testing. Dead cultures do not fluoresce well but provide suitable material for co-agglutination procedures.

Sensitivity Testing

Detection of penicillinase production

It is now essential to check all isolates for production of penicillinase, as most current treatment regimes are ineffective against them.

Screening test

The simplest method of detecting the level of resistance imparted by the production of penicillinase is to check for the ability to produce only a very small (less than 20 mm) zone or no zone at all around a 10-unit (6-μg) disc of penicillin placed on a heavily inoculated lawn of the gonococcus under scrutiny. Certain other highly resistant strains may also produce similar zones of inhibition, but the detection of these strains will not prove any disadvantage. These will be revealed when they fail to give positive reactions in confirmatory tests.

Chromogenic cephalosporinase test

A turbid suspension of an overnight culture of the suspect gonococcus is made in saline and to this is added a drop of a solution of cephalosporin 87/312 (Difco) dissolved in dimethyl sulphoxide and phosphate-buffered saline. A red colouration developing within 10 min of mixing denotes a positive result. Control positive and negative gonococcal cultures should be tested in parallel, but if these prove difficult to maintain, staphylococci will suffice. The reagent may also be dropped directly onto colonies on clear media [e.g. New York City agar (NYC), Oxoid]. Colonies of penicillinase-producing organisms will turn red.

Iodometric Test

This test relies upon the ability of breakdown products of penicillin to compete in the colour reaction that iodine produces with starch. Phosphate-buffered saline (PBS) containing 600 μg of penicillin is placed in wells of a Microtitre plate. Turbid suspensions of gonococci are added and allowed to incubate at room temperature for 1 h, after which 2 drops of starch solution followed, after thorough mixing, by a drop of iodine solution will result in the production of a blue colouration. If penicillinase is present this will rapidly fade, but if it persists for 10 min the result is negative. Premature addition of the iodine will produce a spurious negative result.

Heart plate test

This demonstrates the production of penicillinase dramatically on a culture plate. A lawn of penicillin-sensitive staphylococci is inoculated on a chocolate agar plate. A streak inoculation of the suspect gonococcus is run across the plate from one side to a centrally placed 0.6-μg disc of penicillin.

A similar streak of a sensitive control gonococcus is run to the disc from the opposite side. Incubation of the plate will produce a heart-shaped zone of inhibition with growth of the sensitive staphylococcus right up to the disc where it adjoins the streak of the penicillinase producing gonococcus. Use of control penicillinase-producing and non-producing staphylococci streaked perpendicular to the gonococcal strains will control the test better, but the result is less pleasing to the eye.

Antibiotic sensitivity testing

The necessity of testing all isolates is questionable. Treatment will have to begin long before results are available, but tests on treatment failures are obviously of immediate value and monitoring of the levels of resistance must be carried out, at least at intervals, to detect increases which may dictate alterations in therapeutic practice.

Conventional disc diffusion tests will suffice for most antibiotics—especially those with clearly bimodal distributions of sensitivity, such as streptomycin and spectinomycin—and for other antibiotics where there is sufficient leeway between therapeutically attainable levels of antibiotic and those required for eradication of the organism. In the case of the penicillins and cephalosporins, the margins have become critical for many strains and the distribution of sensitivities is not bimodal. For these, accurate assessment of activity is required. This cannot be achieved with a disc diffusion technique, however well controlled, because differences of growth rates of individual gonococci will produce widely differing zone sizes for organisms of identical sensitivity. Shannon *et al.* (1975) have overcome this problem using a well diffusion technique, but this offers no distinct advantage over the widely used agar dilution method.

Disc diffusion tests

Recommended antibiotic disc contents are: penicillin (for penicillinase screening), 6 μg; and the following, for susceptibility determination: streptomycin, 25 μg; Kanamycin, 30 μg; tetracycline, 10 μg; and spectinomycin, 25 or 100 μg.

Plates should be inoculated with faintly turbid suspensions of organisms using cotton wool swabs. A Stokes technique (Stokes & Waterworth 1972) may be employed, ideally using gonococcal controls. If staphylococci are used in their place, then their more rapid growth will produce smaller control zones with a spurious indication of increased susceptibility of the gonococci. For screening for penicillase producers, a heavier inoculum should be used.

Agar dilution method

This should be the reference method for all antibiotics and should be used for all tests to detect penicillin susceptibility. A series of plates of any suitable nutrient medium containing twofold or fourfold dilutions of antibiotic are poured. It is usually possible to arrange to keep a supply of aliquots of appropriately diluted antibiotics frozen, but in the smaller laboratory an occasional series of plates can be prepared using antibiotic impregnated paper rings produced by Diamed Diagnostics Ltd (Mast House, Denley Road, Merseyside, L20 1EA).

Care must be taken to establish the exact content of ampoules of antibiotics used and the maker's "overage" determined. Certain antibiotics are active only in metabolised form. Thus chloramphenicol base must be employed. Others may prove difficult to dissolve. Rapid dilution of suspensions has yielded amazingly accurate results in the authors' hands, but it is best to employ an effective solute or procure a soluble form of the drug wherever possible. Thus chloramphenicol should be dissolved initially in ethanol, and sulphamethoxazole in alkaline hot distilled water. Spectinomycin powder for assay is available from the manufacturers of the drug; this is soluble in distilled water but the ethical preparation is not.

Suspensions of young cultures containing 10,000 c.f.u. per drop are inoculated onto each of the series of plates. A suitable density can be obtained by making a suspension of organisms of such turbidity that newsprint can just be seen through a 0.5-in. test tube containing it. This should be diluted one hundredfold before use. Plates should be examined after overnight incubation. The minimum inhibitory concentration (MIC) is taken as the lowest plate dilution which shows no growth to the naked eye. Up to five discrete colonies and faint hazes of inoculum should be ignored.

Controls should be included in every run. A sensitive staphylococcus (MIC, 0.03 mg/litre) must be used, and it is good practice to include a highly susceptible *Micrococcus lutea* (MIC, 0.001 mg/litre) and three gonococcal strains whose MIC values are known.

Epidemiological Markers

Gonococci can be divided into sub-groups based on their nutritional requirements. This has become the most accepted method for differentiating strains. Improved knowledge of the composition of the gonococcal outer membranes has allowed the development of reproducible serotyping methods and in the case of penicillin-producing strains subdivision can be made based on the size of plasmids carried. Recently a method has been established for determining the isoenzyme patterns of bacteria, but at present it is possible only to distinguish between species of *Neisseria* rather than within species (Braude *et al.* 1983). Nevertheless, by widening the range of enzymes studied, intra-species variation may be found.

Auxotyping

In 1973, Catlin (1973) developed a defined medium which was capable of supporting the growth of all gonococcal strains. She found that there are variants which are unable to grow in the absence of one or more amino acids, nucleic acid bases or vitamins incorporated. Catlin's medium is complex and has been simplified by others. A simplified medium described by Copley & Egglestone (1983) is shown in Table 1. More commonly used methods are based on requirements in defined media for proline, arginine, hypoxanthine and uracil, but in addition we usually test for the ability of ornithine to replace an arginine requirement.

The set of typing media is prepared by omitting the appropriate factor

TABLE 1

Composition of auxotyping medium (CGC)

Substance	(μg/ml)	Substance	(μg/ml)
Amino acids		Bases	
L-Alanine	50	Hypoxanthine	5
L-Arginine	100	Uracil	10
L-Cysteine $HCl \cdot 2H_2O$	260	Carbohydrates	
L-Glutamic acid	1000	Glucose	2500
L-Glutamine	50	Glycerol	1000
L-Histidine	20	Sodium oxaloacetate	100
L-Isoleucine	30	Sodium acetate	1000
L-Leucine	50	Sodium lactate	250
L-Lysine HCl	50	Vitamins	
L-Methionine	20	Biotin	1
L-Proline	50	Calcium pantothenate	1
L-Seline	50	Pyridoxal phosphate	1
L-Threonine	50	Thiamine HCl	1
L-Tryptophan	50	Thiamine pyrophosphal HCl	1
L-Tyrosine	50	NAD	2.5
L-Valine	60	Others	
Inorganic salts		Soluble starch	1000
$Fe(NO_3)_3$	4	Agar	10,000
K_2HPO_4	4300		
KH_2PO_4	1000		
$MgSO_4 \cdot 7H_2O$	400		
$(NH_4)_2SO_4$	500		
NaCl	5000		

from the complete medium. The ability of ornithine to replace an arginine requirement is tested by adding ornithine 100 μg/ml (w/v) to arginine-free medium. Media solidified with agarose are used for retesting strains that grow poorly on agar-based media. These media are inoculated with a suspension prepared from an 18–24 h culture of *N. gonorrhoeae* grown on chocolate agar plates. Briefly, the growths from these plates are resuspended in phosphate-buffered saline to an optical density of about 0.8 measured by a Unicam SP500 spectophotometer with a wavelength of 600 nm and pathlength of 1 cm. The suspension is applied to each pre-dried auxotyping plate with a sterile cotton wool swab. The plates are then incubated at 36°C in a humid atmosphere containing 5% CO_2 and the growth recorded at 24 and 48 h.

The relative frequency of auxotypes in the Avon area of the UK detected by this method is shown in Table 2. Comparison of isolates from pairs of contacts and larger groups of contacts (Copley *et al.* 1983) showed that with the exception of one contact group which had variable proline requirements, the auxotypes were stable during natural transmission and during laboratory passage.

Plasmid analysis

In 1976, a number of laboratories simultaneously reported the emergence of plasmid-mediated penicillin-resistant strains of gonococci. These strains carried either a 3.2- or 4.4-MD β-lactamase-specifying R plasmid in addition to the 2.6-MD and 24.5-MD plasmids which may be present in other gonococci. Initially it was noted that strains isolated from patients returning from Africa usually carried the 3.2-MD plasmid whereas Asian contacts carried the 4.4-MD plasmid which was frequently associated with the 24.5-MD plasmid. Now both strains are indigenous in England.

It is useful for epidemiological surveillance to be able to identify the type of plasmid carried and a number of methods have been used. Two are described here. The first involves isolation of the plasmid DNA before separation and is based on the method described by Sox *et al.* (1978). The isolated plasmid DNA is then identified electrophoretically using the procedure of Meyers *et al.* (1976). The second, more rapid, method is based on the technique described by Eckhardt (1978) and involves direct lysis of the bacteria in the slots of an agarose gel before separation of plasmid and chromosomal DNA by electrophoresis.

Method 1

One well-dried "chocolate" agar plate (containing ampicillin for β-lactamase-producing strain) is flooded with the suspension of *Neisseria gonor-*

TABLE 2

Auxogroups of 500 consecutive isolates of Neisseria gonorrhoeae*

Requirement†	Number	Percent of isolate
Zero (prototrophic)	153	30.6
Proline	120	24.0
Arginine	56	11.2
Proline and arginine	22	4.4
Arginine, hypoxanthine and uracil	104	20.8
Proline, arginine° and uracil	30	6.0
Other	15	3.0

* Data supplied by C. G. Copley.

† Arginine, arginine requirement satisfied by ornithine; Arginine°, arginine requirement not satisfied by ornithine.

rhoeae and the excess fluid is drained off with a Pasteur pipette. The culture is incubated for 18 h at 36°C in a humid atmosphere containing carbon dioxide. The plates are checked for contamination, 1 ml of glucose broth is added, and the confluent growth is scraped off. This is centrifuged at 10,000 **g** for 5 min, and after removing the supernatant the pellet is kept on ice for the next stage.

After resuspending the pellet in 0.3 ml sucrose–Tris–EDTA, 50 μl lysozyme (5 μg/ml in 0.25 *M* Tris, pH 8.0) is added and the mixture shaken. After 5 min, 50 μl 10% of sodium dodecylsulphate (SDS) is added and the mixture left for 15 min before the addition of 0.1 ml of 5 *M* NaCl. The mixture is stored overnight at 4°C to allow the chromosomal DNA to precipitate. (This can be reduced to 4–5 h if necessary.) On the next day the suspension is centrifuged at 14,000 **g** and 4°C and the pellet is discarded. The supernatant is held at 37°C for 30 min after adding 0.4 ml distilled water and 20 μl RNAse (2 mg/ml in 50 m*M* sodium acetate, pH 5, which has been heated at 80/90°C for 10 min and stored at −20°C). To precipitate the protein, 1 ml Tris-saturated phenol is added and mixed well before centrifuging at 10,000 **g** for 30 min at room temperature. The clear aqueous upper phase (if not clear the phenol extraction must be repeated) is removed and 100 μl of 3 *M* sodium acetate added to it followed by 2 ml of 95% ethanol. The plasmid DNA is allowed to precipitate overnight at −20°C.

On the next day it is centrifuged at 12,000 **g** at −20°C for 20 min, the supernatant poured off and tubes inverted to allow excess fluid to drain away. The deposit is resuspended in 50 μl of Tris–EDTA. Often there is very little visible deposit so it is important to scrape the sides and bottom of the tube. The resuspended deposit is mixed with 0.2 volumes of stopping mixture (Ficoll, 4 g; bromophenol blue, 10 mg; 0.2 *M* ethylenediaminetetraacetic acid, 5 ml; distilled water, 5 ml). Then 30 μl of this mixture is loaded into a 0.7% agarose horizontal slab gel for electrophoresis for 60 min at 2 mA and then for 5–6 h at 5–10 mA and stained with a 1 μg/ml ethidium bromide solution. Viewing with a long wave u.v. transilluminator will reveal characteristic bands of DNA. Comparison with standards and known strains will allow identification.

Method 2

A vertical slab gel (110 × 140 × 6 mm) with 12 slots (dimension 6.5 × 15 × 6 mm) is prepared using 0.7% agarose in electrophoresis buffer and 25 μl lysozyme mixture is added to the empty slots. Using the flat end of a toothpick, four or five colonies from an 18-h culture of *N. gonorrhoeae* are suspended in the 25 μl lysozyme to produce a turbid suspension. After 5 min, 35 μl of the 10% SDS mixture is carefully layered onto the bacterial suspension. This is gently mixed by moving the toothpick from side to side (no more than 4 or 5 times) until the suspension becomes viscous. Finally, 50 μl of 5% SDS mixture is added to each well without disturbing the lower layer. The slots are sealed with 0.7% agarose and electrophoresed for 60 min at 2 mA and then 3.5–5 h at 40 mA. The gels are stained and examined as in method 1.

Serotyping

Several workers have examined the gonococcal outer surface proteins and lipopolysaccharides in the hope of finding a basis of serotyping schemes. Using the co-agglutination method (CoA) Sandström & Danielsson (1980) recently classified gonococci and proposed a multifactor model to explain the antigenic composition of the gonococcal outer membrane. This model was based on three classes of antigens which they tentatively named W, J and M. The antigens of class W are divided into three groups, I, II and III, and are protein by nature. Class J antigens are partially sensitive to pronase and are only shared by closely related species. The antigens of class M are polysaccharide and are divided in seven groups, *a* through *g*.

We have used anti-sera directed against W and M antigens. To prepare these, hyperimmune rabbit sera raised against several of Johnston's major outer membrane protein (MOMP) strains were selectively absorbed with other MOMP strains as indicated in Table 3. These absorbed anti-sera are coated on protein A-bearing staphylococci which had been prepared accord-

TABLE 3

Details of gonococcal outer membrane protein reference strain used for immunisation and absorption in the preparation of co-agglutination reagents

Class	Anti-sera	Absorbing strain
W I	Anti D4	A1:C3
W I	Anti E5	N10:C3
W II	Anti V15	N10:C3
W II	Anti N10	D4:D5
W II	Anti S12	A1
W II	Anti U14	A1:B2
W III	Anti F6	B2:U14
M/a	Anti F6	A1
M/b	Anti D4	B2:A1
M/c	Anti V15	D4:E5
M/d	Anti U14	S12:C3
M/e	Anti S12	N10:H8
M/f	Anti E5	R11:D4

ing to the method of Kessler (1975). Staphylococci coated with normal rabbit serum are used as controls.

Method

Test strains of *N. gonorrhoeae* are prepared by boiling a 1% (w/v) suspension for 1 h. One drop of the coated staphylococci is mixed with an equal volume of the boiled gonococci on a microscope slide and rocked gently for 2 min and examined for agglutination. With this sero-grouping method we have found that the lipopolysaccharide M antigens are readily lost and gained during natural transmission. Protein W antigens on the other hand are stable markers and are thus useful for epidemiological studies.

References

AARDOOM, H. A., DE HOOP, D., ISERIEF, C. O. A., MICHEL, M. F. & STOLZ, E. 1982 Detection of *Neisseria gonorrhoeae* antigen by a solid-phase enzyme immunoassay. *British Journal of Venereal Diseases* **58**, 359–362.

AMIES, C. R. 1967 A modified formula for the preparation of Stuart's transport medium. *Canadian Journal of Public Health* **58**, 296–300.

BRAUDE, A. I., MCCUTCHAN, J. A., ISON, C. & SARGEAUNT, P. R. 1983 Differentiation of Neisseriaceae by isoenzyme electrophoresis. *Journal of Infectious Diseases* **147**, 247–251.

CARY, S. G. & BLAIR, E. B. 1964 New transport medium for shipment of clinical specimens. *Journal of Bacteriology* **88**, 96–98.

CATLIN, B. W. 1973 Nutritional profiles of *Neisseria gonorrhoeae, Neisseria meningitidis* and *Neisseria lactamica* in chemically defined media and the use of growth requirements for gonococcal typing. *Journal of Infectious Diseases* **128**, 178–194.

COPLEY, C. G. & EGGLESTONE, S. I. 1983 Auxotyping of *Neisseria gonorrhoeae* isolated in the United Kingdom. *Journal of Medical Microbiology* **16**, 295–302.

COPLEY, C. G., CHISWELL, C. P. & EGGLESTONE, S. I. 1983 *Neisseria gonorrhoeae:* Stability of typing markers after natural transmission. *British Journal of Venereal Diseases* **59**, 237–241.

DANIELSSON, D. & KRONVALL, G. 1974 Slide agglutination method for the serological identification of *Neisseria gonorrhoeae* with anti-gonococcal antibodies adsorbed to protein A-containing staphylococci. *Applied Microbiology* **27**, 368–374.

DANIELSSON, D. & MOLIN, L. 1971 Demonstration of *Neisseria gonorrhoeae* in prostatic fluid after treatment of uncomplicated gonorrhoeal urethritis, *Acta Dermato-Venereologica* **51**, 73–76.

DEACON, W. E., PEACOCK, W. L., FREEMAN, E. M. & HARRIS, A. 1959 Identification of *Neisseria gonorrhoeae* by means of fluorescent antibodies. *Proceedings of the Society for Experimental Biology and Medicine* **101**, 322–325.

DIENA, B. B., WALLACE, R., KENNY, C. P. & GREENBERG, L. 1970 Dubos oleic acid agar medium in the differentiation of meningococci & gonococci. *Applied Microbiology* **19**, 1025.

ECKHARDT, T. 1978 A rapid method for the identification of plasmid desoxyribonucleic acid in bacteria. *Plasmid* **1**, 584–588.

FLYNN, J. & WAITKINS, S. A. 1972 A serum free medium for testing fermentations of *Neisseria gonorrhoeae. Journal of Clinical Pathology* **25**, 525–527.

KAHN, G. & DANIELSSON, D. 1969 Septic gonococcal arthritis. *Archives of Dermatology* **99**, 421–425.

KELLOG, D. S. & TURNER, E. M. 1973 Rapid fermentation confirmation of *Neisseria gonorrhoeae*. *Applied Microbiology* **25,** 550.

KESSLER, S. W. 1975 Rapid isolation of antigens from cells with a staphylococcal protein A-antibody absorbent: parameters of the interaction of antibody–antigen complexes with Protein A. *Journal of Immunology* **115,** 1617–1624.

KORNER, B. 1962 Storage of gonococci in solid Stuart's medium. *Proceedings of the Northern Dermatological Society* pp. 45–48.

LIND, I. 1964 Non-specific reactions by the application of fluorescent antibodies in bacteriological diagnosis. In *Proceedings of the XIVth Scandinavian Congress of Pathology and Microbiology* pp. 255–256.

MARTIN, J. E. & LESTER, A. 1971 Transgrow, a medium for transport and growth of *Neisseria gonorrhoeae* and *Neisseria meningitidis*. *HSMHA Health Reports* **86,** 30–33.

MEYERS, J. A., SANCHEZ, D., ELWELL, L. P. & FAKLOW, S. 1976 Simple agarose gel electrophoretic method for the identification and characterisation of plasmid deoxyribonucleic acid. *Journal of Bacteriology* **127,** 1529–1533.

MURRAY, E. S., BENFARI, M. J., COPPOLA, S. R., HUGHES, M. D., FENG, W. C., MEDEIROS, R. M. & KUNZ, L. J. 1979 New options for diagnosis and control of gonorrhoeal urethritis in males using uncentrifuged first voided urine (FVU) as a specimen for culture. *American Journal of Public Health* **69,** 596–598.

PEIZER, L. R. & STEFEN, G. L. 1943 Improved method of transportation of material from gonococcal infections. *Journal of Laboratory and Clinical Medicine* **28,** 1121–1126.

PHILLIPS, I., HUMPHREY, D., MIDDLETON, A. & NICOL, C. S. 1972 Diagnosis of gonorrhoea by culture on a selective medium containing vancomycin, colistin, nystatin and trimethoprim. *British Journal of Venereal Diseases* **48,** 287–292.

REYN, A. 1965 Laboratory diagnosis of gonococcal infections. *Bulletin of the World Health Organization* **32,** 449–469.

REYN, A. 1973 Bacteriologisk undersøgelse for gonorré. *Ugeskrift for Laeger* **135,** 909–913.

SANDSTRÖM, E. & DANIELSSON, D. 1980 Serology of *Neisseria gonorrhoeae* classification by coagglutination. *Acta Pathologica et Microbiologica Scandinavica, Section B: Microbiology* **88B,** 27–38.

SCHROETER, A. L. & PÁZIN, G. J. 1970 Gonorrhoea. *Annals of Internal Medicine* **72,** 553–559.

SHANNON, R., HEDGES, A. J. & EDWARDS, R. J. 1975 Distribution of levels of penicillin resistance amongst freshly isolated strains of *Neisseria gonorrhoeae*. Application of a novel sensitivity assay. *British Journal of Venereal Diseases* **51,** 246–250.

SOMMERVILLE, R. G. 1968 An improved method for the direct immunofluorescent identification of *Neisseria gonorrhoeae*. *Bulletin of the World Health Organization* **39,** 942–943.

SOX, T. E., MOHAMMED, W., BLACKMAN, E., BISWAS, G. & SPARLING, F. 1978 Conjugative plasmids in *Neisseria gonorrhoeae*. *Journal of Bacteriology* **134,** 278–286.

STOKES, E. J. & WATERWORTH, P. M. 1972 *Antibiotic Sensitivity Tests by Diffusion Methods*. Broadsheet No. 55 (Revised). London: Association of Clinical Pathologists.

STUART, R. D., TOSHACH, S. R. & PATSULA, T. M. 1954 The problem of transport of specimens for the culture of gonococci. *Canadian Journal of Public Health* **45,** 78–83.

THAYER, J. D. & MARTIN, J. E. 1966 Improved medium selective for cultivation of *Neisseria gonorrhoeae* and *Neisseria meningitidis*. *Public Health Reports* **81,** 559–562.

YOUNG, H. 1978 Cultural diagnosis of gonorrhoea with modified New York City (MNYC) medium. *British Journal of Venereal Diseases* **54,** 36–40.

Differentiation of Staphylococci

K. ROBINSON AND ELAINE SAUVÉ

Division of Agricultural Bacteriology, School of Agriculture, Aberdeen, UK

Precise identification of the staphylococci, and their differentiation from micrococci, has never been easy—a fact recognised by Baird-Parker (1966), who created flexible sub-groups which could be modified in the light of new evidence. Since the publication of the eighth edition of *Bergey's Manual* (Buchanan & Gibbons 1974), the possible medical importance of coagulase-negative staphylococci has prompted an examination of the natural inter-relationships of human isolates using a comprehensive test system (Kloos & Schleifer 1975a; Schleifer & Kloss 1975), which was later condensed into a scheme suitable for use by the routine clinical laboratory (Kloos & Schleifer 1975b). The new species await official approval (Report 1976), presumably because their exact relationships remain uncertain (Baird-Parker 1979) and they may be overclassified (Heddaeus *et al.* 1978).

The probability of achieving adequate identification in a routine laboratory, and agreement between laboratories, is low for a variety of reasons, such as:

(1) The uncertain taxonomic status of the genus.
(2) The limited capacity of simplified conventional biotyping schemes to differentiate between closely related organisms.
(3) The unsuitability of complex chemical techniques, e.g. cell-wall analysis for routine laboratory use.
(4) The low requirement for small laboratories, with limited resources and an infrequent need for identification, to prepare and maintain stocks of media for identification purposes.

The purpose of this chapter is to show how some of the API miniaturised biotyping systems (API Laboratory Products, Basingstoke, England) can be used as an alternative to conventional techniques for the differentiation of

Isolation and Identification of
Micro-organisms of Medical and
Veterinary Importance

Copyright © 1985 by the Society for Applied Bacteriology
All rights of reproduction in any form reserved
0-12-181460-2

staphylococci and how iso-electric focussing of cell-free proteins can be used as a laboratory tool for differentiation at species and strain level.

API Miniaturised Biotyping Systems

Major disadvantages of conventional biotyping methods are laborious media preparation, limited storage life and a requirement for a large amount of incubator space. Miniaturised systems, on the other hand, have none of these problems and have the advantage that the use of heavy inocula reduces the time required to achieve a result and the occurrence of reactions due to contaminants. Both suffer from the disadvantage that their capacity to differentiate depends on the number and choice of tests.

The value of miniaturised systems is well proven, particularly with Enterobacteriaceae (Bale & Matson 1981) and other Gram negative bacteria, and the range of systems is gradually being extended to permit differentiation of Gram positive organisms, including staphylococci, for general differentiation between the major genera, and for screening for enzymes.

Preliminary screening of isolates

All isolations and purifications necessarily involve the use of a simple set of "recognition" tests such as catalase, oxidase, and colonial and cell morphology, which give a clue to the possible identity of the organisms and aid the selection of tests which might be useful for further differentiation or identification.

In the case of Gram positive, catalase-positive cluster-forming cocci, a first essential is to differentiate between micrococci and staphylococci. Tests for this purpose, such as sensitivity to lysostaphin and lysozyme, should be selected from the two currently available schemes for identification by Kloos & Schleifer (1975b) and Baird-Parker (1979) before proceeding to more comprehensive tests for identification.

API systems

The systems to be described, API STAPH, API 20B and API ZYM, were selected because they were designed specifically for the identification of staphylococci or were likely to be of general laboratory use.

Methodology

Organisms should be grown on P agar (Kloos *et al.* 1974), the galleries inoculated according to the manufacturers' instructions, and the results read after 24 h incubation at 37°C.

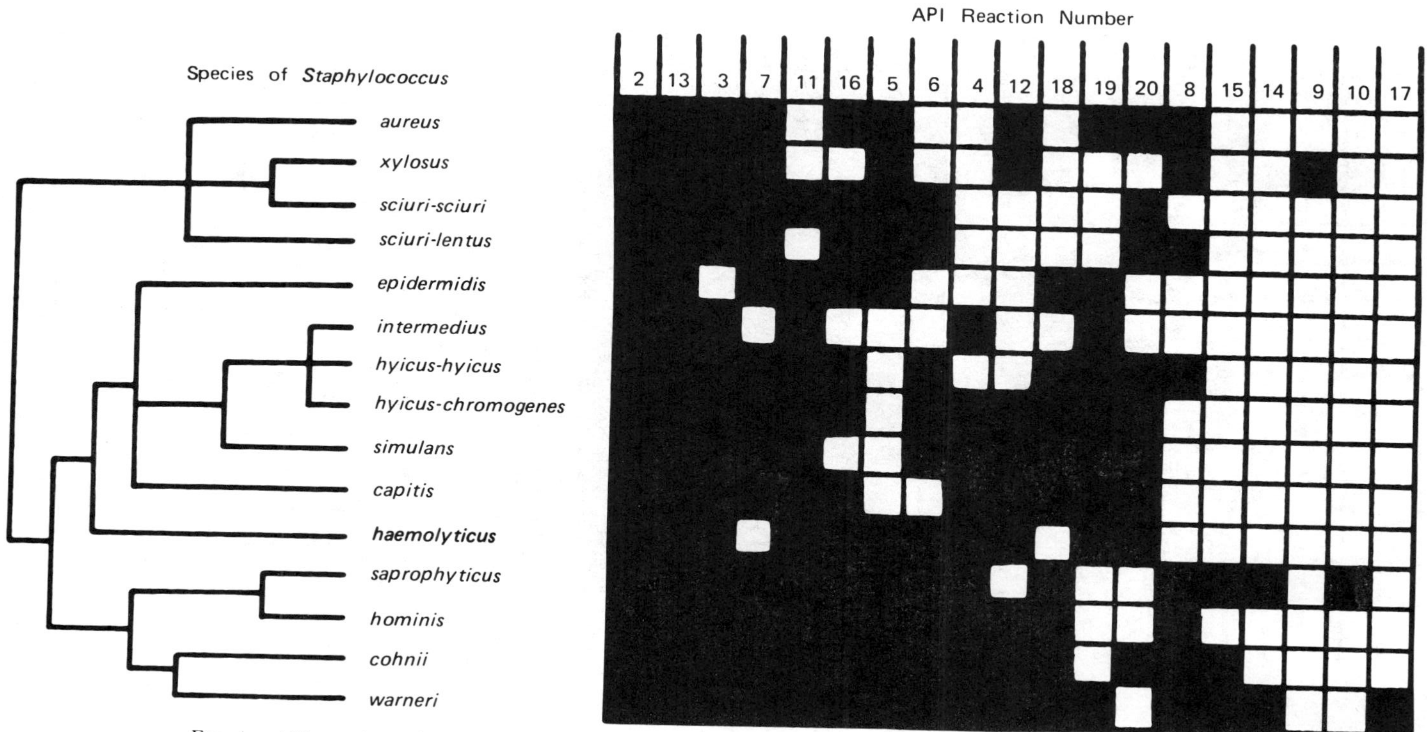

FIG. 1. API reactions of the type species of *Staphylococcus*: (a) API Staph, (b) API 20B, (c) API Zym (continued).

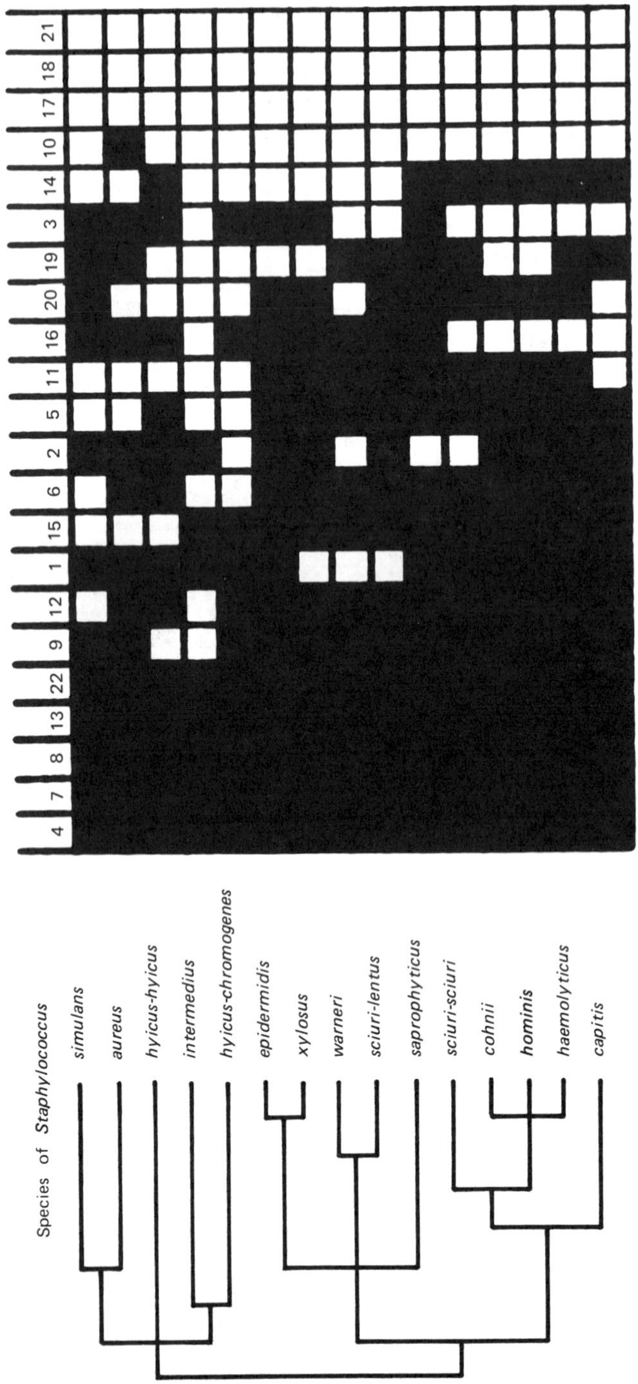

FIG. 1B. (API 20B.)

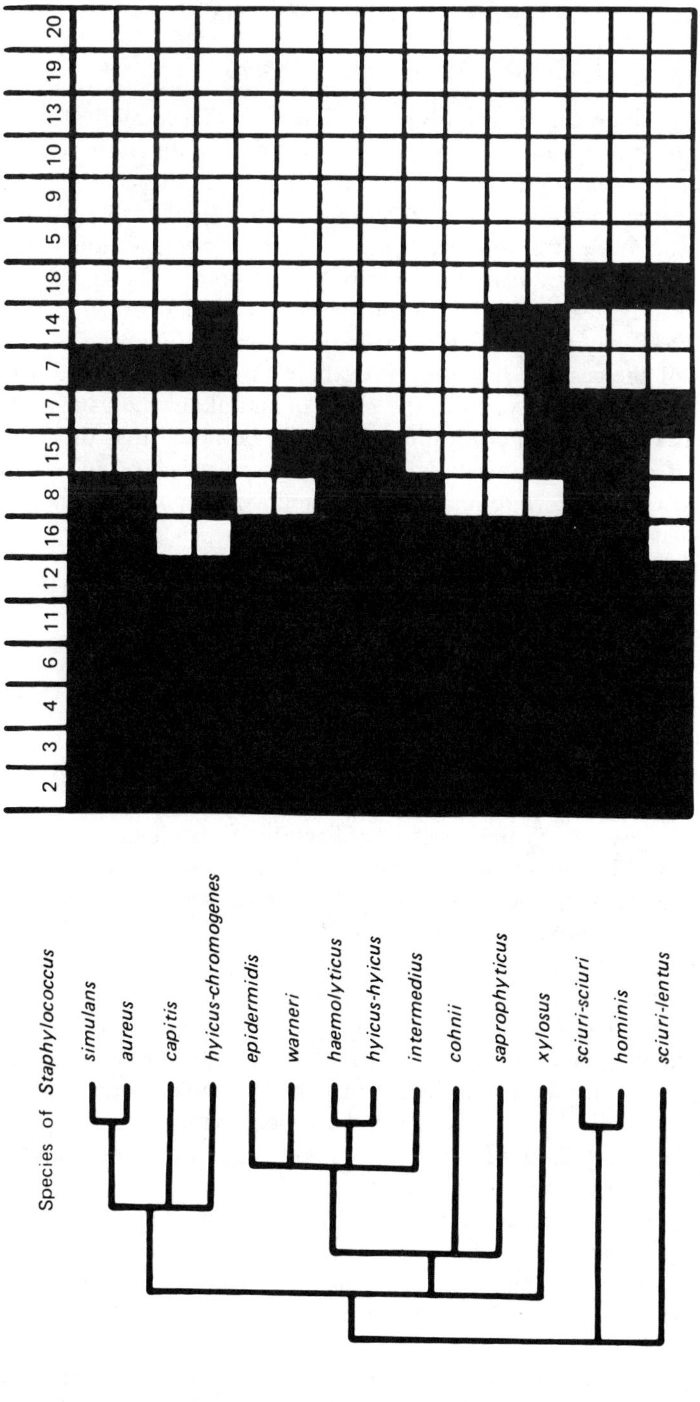

Fig. 1C. (API ZYM.)

Interpretation of results

An identification can be achieved by calculation of a numerical profile and comparing it with those provided in an index. Unfortunately, this method may be unsatisfactory for a variety of reasons.

Firstly, the index may not be sufficiently comprehensive (API Staph); secondly, it may not have been designed to provide species identification of staphylococci (API 20B); and thirdly, the system may not have been intended to be used for identification purposes (API Zym).

Examination of the type species using each system, followed by numerical analysis of the results, has been used here to produce a diagram (Fig. 1) which can, by comparison with the result of an unknown organism, be used to obtain a probable identification. It should be noted that there is a high degree of similarity between many of the type species and that overlap between strains in different clusters may reduce the possibility of achieving an adequate identification.

Helpful key reactions are

(1) API Staph	Gallery nos. 4, 5, 8
(2) API 20B	Gallery nos. 5, 6, 11, 14, 16
(3) API Zym	Gallery nos. 7, 15, 17, 18

Iso-electric Focussing of Cell-free Proteins in Polyacrylamide Gels

Electrophoresis of cell-free extracts of bacteria in starch and polyacrylamide gels followed by detection of iso-enzymes or proteins has been valuable for differentiation of species and strains of several genera of bacteria including *Staphylococcus* (Baillie & Norris 1963; Lund 1965; Cann & Willox 1965; Robinson 1966; Zimmerman & Kloos 1976) and as a technique for the primary screening of ecological isolates (Moore *et al.* 1980).

Iso-electric focussing (IEF) in thin layers of polyacrylamide gel is a technique which separates proteins according to their iso-electric points. Particular advantages of this technique are that the separated proteins are retained within the confines of the gel, that the concentration of proteins into very narrow bands at their pI produces a high degree of resolution, and that variations in experimental conditions within the limits of experimental error do not significantly affect the distribution of the proteins, i.e. there is good inter-gel comparability. In recent years a rapid improvement has taken place in the quality and availability of equipment, materials and technique, and the method is now sufficiently developed for use as a standard routine procedure.

The procedures described here for electro-focussing are based on methods described by Radola (1980), Frey & Radola (1982), Merril *et al.* (1982) and by manufacturers of materials (LKB, Sweden; Pharmacia, Sweden) and have been adopted for routine laboratory use because they provide a relatively rapid method of differentiating bacteria. There is little doubt that the procedures could be improved and/or modified to suit the needs of the taxonomist rather than those of the routine bacteriologist. The method has been shown to be particularly effective for organisms with a limited spectrum of biochemical activity or low optimum growth temperatures and for groups of organisms which are inadequately differentiated by conventional biotyping methods.

Preparation of cell-free extracts

Growth, harvesting and disintegration of organisms

Blood agar base no. 2 (Oxoid, London) in 1-litre Roux flasks is inoculated with a 24-h broth culture and incubated for 24 h at 37°C.

The growth is washed from the agar surface with distilled water, centrifuged at 5000 g for 30 min, and the cell pellet washed twice with distilled water. The cells (approximately 0.5 g wet weight) are suspended in 3 ml distilled water and disintegrated ultrasonically (Robinson 1968). Insoluble cell debris is removed by centrifugation at 4°C and 30,000 **g** for 1 h. The extract is desalted by shaking it gently with Sephadex G-10 or G-25 (Pharmacia, Sweden) or filtering through a short (2 × 1 cm) column of Sephadex. (The latter is preferable for Gram positive bacteria.) The extracts are used immediately or dispensed into freeze-drying ampoules and stored at −20°C.

Composition and preparation of the gel

The following materials are used: acrylamide; *NN'*-methylene bisacrylamide (Bis); ammonium persulphate; Pharmalyte, pH 3–10, and Pharmalyte, pH 4–6.5 (Pharmacia, Sweden); Amberlite MB-1 (BDH, Poole, England) (all chemicals AR grade or "for electrophoresis"); glass moulding plates (LKB, Sweden); polyacrylamide gel moulding sheets (LKB, Sweden); platinum ribbon electrodes (Bio-rad, Watford, England). Most of the chemicals are hazardous and should be handled according to manufacturer's instructions.

Acrylamide stock solution

This is prepared weekly as a stock solution containing: acrylamide 9.7 g; Bis 0.3 g; Amberlite MB-1 1.0 g; distilled water 100 ml. It is stirred gently for 1 h, filtered through filter paper to remove resin beads and stored in the dark at 4°C.

Ammonium persulphate solution

1% solution w/v in distilled water is prepared as required.

Note: Failure of the gel to polymerise has, in our experience, been due to this reagent. When it is not in frequent use it is standard practice to remove and discard a 1- to 2-cm layer at the surface and mix the remainder thoroughly; in addition a new batch of reagent is used at 3-month intervals.

Gel compostion

The following reagents are mixed and degassed: acrylamide stock solution, 10 ml; Pharmalyte, pH 3–10 (or 4–6.5), 1.5 ml; 50% v/v glycerol, 4.2 ml; distilled water, 4.3 ml. When using a high vacuum pump, care should be taken to avoid loss of water by evaporation.

After degassing, 0.5 ml 1% w/v ammonium persulphate solution is added below the surface and mixed gently to avoid reaeration.

Casting the gel

Thin-layer gels (0.5 mm) can be cast on gel moulding sheets using glass gel casting plates by careful injection of the gel solution into the space between the plates according to manufacturer's instructions (LKB, Sweden). It has been found convenient to cast gels the day before they are required and store them overnight at 4°C with the ends of the casting frames sealed with plastic film.

The gel on its support film is removed from the casting frame immediately prior to use.

Electro-focussing

Conditions

Buffers	pH 3–10:	Anolyte 1 mole/litre orthophosphoric acid Catholyte 1 *M* NaOH
	pH 4–6.5:	Anolyte 0.5 mole/litre acetic acid Catholyte 0.5 mole/litre NaOH
Coolant temperature		8°C
Power		Voltage (maximum) 1500 V; power (initial), 15 W
Inter-electrode distance		10 cm
Pre-focussing		45 min
Focussing		1 h

Pre-focussing

The gel is placed on the cooling platten of the electrophoresis tank with a few millilitres of water between them to provide good thermal contact and is blotted to remove the surplus water which emerges at the edges of the support film.

The buffer wicks (LKB, Sweden) are soaked in buffer and blotted to remove surplus liquid before placing on the gel. They are trimmed to the length of the gel.

The electrodes are placed on the wicks, ensuring that they make firm contact, and power is applied for 45 min.

Pre-focussing is necessary to establish the pH gradient in the gel. Protein separation can be achieved without prefocussing or by prefocussing for much shorter times, e.g. 5–10 min. Resolution is poorer, however, and wavy protein bands may be produced due to the simultaneous movement of ampholytes during the formation of the pH gradient with proteins, salts and other substances in the cell-free extract.

Application of samples

After pre-focussing, sample strips 1 × 0.5 cm (LKB, Sweden) are placed in a line 2 cm from the cathode and separated from each other by not more than 1 mm. Ten microliters of sample are placed on the strip using disposable micro-capillary tubes.

Positioning the sample strips as described has two advantages. Firstly, proteins in adjacent samples with the same pI join to form a continuous line, and secondly, wave motions in the gel, which occur with larger inter-sample distances, are controlled.

The quantity of sample applied represents approximately 75% of the absorptive capacity of the strip. It should be noted that other types of strip have different absorptive capacities and also different protein adsorptive capacities (Wadström & Smyth 1975). Over-application of liquid has been found to produce a streaming effect due to movement of the surplus from the sides of the strip as well as the leading edges.

Focussing

Power should be re-applied to the gel immediately following sample application. Sample strips should be removed after 10 min of focussing.

Fixation, staining and destaining

The purpose of this stage is to remove ampholytes from the gel, to stain the proteins and to remove surplus stain from the gel background. A variety of methods are described in the literature and by the manufacturers of electrofocussing materials. The methods described here are, with some modifications, based on those of Radola (1980) and Frey & Radola (1982).

(a)

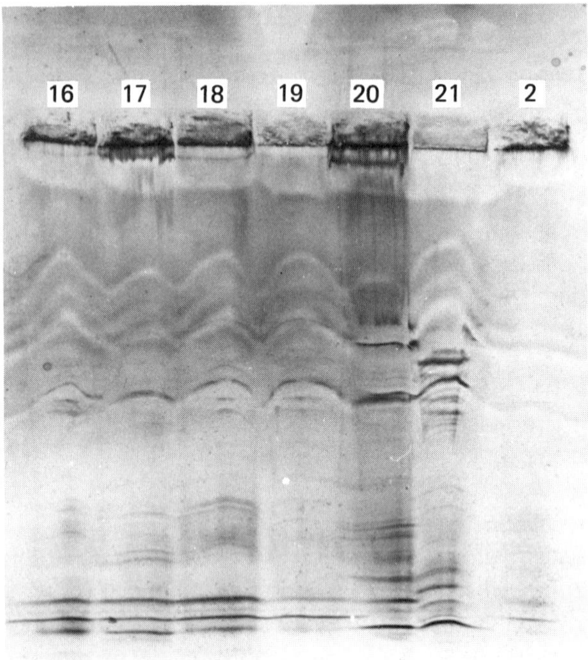

(b)

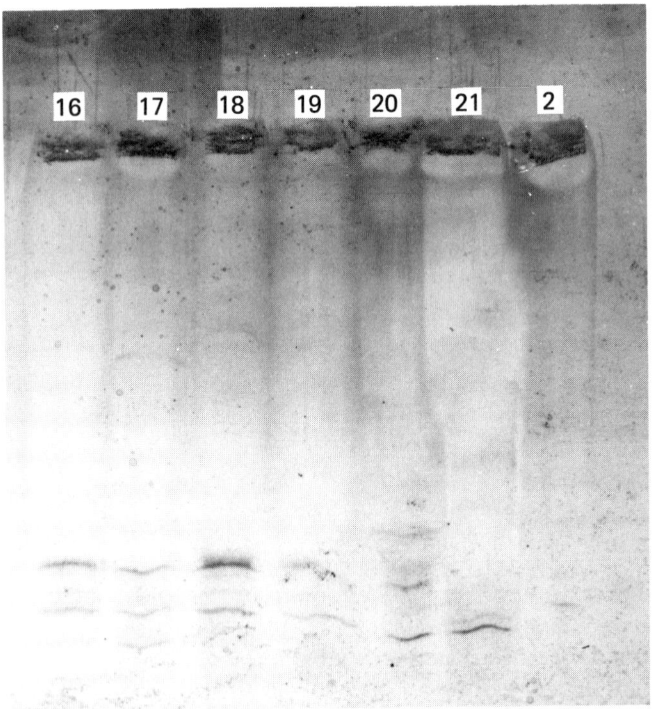

Method

(1) The gel is fixed in 20% w/v trichloroacetic acid with shaking for 5 min. The fixer is then changed and fixing continued for 10 min.
(2) It is then rinsed for 15–30 s in Universal Solvent (methanol:water:glacial acetic acid, 25 : 65 : 10) prepared freshly from methanol and water:glacial acetic acid (65 : 10).
(3) It is stained with 0.05% w/v Serva blue W (Serva, Cambridge, England) in distilled water (prepared from a 1% stock solution) until bands are visible.
(4) Then the gel is destained with distilled water until the background is clear and the protein bands are clearly visible. (Protein bands fade with prolonged immersion in water and therefore recording and preservation should be done as soon as possible.)

Recording and preservation

Photography

The wet gel can be photographed on Technical Pan 2415 film (Kodak, London, England) with the aid of a green filter and transmitted light. Develop in DG 10 (Kodak) and print on a suitable high-contrast paper.

Drying the gel

A permanent preparation can be made by drying the gel, at room temperature, firmly clamped at the edges of the support film to a sheet of glass. After drying, protect the surface of the gel with a sheet of transparent plastic.

Interpretation of results

Differentiation can be obtained by cutting out the protein profiles from the photographic print and visually matching the profiles with those of known organisms.

The protein profiles at pH 3–10 of a number of *Staphylococcus epidermidis* strains from a variety of sources are shown in Fig. 2*a* and demonstrate the high degree of similarity occurring within a species. The pH 4–6.5 profiles of the same organisms show how they can be differentiated further (Fig. 2*b*).

FIG. 2. Separation of cell-free proteins of *Staphylococcus epidermidis* spp. by iso-electric focussing in thin-layer polyacrylamide gels: (a) pH 3–10, (b) pH 4–6.5. Strains are numbered as 2, ATCC 14990 (human); 16, K206 (food); 17, K208 (pig); 18, K282 (dog); 19, NCD0 (dairy); 20, K341 (clinical); 21, ATCC 14990. Cultures were received from other laboratories as follows: 2, Professor K. H. Schleifer, Technical University, Munich; and 16–21, Dr. R. K. A. Feltham, Department of Microbiology, University of Leicester.

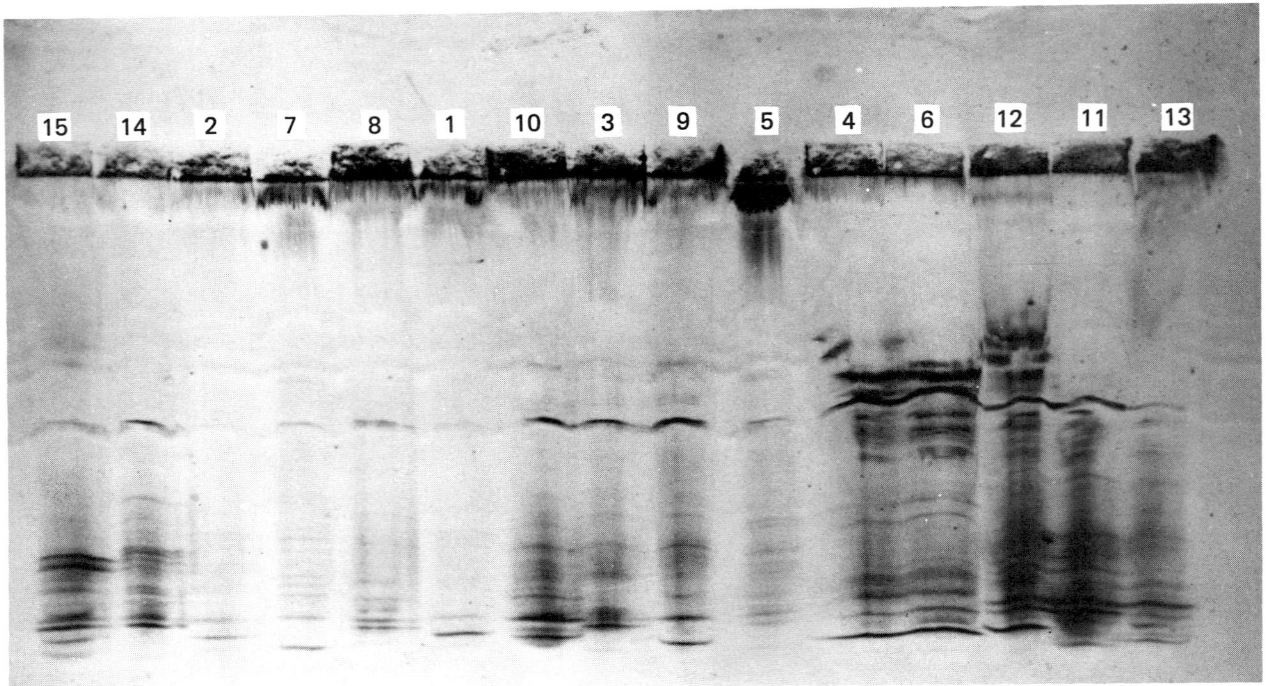

FIG. 3. Separation of the cell-free proteins of the type species of *Staphylococcus* by iso-electric focussing in thin-layer polyacrylamide gels at pH 3–10. Species are numbered as 1, *S. simulans;* 2, *S. epidermidis;* 3, *S. aureus;* 4, *S. sciuri* ssp. *sciuri;* 5, *S. cohnii;* 6, *S. hominis;* 7, *S. saprophyticus;* 8, *S. capitis;* 9, *S. haemolyticus;* 10, *S. warneri;* 11, *S. intermedius;* 12, *S. sciuri* ssp. *lentus;* 13, *S. xylosus;* 14, *S. hyicus* ssp. *hyicus;* 15, *S. hyicus* ssp. *chromogenes*. All cultures with the exception of 3 (NCTC 8532) were received from Professor K. H. Schleifer, Technical University, Munich.

The pH 3–10 profiles of the type species are illustrated in Fig. 3.

Ideally, differentiation and/or identification should be achieved using computerised pattern-matching techniques of densitometric tracings. This, however, would require the availability of a suitable densitometer, the incorporation of standards to "normalise" the profile (Kersters & de Ley 1975), and access to a computer containing a reference file of profiles for known species. Further developments are needed to provide such facilities.

Hazards

Many of the chemicals used for gel preparation, fixing, staining etc. are hazardous and care should be taken that they are handled and disposed of with suitable safety precautions. In addition it is essential to use electrophoresis equipment which possesses safeguards to prevent access to live electricity. It is recommended that the power supply unit and mains electricity should be switched off as part of the routine adopted prior to opening the electrophoresis tank.

General Observations

Essential requirements of the IEF technique used for identification purposes are that it should be simple to perform, be reasonably rapid, be capable of dealing with large numbers of samples, and give good inter-gel reproducibility. The two most probable sources of error are gel composition and focussing time. Consideration should be given to the purchase of equipment which, by measuring volt-hours, would compensate for differences in resistance between gels and help to achieve more standardised conditions.

In our experience decreasing or increasing electro-focussing time either compresses or extends the protein profile. Despite these changes, visual pattern matching is still possible.

There appears to be little advantage in reducing electro-focussing time to less than 1 h. This time allows the staff to undertake other tasks without being unduly rushed.

Alternative staining techniques are available, of which silver staining with its much greater sensitivity can be recommended. Silver staining is more complicated to perform, and the profile is more complex and would be preferable for taxonomic rather than routine purposes. A number of staining techniques are available (Ochs *et al.* 1981), and the method by Merril *et al.* (1982) has been successfully applied in this laboratory to thin-layer gels which have been stained previously with Coomassie blue (Anon. 1982) but not after staining with Serva blue W.

The ampholyte initially selected (i.e. Pharmalyte) was chosen at random from those available and has continued to be used. Lååas & Olsson (1981) have found that Pharmalyte focussed faster than other amphyolytes, and it is possible therefore that under the conditions described here, the profiles formed using other ampholytes may not be exactly comparable.

References

ANON. 1982 *The Bio-rad Silver Stain*. Technical Bulletin No. 1089. Watford, England: Bio-rad Laboratories Ltd.

BAILLIE, A. & NORRIS, J. R. 1963 Studies on enzyme changes during sporulation in *Bacillus cereus* using starch gel electrophoresis. *Journal of Applied Bacteriology* **26,** 102–106.

BAIRD-PARKER, A. C. 1966 Methods for studying staphylococci and micrococci. In *Identification Methods for Microbiologists: Part A,* ed. Gibbs, B. M. & Skinner, F. A., pp. 59–64. Society for Applied Bacteriology Technical Series No. 1. London & New York: Academic Press.

BAIRD-PARKER, A. C. 1974 Genus *Staphylococcus*. In *Bergey's Manual of Determinative Bacteriology 8th ed,* ed. Buchanan, R. E. & Gibbons, N. E., pp. 483–489. Baltimore, Maryland: Williams & Wilkins.

BAIRD-PARKER, A. C. 1979 Methods for identifying staphylococci and micrococci. In *Identification Methods for Microbiologists* 2nd edn, ed. Skinner, F. A. & Lovelock, D. W., pp. 201–209. Society for Applied Bacteriology Technical Series No. 14. London & New York: Academic Press.

BALE, M. J. AND MATSON, J. M. 1981 Time–motion and cost comparison study of Micro-ID, API 20E and conventional biochemical testing in identification of *Enterobacteriaceae. Journal of Clinical Microbiology,* 14, 665–670.

CANN, D. C. & WILLOX 1965 Analysis of multimolecular enzymes as an aid to the identification of certain rapidly growing mycobacteria using starch gel electrophoresis. *Journal of Applied Bacteriology* **28,** 165–173.

FREY, M. & RADOLA, B. J. 1982 Rapid staining of proteins in ultra thin iso-electric focussing in polyacrylamide gels. *Electrophoresis* **3,** 27–32.

HEDDAEUS, H., HECZKO, P. B., MORDARSKI, M., GOODFELLOW, M. & PULVERER, G. 1978 Classification of coagulase-negative staphylococci according to recently proposed taxonomic schemes. *FEMS Microbiology Letters* **3,** 133–135.

KERSTERS, K. & DE LEY, J. 1975 Identification and grouping of bacteria by numberical analysis of their electrophoretic protein patterns. *Journal of General Microbiology* **87,** 333–342.

KLOOS, W. E. & SCHLEIFER, K. H. 1975*a* Isolation and characterisation of staphylococci from human skin. II. Descriptions of four new species: *Staphylococcus warneri, Staphylococcus capitis, Staphylococcus hominis,* and *Staphylococcus simulans. International Journal of Systematic Bacteriology* **25,** 62–79.

KLOOS, W. E. & SCHLEIFER, K. H. 1975*b* Simplified scheme for the routine identification of human *Staphylococcus* species. *Journal of Clinical Microbiology* **1,** 82–88.

KLOOS, W. E., TORNABEENE, T. G. & SCHLEIFER, K. H. 1974 Isolation and characterisation of micrococci from human skin including two new species: *Micrococcus lylae* and *Micrococcus kristinae. International Journal of Systematic Bacteriology* **24,** 79–101.

LÅÅS, T. & OLSSON, I. 1981 pH-Gradient development and focussing speed in thin-layer polyacrylamide gel isoelectric focussing: A comparison between Pharmalyte, Ampholine and Servalyt. *Electrophoresis* **2**, 235–239.

LUND, B. M. 1965 A comparison by the use of gel electrophoresis of soluble protein components and esterase enzymes of some Group D streptococci. *Journal of General Microbiology* **40**, 413–419.

MERRIL, C. R., GOLDMAN, D. & VAN KEUREN, M. L. 1982 Simplified silver protein detection and image enhancement methods in polyacrylamide gels. *Electrophoresis* **3**, 17–20.

MOORE, W. E. C., HASH, D. E., HOLDEMAN, L. V. & CATO, E. P. 1980 Polyacrylamide slab gel electrophoresis of soluble proteins for studies of bacterial floras. *Applied and Environmental Microbiology* **39**, 900–907.

OCHS, D., MCCONKEY, E. H. & SAMMONS, D. W. 1981 Silver stains for proteins in polyacrylamide gels: A comparison of six methods. *Electrophoresis* **2**, 304–306.

RADOLA, B. J. 1980 Ultra-thin iso-electric focussing in 50–100 μm polyacrylamide gels on silanised glass plates or polyester films. *Electrophoresis* **1**, 43–46.

REPORT 1976 ICSB Subcommittee on the Taxonomy of Staphylococci and Micrococci. In *Staphylococci and Staphylococcal Diseases*, ed. Jeljaszewicz, J. Stuttgart & New York: Gustav Fisher Verlag. (ISBN 3-437-10436-5.)

ROBINSON, K. 1966 Some observations on the taxonomy of the genus *Microbacterium*. II. Cell wall analysis, gel electrophoresis and serology. *Journal of Applied Bacteriology* **29**, 616–624.

ROBINSON, K. 1968 The use of cell wall analysis and gel electrophoresis for the identification of coryneform bacteria. In *Identification Methods for Microbiologists: Part B*, ed. Gibbs B. M. & Shapton, D. A., pp. 85–92. Society for Applied Bacteriology Technical Series No. 2. London & New York: Academic Press.

SCHLEIFER, K. H. & KLOOS, W. E. 1975 Isolation and characterisation of bacteria from human skin. I. Amended descriptions of *Staphylococcus epidermidis* and *Staphylococcus saprophyticus* and descriptions of three new species *Staphylococcus cohnii*, *Staphylococcus haemolyticus* and *Staphylococcus xylosus*. *International Journal of Systematic Bacteriology* **25**, 60–61.

WADSTRÖM, T. & SMYTH, C. J. 1975 Some applications of a zymogram method in combination with isoelectric focussing in polyacrylamide gels. In *Isoelectric Focussing*, ed. Arbuthnott, J. P. & Beeley, J. A. London: Butterworths.

ZIMMERMAN, R. J. & KLOOS, W. E. 1976 Comparative zone electrophoresis of esterases of *Staphylococcus* species isolated from mammalian skin. *Canadian Journal of Microbiology* **22**, 771–779.

Streptococcus mutans and Other Streptococci from the Oral Cavity

D. BEIGHTON

Dental Research Unit, Royal College of Surgeons of England, West Hill, Downe, Near Orpington, Kent, UK

Streptococcus mutans was first isolated by Clarke (1924) from early carious lesions on human teeth. He suggested that it was implicated in the caries process, but his finding was not followed up for over three decades because of the vigorous efforts at the time to incriminate lactobacilli as the major aetiological agent of tooth decay. Hemmens *et al.* (1946), in a longitudinal study, clearly demonstrated that lactobacilli were not significantly involved in the initiation of tooth decay but that they colonised carious lesions after cavity formation. The demonstration that streptococci similar to those able to cause tooth decay in rodents were present in human carious lesions (Jablon & Zinner 1966) and had physiological properties similar to *Strep. mutans* (Carlsson 1968) led to an explosion of research into the role of this organism in human dental caries. *Streptococcus mutans* is now considered to play a major role in the aetiology of tooth decay.

The isolation of *Strep. mutans* from the oral cavity is not difficult. Most studies, however, are concerned with quantifying the level of *Strep. mutans*, and usually other bacterial genera, in dental plaque, and this is subject to many technical difficulties (Hardie & Bowden 1976*a*). *Streptococcus mutans* has frequently been identified as the causative organism in bacterial endocarditis (Parker & Ball 1976). The *Strep. mutans* strains from blood cultures are not different from oral strains (Facklam 1974), although their identification and differentiation from *Strep. bovis* may prove difficult.

Handling of Specimens

Most dental studies are concerned with determining the total number or proportion of *Strep. mutans* in a sample. The major obstacle to this is ade-

Isolation and Identification of
Micro-organisms of Medical and
Veterinary Importance

Copyright © 1985 by the Society for Applied Bacteriology
All rights of reproduction in any form reserved
0-12-181460-2

quate dispersion of the sample, especially dental plaque. The most common methods of dispersing plaque samples are sonication and homogenization in a tissue grinder (Hardie & Bowden 1976*a*). Mucosal surfaces may be sampled with alginate swabs which can be dissolved in diluent supplemented with 1% hexametaphosphate (van Houte *et al.* 1970). Alternatively, cotton wool swabs may be used and adherent bacteria removed by vigorously vortexing in the presence of sterile glass balls.

Streptococcus mutans can often be isolated from saliva, but it is best to disperse saliva samples before dilution. It is present in low numbers in human faeces (Unsworth 1980). The faecal samples should be adequately dispersed by shaking with glass balls (Walker 1981) or a similar device before dilution and plating.

Depending on the size of the plaque sample, the concentration of *Strep. mutans* may be determined as a percentage of the total anaerobic colony count, colony-forming units (c.f.u.) per milligram of plaque, or c.f.u. per plaque parameter (e.g., μg DNA or μg protein). In saliva *Strep. mutans* are often enumerated as c.f.u. per millilitre of stimulated saliva, while *Strep. mutans* in faeces are usually determined as c.f.u. per gramme.

Dental plaque

If *Strep. mutans* are the sole object of study then dispersed plaque samples should be plated on to trypticase–yeast–cystine (TYC) medium and mitis salivarius–bacitracin (MSB) agar (or perhaps TYCSB medium) and incubated for 3 days anaerobically. If other streptococci are to be studied, mitis salivarius (MS) agar should be used, as the recovery of oral streptococci other than *Strep. mutans* is greater on MS agar than on TYC medium (Beighton *et al.* 1982).

Mucosal surfaces

The concentration of *Strep. mutans* on the tongue, cheeks and gingiva is so low as to necessitate the use of MSB agar or TYCB medium for their recovery.

Saliva

The level of *Strep. mutans* in saliva varies in relation to the plaque level (van Houte & Green 1974). TYC medium and the more selective *Strep. mutans* media should be employed.

Faeces

Examination of faeces for *Strep. mutans* is very difficult, and the carriage rate is low. A variety of media has been employed, including MS agar, MSB agar and Columbia agar (Oxoid) supplemented with polymixin B (50 units/ml), nalidixic acid (20 μg/ml) and crystal violet (0.8 μg/ml) (Unsworth 1980).

Isolation

The techniques used to isolate *Strep. mutans* from a sample will depend greatly on the nature of the sample and the site from which it was collected. The proportion of *Strep. mutans* on the smooth surfaces of the teeth is considerably lower than on the occlusal surfaces (Ikeda *et al.* 1973), while very few of these organisms are isolated from the mucosal surfaces of the oral cavity (Carlsson 1967*a*). They may also be isolated from saliva, although originally resident on the teeth; the salivary *Strep. mutans* concentrations correlate well with the plaque levels (van Houte & Green 1974). If the concentration of *Strep. mutans* in dental plaque is low, however, then its isolation from saliva may be difficult. *Streptococcus mutans* may be isolated from faeces, but as the bacterial populations of the oral cavity and faeces are considerably different, selective media used for the isolation of *Strep. mutans* from oral samples may not be suitable for their enumeration in faeces. The isolation of *Strep. mutans* from blood cultures presents no problems provided adequately enriched media are employed.

Growth of Streptococcus mutans *on streptococcal-selective media*

The two main streptococcal-selective media employed in the isolation of *Strep. mutans* from dental plaque are MS agar (Chapman 1946) and TYC agar (de Stoppelaar *et al.* 1967), both of which are available commercially. MS agar is selective by virtue of the presence of potassium tellurite (10 μg/ml), crystal violet (0.8 μg/ml) and Trypan blue (75 μg/ml). Linke (1977) has shown Trypan blue at 50 μg/ml inhibits certain *Strep. mutans* strains, and others (Liljemark *et al.* 1976; Staat 1976) have shown that the recovery of *Strep. mutans* on MS agar is inhibited compared with the recovery on blood agar; the degree of inhibition depends on the source of the MS agar. In our hands, MS agar (Oxoid) gives recoveries of stock *Strep. mutans* strains equivalent to those obtained on horse blood agar.

TYC agar is inhibitory because of the inclusion of sodium acetate, and, as it is transparent, colonies are more clearly visible than those on MS agar. The recovery of *Strep. mutans* from plaque samples on TYC agar is better than on MS agar (Beighton *et al.* 1982). Both media are incubated anaerobically in an atmosphere of 10% CO_2 and 90% H_2 for 3 days. This allows the colonies of all the streptococci to grow to a reasonable size so that, provided the plates have less than 100 colonies, a differential colony count can be made without the need for microscopic examination. It is suggested that *Strep. mutans* forms characteristic colonies of each of these media. To some extent this is true, but on MS agar it is very unwise to attempt to distinguish *Strep. mutans* by "characteristic colony morphology" as it can easily be confused with some *Strep. sanguis* isolates. On TYC agar some *Strep. mutans* serotype *c* strains produce a yellow pigment (Woltjes *et al.* 1980), while colonies of *Strep. mutans* serotypes *d*, *g* and *h* are surrounded by a white halo (Beighton *et al.* 1981; van Palenstein Helderman *et al.* 1983). Again it is often difficult to distinguish *Strep. mutans* colonies from *Strep. sanguis* colonies, and the identity of isolates must ultimately rest on the results of biochemical tests.

Both MS agar and TYC agar are highly selective, and only occasionally are extraneous organisms isolated. Care must be taken not to add potassium tellurite to the MS agar while it is too hot (over 50°C) since this will cause the tellurite to decompose and allow Gram positive filamentous organisms to grow. This same problem arises if the medium is inadequately mixed after the addition of the potassium tellurite (1 ml/litre of a 1% solution). If the proportion of veillonellas in a plaque sample is very high, then these too will be isolated on the MS agar, as will the occasional lactobacillus or yeast. On TYC agar actinomyces may grow, but their colonies are usually too small to be a problem. Catalase-positive cocci will grow on TYC medium but their large (3 mm) low convex white colonies with an entire edge are easily distinguished from those of *Strep. mutans*.

Growth of Streptococcus mutans *on specific selective media*

Many media have been described for the selective cultivation of *Strep. mutans* from the oral cavity. Sulphonamide was added to MS agar (Carlsson 1967*b*) to recover *Strep. mutans* selectively from plaque, but Mejare & Edwardsson (1975) used this medium to isolate *Strep. milleri* from the gingival crevice, and it has been suggested that *Strep. mutans* serotypes *d* and *g* are inhibited by sulphonamide (Emilson & Bratthall 1976). *Streptococcus mutans* and a few other streptococci are able to grow in the presence of high levels of sucrose. Ikeda & Sandham (1972) added 40% w/v sucrose to MS agar to make it selective for *Strep. mutans*, but *Strep. salivarius* strains will also grow in the presence of high sucrose concentrations (Gold *et al.* 1973). Colman & Williams (1972) showed that *Strep. mutans* strains are resistant to bacitracin while most other oral streptococci are inhibited. Gold *et al.* (1973) capitalized on these observations and formulated the most widely used medium for selectively isolating *Strep. mutans* from dental plaque by adding sucrose to a final concentration of 20% w/v and 0.2 units/ml bacitracin to MS agar (MSB agar). *Streptococcus mutans* serotype *a* (Staat 1976) and serotype *h* (Beighton *et al.* 1981) are inhibited on MSB agar, but neither serotype is normally present in human dental plaque. Although widely used, MSB agar is partially inhibitory for *Strep. mutans* because potassium tellurite and Trypan blue are present in the basal medium: both these antibacterials may inhibit *Strep. mutans* isolates (Linke 1977; Little *et al.* 1977). In our hands MSB agar will permit the growth of other oral streptococci, especially *Strep. milleri*, so that it is necessary to confirm the identity of colonies growing on the MSB agar with a short set of biochemical tests.

Linke (1977) devised a medium using mannitol, sorbitol, basic fuchin and sodium azide (MSFA medium) on which *Strep. mutans* strains produce characteristic colonies, but other mannitol-fermenting strains will also grow. This medium has not been widely used; its general suitability for isolating *Strep. mutans* from dental plaque warrants further investigation.

Recently van Palenstein Helderman *et al.* (1983) modified TYC medium by increasing the sucrose concentration to 20% w/v and by adding bacitracin at 0.1 units/ml (TYCSB medium). This medium may be an improvement on MSB agar, but *Strep. mutans* serotypes *a* and *h* will still be inhibited. However, an advantage of the medium is that, as on TYC medium, *Strep. mutans* serotypes *c, e* and *f* form colonies distinctly different from those formed by serotypes *d, g* and *h.*

Culture media

TYC medium

This medium (de Stoppelaar *et al.* 1967) contains (g/litre): trypticase, 15; yeast extract, 5; L-cystine, 0.2; Na_2SO_3, 0.1; NaCl, 1; $Na_2HPO_4 \cdot 12H_2O$), 2; $NaHCO_3$, 2; sodium acetate trihydrate, 20; sucrose, 50; agar, 12; pH 7.3.

TYC medium is available commercially from LabM (London, UK).

TYC medium tends to inhibit oral streptococci other than *Strep. mutans* and may be made more selective for *Strep. mutans* by the addition of an additional 150 g/litre of sucrose and 100 units/litre of filter-sterilized bacitracin to the autoclaved and cooled medium (van Palenstein Helderman *et al.* 1983). TYC medium is useful for determining polysaccharide production.

MS agar

Although this medium was designed by Chapman (1946) to isolate streptococci from faeces, it has been very widely used by dental microbiologists to isolate streptococci from the oral cavity. It is available from several commercial companies (Sterilin, Difco, BBL).

MS agar may be modified to increase its selectivity for *Strep. mutans* by the addition of 150 g/litre of sucrose before autoclaving and 200 units/litre of filter-sterilized bacitracin to the sterile and cooled basal medium (Gold *et al.* 1973).

As the source of MS agar can significantly affect the recovery of *Strep. mutans* (Staat 1976), it is suggested that recoveries of stock *Strep. mutans* strains on, say, horse blood agar (HBA) and MS agar be compared.

Fermentation-based medium

The basal medium used by Shklair & Keene (1974) has the following composition (g/litre): thioglycollate medium without dextrose or indicator (Difco), 24; purple broth base (Difco), 16. Carbohydrates are added as filter-sterilized solutions to the cooled base. We find, however, that autoclaving mannitol and sorbitol in this medium does not influence the outcome of fermentation reactions by *Strep. mutans*. Other media for the identification of *Strep. mutans* and other oral streptococci are described by Colman (1976).

Identification

The *Strep. mutans* strains first isolated by Clarke in 1924 were Gram positive cocci growing in chains; they produced acid from mannitol and raffinose, and they were nonhaemolytic. An unusual feature of *Strep. mutans* was that under certain growth conditions, especially low pH, the coccal cells tend to elongate to form short slightly irregular rods which might be mistaken for coryneform bacteria.

It is difficult to identify *Strep. mutans* colonies reliably on any of the selective media, though glucan-forming colonies on MSB agar or TYCB medium are very likely to be those of *Strep. mutans*. It is therefore necessary to use a number of biochemical tests in order to distinguish them from other oral streptococci. Several taxonomic studies have clearly established *Strep. mutans* as a fairly homogeneous group of organisms (Colman 1968; Carlsson 1968; Guggenheim 1968; Hardie and Bowden 1976*b*). A scheme for identifying the major oral streptococci is shown in Table 1.

In our laboratory, single colonies are picked up from the streptococcal selective media directly into 20-ml volumes of Todd–Hewitt broth and grown at 37°C for 2 days. These cultures are checked for purity by plating on HBA and in the majority of instances the cultures are pure. Failure to grow in Todd–Hewitt broth is good indication that the isolates are not streptococci. Actinomyces growing on the streptococcal-selective media fail to grow in

TABLE 1

Differentiation of major human oral streptococci

Reaction	*Streptococcus* species				
	mutans	*sanguis*	*mitior*	*milleri*	*salivarius*
Fermentation of					
Mannitol	+	−	−	−	−
Sorbitol	+/−[†]	−	−	−	−
Hydrolysis of					
Arginine	−[*]	+[‡]	−	+	−
Aesculin	+/−[†]	+[‡]	−	+	+
Production of					
Acetoin	+	−	−	+	−
Glucan from sucrose	+	+[‡]	+	−	+/−
Fructan from sucrose	NT	NT	NT	−	+
H_2O_2	−/+[†]	+	+	−	−

[*] Serotype *b* strains hydrolyse arginine.
[†] Serotypes *d*, *g* and *h* strains may give the less common result.
[‡] Negative results often found in one of these tests.
NT, not testable.

Todd–Hewitt broth, but it is necessary to check the identity of the subcultured organisms that fail to grow by examining Gram stained films of other representatives of the same colony type. The Todd–Hewitt broth cultures are used to inoculate test media, then centrifuged to pellet the bacteria, which are used, without extraction, to determine the serotype of the *Strep. mutans* isolate by immunodiffusion.

Streptococcus mutans can be differentiated from other oral streptococci primarily on the basis of acid production from mannitol and sorbitol and glucan formation from sucrose. The media and methods described by Colman (1976) are reliable although we have used the basal medium described by Shklair & Keene (1974) for the carbohydrate fermentations. In their method bromocresol purple rather than phenol red (Colman 1976) is used as the pH indicator, and this produces very satisfactory and unequivocal fermentation results, while phenol red sometimes gives an indifferent colour change. Glucan production in broth can sometimes be difficult to demonstrate if the bacteria produce an excess of water-insoluble glucan (mutan; Guggenheim 1970). To obviate this problem, examination of colonies on TYC medium for glucan production is recommended.

In our hands the media used by Colman (1976) and the basal medium for carbohydrate fermentations described by Shklair & Keene (1974) can be used in a miniaturized system as suggested for other organisms by Jayne-

Williams (1975). Approximately 250 μl of each test medium, dispensed into a well of a sterile 96-well Microtitre tray, is inoculated with a drop of Todd–Hewitt broth culture. Flat-bottomed wells are more suitable as they have a greater capacity than U- or V-shaped wells and suffer less from optical distortion, which greatly hinders the filling of wells. The inoculated trays are incubated in candle jars for 2 days before recording the results. The identification of stock strains by this method is identical with that obtained using conventional methods.

Sub-divisions within Streptococcus mutans

Although *Strep. mutans* strains form a particularly homogeneous group a degree of serological diversity has been demonstrated. Jablon & Zinner (1966) showed that *Strep. mutans* strains from hamsters and rats were serologically distinct. Later, Bratthall (1970) described five serotypes (*a*, *b*, *c*, *d*, *e*) amongst *Strep. mutans* strains; Perch *et al.* (1974) demonstrated the existence of two additional serotypes (*f* and *g*), and we recently isolated strains classified as serotype *h* from the dental plaque of monkeys (Beighton *et al.* 1981).

On the basis of the DNA base composition (Coykendall 1977; Beighton *et al.* 1984), DNA homology studies (Coykendall 1974), and differences in the protein profiles of whole-cell extracts subjected to SDS-PAGE (Russell 1976; Beighton *et al.* 1981; Beighton *et al.* 1984), *Strep. mutans*-like strains can be divided into six species. Three of these species (*Strep. mutans*, *Strep. ferus* and *Strep. macacae*) have the serotype *c* polysaccharide antigen in common. In Table 2 the subdivision of *Strep. mutans* into species and the host range of each is presented. Only serotypes *c* and *e* (*Strep. mutans*, after the original description in Clarke 1924) and *d* and *g* are commonly isolated from the oral cavity of humans.

Shklair & Keene (1974, 1976) subdivided *Strep. mutans* serotypes into five biotypes based on the fermentation of mannitol (with and without bacitracin), sorbitol, raffinose and melibiose and production of ammonia from arginine. According to Shklair & Keene (1974) a key characteristic in differentiating serotype *c* from serotype *e* (biotypes I and V respectively) is melibiose fermentation, but a study of 137 clinical isolates of serotype *c* and *e* failed to demonstrate the suitability of melibiose fermentation for differentiating between these two serotypes (Hamada *et al.* 1979). Kral & Daneo-Moore (1981) showed that acid production from starch and dextrin by *Strep. ferus* differentiated it from *Strep. mutans* serotype *c* strains. In our laboratory we identified *Strep. macacae*, which also possesses the serotype *c* antigen, and found that failure to produce acid from inulin and bacitracin sensitivity distinguished it from *Strep. mutans* serotype *c*. Although biotyping does not permit the reliable designation of serotype to *Strep. mutans* strains it does enable strains to be placed into the appropriate genetic group (Table 2).

TABLE 2

A scheme for subdividing Streptococcus mutans *strains and differentiating them from* Strep. ferus *and* Strep. macacae

Serotype	DNA base content (mole % G & C)	Suggested species name	Principal host	Fermentation of					Resistance to bacitracin	H_2O_2 production	Arginine hydrolysis	White halo on TYC medium
				Mannitol	Sorbitol	Inulin	Dextrin	Raffinose				
a	42–44	*cricetus*	Hamster	+	+	+	–	+	–	–	–	–
b	41–44	*rattus*	Rat	+	+	+	–	+	+	–	+	–
c,e,f	36–38	*mutans*	Humans and monkeys	+	+	+	–	+	+	–	–	–
d,g,h	44–46	*sobrinus*	Humans and monkeys	+	+/–	+	–	–	+/–	+/–	–	+
c	43–45	*ferus*	Rat	+	+	+	+	–	–	–	–	–
c	35–36	*macacae*	Monkeys	+	v	–	–	+	–	–	–	–

Data combined from Shklair & Keene (1974); Kral & Daneo-Moore (1981); Beighton *et al.* (1981); Coykendall (1977); van Palenstein Helderman *et al.* (1983); Beighton *et al.* (1984).

For the most part, human dental studies are concerned with assigning strains to the serotype *c, e, f* or to the serotype *d, g, h* genogroups, i.e. *Strep. mutans* and *Strep. sobrinus*, respectively. If specific antisera (Bratthall 1970) are available, the isolates can be studied further. This should be done for epidemiological studies, as different *Strep. mutans* serotypes do not all display the same degree of cariogenicity (Hamada *et al.* 1978; Beighton *et al.* 1983) and different serotypes may be associated with caries on different tooth surfaces (Huis in 't Veld *et al.* 1978).

Identification of Streptococcus mutans *from clinical specimens*

Streptococcus mutans can be readily identified from other oral streptococci in samples taken from the oral cavity. It may be confused, however, with other streptococci present in specimens collected from non-oral sites. This problem may be exacerbated if commercial kits for the identification of streptococci are used. The API-20 strep kit (API Laboratory Products Ltd., Basingstoke, Hants.) is a convenient and reliable method for identifying streptococci (Ball & Colman 1982), but the differentiation between *Strep. mutans*, *Strep. bovis* I and *Strep. salivarius* rests on only four fermentation reactions (Table 3). These three species are clinically important, as they are often isolated from blood cultures (Parker & Ball 1976).

To increase the reliability of the identification of these bacteria, several

TABLE 3

Differentiation of Streptococcus mutans *from certain other streptococci isolated from clinical specimens*

	Streptococcus species			
	mutans			
Reaction	*c,e,f*	*d,g,h*	*salivarius*	*bovis I*
Fermentation of				
Mannitol*	+	+	−	+
Sorbitol*	+	+/−	−	−
Starch*	−	−	+	+
Glycogen*	−	−	−	+
Lancefield group D	−	−	−	+
Halo on TYC medium	−	+	−	−
Large mucoid colony on TYC medium	−	−	−	+
Major polysaccharide from sucrose	Glucan	Glucan	Glucan or fructan	Glucan

* Tests used in API-20 strep.

additional tests should be performed. Only *Strep. bovis* I possesses the Lancefield group D antigen, and on TYC medium *Strep. bovis* I forms large (>5 mm in diameter) mucoid colonies while *Strep. mutans* forms erose, white or yellow colonies 1–2 mm in diameter and *Strep. salivarius* may form either large, very erose, white colonies, gumdrop- or draughtsman-shaped colonies or colonies similar to those of *Strep. mutans*, depending on the amount of fructan formed from sucrose.

Examination of isolates for the Lancefield group D antigen and colonial morphology on TYC medium (or other similar sucrose-containing agar) will increase the reliability of the identification, especially of *Strep. bovis* I.

Streptococcus mutans is a normal component of the bacterial plaque present on human teeth. Given the appropriate environment it becomes involved in the initiation of tooth decay but it does not act alone. As a member of a complex microbial community it is subjected to various interactions with both the host and other constituents of plaque. These interactions have been studied extensively and provide a fascinating insight into the ecology, not only of *Strep. mutans*, but also of other members of the plaque flora. A comprehensive review of *Strep. mutans* biology has been written by Hamada & Slade (1980), and the book by Marsh & Martin (1984) provides an excellent introduction to the subject of dental plaque ecology.

I should like to thank Dr. G. Colman and Mrs. H. Hayday for their assistance in the preparation of this chapter.

References

Ball, L. C. & Colman, G. 1982 A comparison of conventional methods and API galleries for the identification of streptococci. In *Basic Concepts of Streptococci and Streptococcal Diseases*, ed. Holm, S. E. & Christensen, P., pp. 41–42. Chertsey, England: Reedbooks Ltd.

Beighton, D., Russell, R. R. B. & Hayday, H. 1981 The isolation and characterization of *Streptococcus mutans* serotype *h* from dental plaque of monkeys (*Macaca fascicularis*). *Journal of General Microbiology* **124**, 271–279.

Beighton, D., Hayday, H. & Walker, J. 1982 The acquisition of *Streptococcus mutans* by infant monkeys (*Macaca fascicularis*) and its relationship to the initiation of dental caries. *Journal of General Microbiology* **128**, 1881–1892.

Beighton, D., Cohen, B. & Hayday, H. 1983 The association between dental caries incidence and two *Streptococcus mutans* populations in monkeys, *Macaca fascicularis*. *Caries Research* **17**, 125–130.

Beighton, D., Hayday, H., Russell, R. R. B. & Whiley, R. A. 1984 *Streptococcus macacae* sp. nov., a "mutans-like" species from the dental plaque of monkeys (*Macaca fascicularis*). *International Journal of Systematic Bacteriology* **34**, 332–335.

Bratthall, D. 1970 Demonstration of five serological groups of streptococcal strains resembling *Streptococcus mutans*. *Odontologisk Revy* **21**, 181–196.

CARLSSON, J. 1967*a* Presence of various types of non-hemolytic streptococci in dental plaque and in other sites of the oral cavity in man. *Odontologisk Revy* **18**, 55–74.

CARLSSON, J. 1967*b* A medium for the isolation of *Streptococcus mutans*. *Archives of Oral Biology* **12**, 1657–1658.

CARLSSON, J. 1968 A numerical taxonomic study of human oral streptococci. *Odontologisk Revy* **19**, 137–160.

CHAPMAN, G. H. 1946 The isolation and testing of fecal streptococci. *American Journal of Digestive Disease* **13**, 105–107.

CLARKE, J. K. 1924 On the bacterial factor in the aetiology of dental caries. *British Journal of Experimental Pathology* **5**, 141–147.

COLMAN, G. 1968 The application of computers to the classification of streptococci. *Journal of General Microbiology* **50**, 149–158.

COLMAN, G. 1976 The viridans streptococci. In *Selected Topics in Clinical Bacteriology*, ed. De Louvois, J., pp. 179–198. London: Baillière Tindall.

COLMAN, G. & WILLIAMS, R. E. O. 1972 Taxonomy of some human viridans streptococci. In *Streptococci and Streptococcal Disease*, ed. Wannamaker, L. & Matson, J. M., pp. 281–299. London & New York: Academic Press.

COYKENDALL, A. L. 1974 Four types of *Streptococcus mutans* based on their genetic, antigenic and biochemical characteristics. *Journal of General Microbiology* **83**, 327–338.

COYKENDALL, A. L. 1977 Proposal to elevate the subspecies of *Streptococcus mutans* to species status, based on their molecular composition. *International Journal of Systematic Bacteriology* **27**, 26–30.

DE STOPPELAAR, T. D., VAN HOUTE, J. & DE MOOR, C. E. 1967 The presence of dextran-forming bacteria, resembling *Streptococcus bovis* and *Streptococcus sanguis*, in human dental plaque. *Archives of Oral Biology* **12**, 1199–1201.

EMILSON, C. G. & BRATTHALL, D. 1976 Growth of *Streptococcus mutans* on various selective media. *Journal of Clinical Microbiology* **4**, 95–98.

FACKLAM, R. R. 1974 Characterization of *Streptococcus mutans* isolated from human dental plaque and blood. *International Journal of Systematic Bacteriology* **24**, 313–319.

GOLD, O. G., JORDAN, H. V. & VAN HOUTE, J. 1973 A selective medium for *Streptococcus mutans*. *Archives of Oral Biology* **18**, 1357–1364.

GUGGENHEIM, B. 1968 Streptococci of dental plaque. *Caries Research* **2**, 147–163.

GUGGENHEIM, B. 1970 Enzymatic hydrolysis and structure of water-insoluble glucan produced by glucosyltransferases from a strain of *Streptococcus mutans*. *Helvetica Odontologica Acta* **14**, 89–108.

HAMADA, S. & SLADE, H. D. 1980 Biology, immunology and cariogenicity of *Streptococcus mutans*. *Microbiological Reviews* **44**, 331–384.

HAMADA, S., OOSHIMA, T., TORII, M., IMANISHI, N., MASUDA, N., SOBUE, S. & KOTANI, S. 1978 Dental caries induction in experimental animals by clinical strains of *Streptococcus mutans* isolated from Japanese children. *Microbiology and Immunology* **22**, 301–314.

HAMADA, S., MASUDA, N. & SHIMAMOTO, T. 1979 Some biological properties of *Streptococcus mutans* isolated from human mouths, with reference to the correlation with serotypes. *Archives of Oral Biology* **24**, 627–631.

HARDIE, J. M. & BOWDEN, G. H. 1976*a* The microbial flora of dental plaque: bacterial succession and isolation considerations. In *Microbial Aspects of Dental Caries*, ed. Stiles, H. M., Loesche, W. J. & O'Brien, R. C., Special Supplement to *Microbiology Abstracts* **1**, 63–87.

HARDIE, J. M. & BOWDEN, G. H. 1976*b* Physiological classification of oral viridans streptococci. *Journal of Dental Research* **55**, 166A–176A.

HEMMENS, E. S., BLAYNEY, J. R., BRADEL, S. F. & HARRISON, R. W. 1946 The microbic flora of the dental plaque in relation to the beginning of caries. *Journal of Dental Research* **25**, 195–205.

HUIS IN 'T VELD, J., BANNET, D., VAN PALENSTEIN HELDERMAN, W., SANPAIO CAMARGO, P. & BACKER DIRKS, O. 1978 Antibodies against *Streptococcus mutans* and glucosyltransferases in caries-free and caries active military recruits. *Advances in Experimental Medicine and Biology* **107**, 369–381.

IKEDA, T. & SANDHAM, H. J. 1972 A medium for the recognition and enumeration of *Streptococcus mutans*. *Archives of Oral Biology* **17**, 601–604.

IKEDA, T., SANDHAM, H. J. & BRADLEY, E. L. 1973 Changes in *Streptococcus mutans* and lactobacilli in plaque in relation to the initiation of dental caries in Negro children. *Archives of Oral Biology* **18**, 555–566.

JABLON, J. M. & ZINNER, D. D. 1966 Differentiation of cariogenic streptococci by fluorescent antibody. *Journal of Bacteriology* **92**, 1590–1596.

JAYNE-WILLIAMS, D. J. 1975 Miniaturized methods for the characterization of bacterial isolates. *Journal of Applied Bacteriology* **38**, 305–309.

KRAL, T. A. & DANEO-MOORE, L. 1981 Biochemical differentiation of certain oral streptococci. *Journal of Dental Research* **60**, 1713–1718.

LILJEMARK, W. F., OKRENT, D. H. & BLOOMQUIST, C. G. 1976 Differential recovery of *Streptococcus mutans* from various mitis-salivarius-agar preparations. *Journal of Clinical Microbiology* **4**, 108–109.

LINKE, H. A. B. 1977 New medium for the isolation of *Streptococcus mutans* and its differentiation from other oral streptococci. *Journal of Clinical Microbiology* **5**, 604–609.

LITTLE, W. A., KORTS, D. C., THOMSON, L. A. & BOWEN, W. H. 1977 Comparative recovery of *Streptococcus mutans* on ten isolation media. *Journal of Clinical Microbiology* **5**, 578–583.

MARSH, P. D. & MARTIN, M. V. 1980 *Oral Microbiology*. Walton-on-Thames, Surrey, England: Thomas Nelson & Sons Ltd.

MEJARE, B. & EDWARDSSON, S. 1975 *Streptococcus milleri* (Guthof): an indigenous organism of the human oral cavity. *Archives of Oral Biology* **20**, 757–762.

PARKER, M. T. & BALL, L. C. 1976 Streptococci and aerococci associated with systemic infection in man. *Journal of Medical Microbiology* **9**, 275–302.

PERCH, B., KJEMS, E. & RAVN, T. 1974 Biochemical and serological properties of *Streptococcus mutans* from various human and animal sources. *Acta Pathologica et Microbiologica Scandinavica, Section B: Microbiology and Immunology* **82B**, 357–370.

RUSSELL, R. R. B. 1976 Classification of *Streptococcus mutans* strains by SDS gel electrophoresis. *Microbios Letters* **2**, 55–59.

SHKLAIR, I. L. & KEENE, H. J. 1974 A biochemical scheme for the separation of the five varieties of *Streptococcus mutans*. *Archives of Oral Biology* **19**, 1079–1081.

SHKLAIR, I. L. & KEENE, H. J. 1976 Biochemical characterization and distribution of *Streptococcus mutans* in three diverse populations. In *Microbial Aspects of Dental Caries*, ed. Stiles, H. M., Loesche, W. J. & O'Brien, T. C., Special Supplement to *Microbiology Abstracts* **3**, 201–210.

STAAT, R. H. 1976 Inhibition of *Streptococcus mutans* strains by different mitis-salivarius agar preparations. *Journal of Clinical Microbiology* **3**, 318–380.

UNSWORTH, P. F. 1980 The isolation of streptococci from human faeces. *Journal of Hygiene* **85**, 153–164.

VAN HOUTE, J. & GREEN, D. B. 1974 Relationship between the concentration of bacteria in saliva and the colonization of teeth in humans. *Infection and Immunity* **9**, 624–630.

VAN HOUTE, J., GIBBONS, R. J. & BANGHART, S. B. 1970 Adherence as a determinant of the presence of *Streptococcus salivarius* and *Streptococcus sanguis* on the human tooth surface. *Archives of Oral Biology* **15**, 1025–1034.

VAN PALENSTEIN HELDERMAN, M., IJSSELDIJK, M. & HUIS IN 'T VELD, J. H. J. 1983 A selective medium for the two major subgroups of the bacterium *Streptococcus mutans* isolated from human dental plaque and saliva. *Archives of Oral Biology* **28**, 599–603.

WALKER, J. 1981 Antibody responses of monkeys to oral and local immunization with *Streptococcus mutans*. *Infection and Immunity* **31**, 61–70.

WOLTJES, J., LEGDEUR-VELTHUIS, H., EGGINK, C. O. & DE GRAAFF, J. 1980 β-Haemolysis and pigment production by the oral bacterium *Streptococcus mutans*. *Archives of Oral Biology* **25**, 279–281.

Listeriosis in Farm Animals in Great Britain

M. Gitter

Ministry of Agriculture, Fisheries and Food, Central Veterinary Laboratory, New Haw, Weybridge, Surrey, UK

During the last decade or so, *Listeria monocytogenes* has been receiving increasing attention as an agent affecting animal and human health, and we have often asked ourselves the question: how many cases have been overlooked and to what extent do our laboratory records reflect the actual incidence of the disease? Our records are based on the reports of the Veterinary Investigation Centres, and we know that listeriosis in farm animals occurs in all parts of the country and involves many species of animals and birds (Table 1). For many years listeriosis in Great Britain was regarded as mainly a disease of sheep, but during the brucella eradication campaign listeria infections were reported more frequently in cattle than in sheep. Progress in the eradication of brucellosis, however, was followed by a gradual rise in the numbers of cases of ovine listeriosis reported, and this culminated in 1981/1982 in a marked preponderance of ovine listeriosis over any other kind of animal listeriosis in Great Britain (Fig. 1).

Clinical Manifestations

Clinical disease is most common in winter and early spring. Until 1982 it was usually sporadic, although occasional outbreaks were recorded in which the mortality was as high as to 10%. Listeriosis shows a wide variety of clinical manifestations in the form of five separate diseases or syndromes which seldom overlap, each of which probably has a separate pathogenesis: (1) infection of the pregnant uterus with subsequent abortion, stillbirth or neonatal death, (2) septicaemia with miliary lesions, mainly in unweaned animals, (3) encephalitis, (4) purulent conjunctivitis and keratitis, and (5) mastitis. The term "typical listeriosis" is therefore quite meaningless.

Isolation and Identification of
Micro-organisms of Medical and
Veterinary Importance

Copyright © 1985 by the Society for Applied Bacteriology
All rights of reproduction in any form reserved
0-12-181460-2

TABLE 1

Incidents of listeriosis in Great Britain reported by Veterinary Investigation Centres

Year	Cattle	Sheep	Pigs	Horses	Goats	Rabbits	Guinea pigs	Dogs	Cats	Any other animal	Fowl	Other species of birds	Total
1975	37	44	1		1					2	1	1	87
1976	58	49	2		1	2					3	1	116
1977	52	39	2	2				1		1	2	1	100
1978	75	70		1		3					1		150
1979	65	86	2		1					1	1	2	158
1980	60	101	3		1		1			1	2	3	172
1981	40	82	3	1	5	2				1	1	1	136
1982	85	195	1		4	3	1				2	2	293
Total	472	666	14	4	13	10	2	1	—	6	13	11	1212

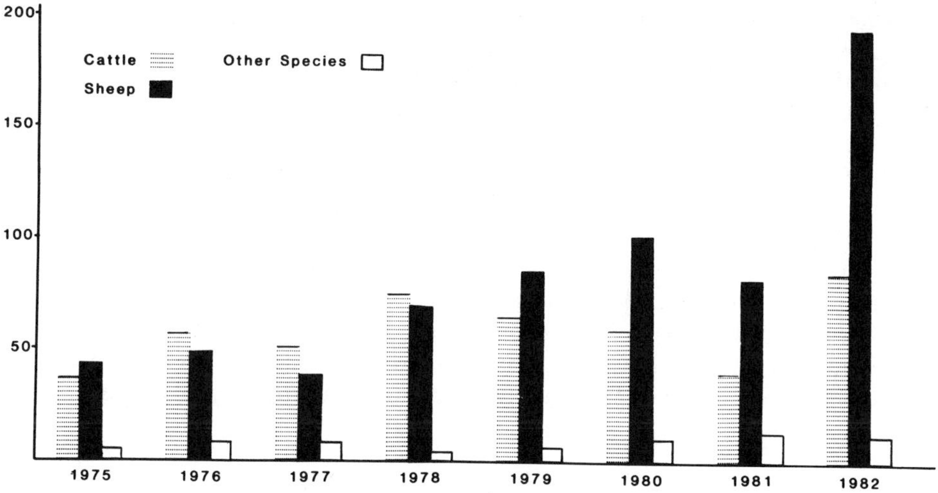

FIG. 1. Incidents of animal listeriosis, Great Britain, 1975–1982.

Abortion in cattle and sheep

Abortion in cattle and sheep occurs during the latter stages of pregnancy. It is usually sporadic, especially in cattle, although in sheep outbreaks involving several animals have been recorded. There is evidence, both field and experimental (Gitter *et al.* 1981), that diagnosed cases of listerial abortion are the proverbial tip of an iceberg, and far more animals are infected than those which abort. The uterine infection is of haematogenous origin and is preceded by asymptomatic bacteraemia. The placenta is quickly invaded by the organisms; the fetus dies and is expelled within 5 days. By this time, autolytic changes may mask minor lesions, and the only tissue alterations may be small foci of hepatic necrosis. Abortions at this stage are not usually accompanied by severe systemic disease of the dam, although on a few occasions we have isolated the organisms from the milk of an aborting cow. The placenta may be retained as a result of mild metritis, but the organisms and associated inflammation are quickly cleared from the non-pregnant uterus. If the fetus is near term when the infection develops, an abnormal parturition can occur with dystokia, severe metritis and septicaemia being the most common sequelae. There is no evidence that simple listerial metritis has any long-lasting effect on the uterus or reproductive function, and subsequent pregnancies are usually uncomplicated.

We have no accurate information about the incidence of listerial abortion in Great Britain. Bovine fetuses are now seldom submitted for laboratory examination, but if serological evidence could be taken as an indication that abortion was associated with *L. monocytogenes* then this widely distributed organism could be regarded as an important cause of bovine abortion (Gitter 1979).

Information on the incidence of listerial abortion in sheep is also inadequate. Abortions, which are often sporadic, occur in the latter stages of pregnancy, often when lambing is under way and the farmer is too busy to take a single fetus to the laboratory, especially when the ewes appear to be healthy.

Diagnosis of listerial abortion or septicaemia is not difficult and the organism can be isolated readily from the internal organs, fetal stomach contents and placenta, but a random placental isolate may be difficult to interpret. In such cases histological examination of the placenta, if it is not too decomposed, and serological examination of paired serum samples taken from the dam may be helpful in assessing the significance of the placental isolate.

Visceral or septicaemic infections

The visceral or septicaemic form, with the characteristic "sawdust" liver lesions, occurs most frequently in the first weeks of life, and the neonatal infection can probably be considered as a continuation of an intrauterine infection. This type of listeriosis generally occurs as outbreaks, and losses can be high. In spite of septicaemia, the cerebral and visceral forms very seldom overlap in the same animal, and the clinical symptoms alone are of limited diagnostic value. The affected lambs just look ill and sometimes scour, but there is nothing specific about the clinical manifestations that could not as easily be attributed to other ailments of young lambs. The septicaemic form is uncommon in young calves and is usually sporadic.

Encephalitis of ruminants

Encephalitis of ruminants has been the most frequently recognised form of animal listeriosis, especially during the last 2 years in sheep, and could prove to be of considerable economic importance. *Listeria monocytogenes* seems to have a remarkable affinity for the brain stem; the lesion is most severe in the medulla and pons, and the evidence is against the haematogenous route (Asahi *et al.* 1957).

In cattle, encephalitis and facial paralysis often occur in the same animal, and can be manifested by a drooping ear, eyelid or lip with consequent dropping of cud. There may be paralysis of the masticatory muscles and of the pharynx. It is often accompanied by blindness, ataxia, dullness and circling. Prompt wide-spectrum antibiotic treatment is claimed to have led to recovery but generally prognosis is poor, with death supervening within 4–10 days of the onset of symptoms, although the animals are often destroyed earlier on humanitarian grounds. The incidence is usually sporadic, and although this form occurs principally in adult animals it has also been reported in calves.

In sheep and goats the disease often runs an acute course with death sometimes supervening within hours of the onset of clinical symptoms. Facial paralysis in sheep is rare. During the last 2 years, numerous outbreaks of encephalitis in ewes, and some in lambs, have been recorded, with the incidence being highest in Scotland and the north of England, lower in Wales and the south-west of England, and lowest in the south-east (Anon. 1983). This represents a change in the pattern of this form of listeriosis, which used to occur only sporadically in Great Britain with outbreaks being very rare.

It is of interest that although most cases of cerebral listeriosis occur during pregnancy or soon after lambing, listerial abortions in such flocks were almost unknown. Recent field reports, however, indicate a further change in the pattern of the disease, with both abortion and encephalitis being diagnosed in the same flocks, although not in the same animals. The picture is also often confused by concurrent disease in the flock or even in the same animal.

Ante-mortem diagnosis of listerial encephalitis is hampered by the lack of satisfactory diagnostic tests. The peripheral monocytosis, often found in mono-gastric animals, is not a characteristic of ruminants. Often the animals, especially sheep and goats, die before a serological response can be detected.

Bacteriological examination of the brain is often difficult and time-consuming, but the histological picture is, at the present stage of our knowledge, pathognomonic. The typical parenchymal lesion is a micro-abscess which may begin as a tiny collection of neutrophils but more usually does so as a minute focus of microglial reaction. The glial nodules may persist, but there is a tendency for them to be infiltrated by neutrophils and for their centre to liquify. There is also heavy perivascular cuffing, with the cuffs composed mainly of lymphocytes and histiocytes with a few admixed neutrophils and eosinophils. These lesions are confined to the brain stem and medulla. Mild meningitis often affects the cerebellum and anterior cervical cord (Jubb & Kennedy 1970). In view of the findings of Grønstøl (1980b) and our own observations concerning the isolation of listeria from sheep brains which did not show lesions, it is essential that histological examination be carried out in order to confirm the diagnosis of listerial encephalitis. The mere isolation of the organism is not a sufficient ground for this diagnosis, and there is also a possibility of a mixed infection which may not be evident on bacteriological examination only.

Infectious bovine keratitis

Isolation of *L. monocytogenes* has been reported in outbreaks of infectious bovine keratitis but the aetiology of this condition is still unknown. A number of potential pathogens have been implicated, however, such as the virus of infectious bovine rhino-tracheitis, *Mycoplasma bovoculi, Moraxella bovis, Branhamella (Neisseria) catarrhalis* and *L. monycytogenes,* with two or three of these micro-organisms being isolated from one eye.

Bovine mastitis

Despite the wide distribution of *L. monocytogenes,* there are few reports in the literature on bovine mastitis caused by this organism. This is surprising in view of the marked increase in the incidence of environmental mastitis but unless especially looked for *L. monocytogenes* may easily escape detection in routine examination of material submitted to a diagnostic laboratory (Gitter *et al.* 1980). The potential hazards of the misidentification of listeria as a diphtheroid contaminant or as streptococcal species are well known. The cases of listerial mastitis reported in the literature varied in severity. Some were acute, some chronic and some subclinical, and the response to antibiotic treatment was generally disappointing. Sometimes only one-quarter of the udder was affected, sometimes the entire udder, and the organism could be excreted for up to 3 years (Donker-Voet 1965).

Mixed infections

The possibility of multifactorial disease must be borne in mind. We have come across the following infections and conditions which occurred concurrently with listeriosis either in the same flock or herd or even in the same animal.

Bovine abortion: *Leptospira hardjo,* bovine virus diarrhoea and *Brucella abortus.*

Ovine encephalitis: Streptococcal meningitis, pasteurella meningitis, pregnancy toxaemia, scrapie and cerebro–cortical necrosis. In one outbreak where five ewes were showing similar clinical symptoms, one was affected with listerial encephalitis, two with cerebro–cortical necrosis, and two with pregnancy toxaemia.

Bovine mastitis: In a case of clinical mastitis where *L. monocytogenes* was isolated from the affected quarter, which did not respond to treatment, histological examination of the udder revealed lesions similar to those of experimental mycoplasma mastitis. Mycoplasmas were not isolated from the milk or the udder (Gitter *et al.* 1980).

Listeriosis in other animals

Pigs

The reported disease incidence is remarkably low. The few cases mention abortion, stillbirth, septicaemia and encephalitis among sucking pigs. The clinical symptoms vary; nervous symptoms in the forms of fits, nystagmus and opisthotonos, or sudden death in a septicaemic form with multiple necrotic foci in the liver have been reported.

Domestic fowl

The recorded incidence is low. As in mammals, it is the young fowl that appears to be most susceptible, but sporadic incidents were reported in older birds. Listerial infection in birds is usually of the septicaemic form although the central nervous system can also be affected. The most conspicuous lesions fairly often encountered are massive areas of myocardial degeneration and necrosis and necrotic foci are also found, with variable frequency in liver, spleen and kidneys.

Experimentally, birds show considerable resistance to infection, and this is borne out by the field experience that outbreaks of listeriosis usually occur as a sequel to other debilitating diseases of poultry.

In surveys on "oven-ready poultry", *L. monocytogenes* was isolated on one occasion from 27 out of 51 birds (53%) (Kwantes & Isaac 1975) and from 10 out of 68 birds (14.7%) on another (Gitter 1976).

Rodents

The disease has been encountered in chinchillas and rabbits in which it caused septicaemia and abortion.

Silage

There are many reports associating silage feeding with listeriosis in sheep. Not only have many strains of listeria been isolated from silage, in particular the surface layers and poor quality silage, but silage seems to increase the susceptibility of sheep to the disease (Grøstøl 1980a). Listeria has been isolated from silage with a pH less than 4 (Grønstøl 1979) and Dijkstra (1975) reported that in 16% of silage samples listerias survived for 10–12 years. Clearly more investigation should be directed into the association between listeriosis and silage feeding.

Isolation

Knowledge of the pathogenesis and clinical manifestations is very helpful, if not essential, for correct diagnostic measures to be applied. Figure 2 illustrates a general scheme for isolating listerias.

Abortion and septicaemia

Viscera and fetal stomach contents

These are plated on 5% sheep blood agar which is incubated for 24–48 h at 37°C. Haemolysis varies from scanty to pronounced especially in *L. monocytogenes* type 5.

Placenta

A loopful of macerated portion of the placenta is plated on blood agar and on blood agar containing nalidixic acid (0.004%) and thallium acetate (0.02%) (NATA agar) and incubated at 37°C for 24–48 h. The rest of the macerated tissue is placed in tryptose phosphate broth, which is then kept at 4°C and subcultured periodically on NATA agar. Suspect colonies are picked up by reflected light (Henry 1933; Gray 1957) and subcultured on blood agar.

Encephalitis

Bacteriological examination. A small portion of the medulla, about 5 mm^3, is macerated in a sterile grinder and plated directly on 5% sheep blood agar, which is then incubated at 37°C and examined as described above. The rest of the macerated portion of the medulla is washed out of the grinder with tryptose phosphate broth, placed at 4°C and subcultured periodically on NATA agar, and suspect colonies are examined in the same way.

Histological examination. After the removal of a portion of medulla for bacteriological examination, the brain is placed in neutral, phosphate-buffered 10% formalin for histological examination. This is often completed before the results of bacteriological examination are known. If the results of histological examination are negative, bacteriological examination can be discon-

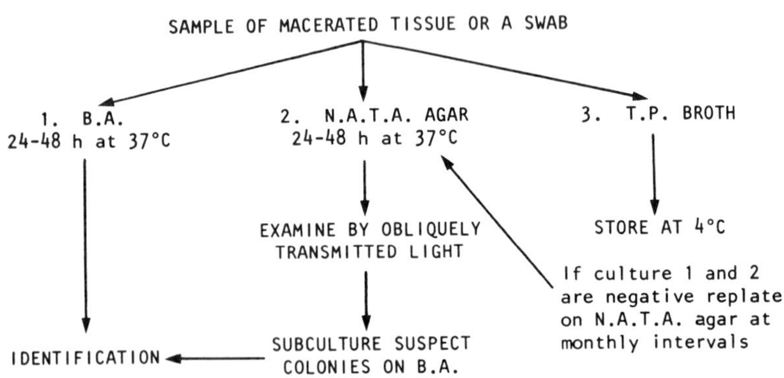

FIG. 2. Examination for *L. monocytogenes*. B.A., blood agar; N.A.T.A., nalidixic acid–thallium acetate; T.P.B., tryptose phosphate broth.

tinued. When the characteristic lesions of listeria encephalitis are present, bacteriological examination can continue if it is desirable to identify the serotype involved; otherwise there is no practical advantage in pursuing the examination. As already mentioned, diagnosis of listerial encephalitis cannot be based entirely on the isolation of listeria from the brain.

Purulent conjunctivitis and keratitis

A swab taken from the conjunctival sac is used to inoculate 5% sheep blood agar (incubated at 37°C) and is then placed in tryptose phosphate broth and left at 4°C. Subsequent examinations are as in the bacteriological examination of placenta and medulla.

Mastitis

The milk is centrifuged and the deposit is cultured on sheep blood agar and NATA agar.

Silage

Examination of silage would proceed along the lines outlined by J. Watkins (this volume, p. 201).

Identification

The isolates are identified by their cellular morphology, colonial appearance, tumbling motility at 22°C, positive catalase reaction, hydrolysis of aesculin and the examination of culture plates by obliquely transmitted light. (See also J. Watkins, this volume, p. 201.)

Serum agglutination test

This is carried out with a trypsinised antigen on mercapto–ethanol-reduced sera (Gitter 1979), but in view of the ubiquity of listerias serological examination alone cannot be regarded as fully diagnostic of the disease. Examination of paired serum samples may be helpful, however, in cases of abortion where fetal specimens are not available or in the case of a random placental isolate.

Comment

Listeriosis is not a notifiable disease in Great Britain. The available data are not truly representative of the extent of its incidence, but they do reflect the trend and pattern of the disease. We know that this elusive pathogen seems to have altered some of its clinical manifestations, and a marked increase in the incidence of the disease has been recorded by the Veterinary Investigation Centres. Outbreaks in marmosets (J. Knight, personal communication) and fruit bats (P. A. Gillespie, personal communication) in zoological gardens and a series of isolates of *L. monocytogenes* from seagulls' droppings (Roberts & Fenlon, personal communication) have also been reported.

If we become too complacent in our new-found knowledge, *Listeria monocytogenes* in all its clinical guises will continue to produce serious and often fatal disease, and if the recently recorded trend continues it may become a source of serious economic losses.

References

ANON. 1983 Listeria infections in farm animals. *Veterinary Record* **112,** 314.

ASAHI, O., HOSODA, T. & AKIYAMA, Y. 1957 Studies on the mechanism of infection of the brain with *Listeria monocytogenes*. *American Journal of Veterinary Research* **18,** 147–157.

DIJKSTRA, R. G. 1975 Recent experiences on the survival times of *Listeria* bacteria in suspensions of brain, tissue, silage, faeces and in milk. In *Problems of Listeriosis*, ed. Woodbine, M., pp. 71–73. Leicester, England: Leicester University Press.

DONKER-VOET, J. 1965 Listeriosis in animals. *Bulletin de l'Office International des Epizooties* **64,** 757–764.

GITTER, M. 1976 *Listeria monocytogenes* in 'oven ready' poultry. *Veterinary Record* **99,** 336.

GITTER, M. 1979 *Listeria monocytogenes* infection in bovine abortion. In *Problems of Listeriosis*, ed. Ivanov, I., pp. 193–205. Sofia, Bulgaria.

GITTER, M., BRADLEY, R. & BLAMPIED, P. H. 1980 *Listeria monocytogenes* infection in bovine mastitis. *Veterinary Record* **107,** 390–393.

GITTER, M., RICHARDSON, C. & BOUGHTON, E. 1981 Experimental infection of pregnant ewes with *L. monocytogenes*. In *Problems of Listeriosis*. Madrid, Spain.

GRAY, M. L. 1957 A rapid method for the detection of colonies of *Listeria monocytogenes*. *Zentralblatt für Bakteriologie, Parasitenkunde, Infektionskrankheiten und Hygiene, Abteilung 1: Originale* **169,** 373–377.

GRØNSTØL, H. 1979 Listeriosis in sheep. Isolation of *Listeria monocytogenes* from grass silage. *Acta Veterinaria Scandinavica* **20,** 492–497.

GRØNSTØL, H. 1980*a* Listeriosis in sheep. *Listeria monocytogenes* in sheep fed hay or grass silage during pregnancy. Immunological state, white blood cells, total serum protein and serum iron. *Acta Veterinaria Scandinavica* **21,** 1–10.

GRØNSTØL, H. 1980*b* Listeriosis in sheep. Isolation of *Listeria monocytogenes* from organs of slaughtered animals and dead animals submitted for post mortem examination. *Acta Veterinaria Scandinavica* **21,** 11–17.

HENRY, B. S. 1933 Dissociation in the genus *Brucella*. *Journal of Infectious Diseases* **52,** 374–402.

JUBB, K. V. F. & KENNEDY, P. C. 1970 *Pathology of Domestic Animals* 2nd ed., Vol. 2, p. 410. New York & London: Academic Press.

KWANTES, M. & ISAAC, M. 1975 *Listeria* infection in West Glamorgan. In *Problems of Listeriosis*, ed. Woodbine, M., pp. 112–114. Leicester, England: Leicester University Press.

Isolation of *Listeria monocytogenes*

J. WATKINS*

Yorkshire Water Authority, South Eastern Division, Divisional Laboratory, Green Lane, Scawthorpe, Doncaster, UK

The problem of listeriosis would appear to be intransigent. *Listeria monocytogenes* is widely distributed in nature (Weis 1975) and is able to survive in the natural environment (Welshimer 1960). There is clear association with warm blooded animals and man, and transient residence in the gut is an unquestionable feature (Kampelmacher & Van Noorle Jansen 1969, 1972) not necessarily associated with the direct contact of faecal material. The organisms have been demonstrated in milk by Hyslop & Osborne (1959) and Gitter *et al.* (1980), in oven-ready poultry by Gitter (1976), and more recently in coleslaw by Schelch *et al.* (1983) in an outbreak of human listeriosis in Canada. The ability of the organism to survive in the environment and in some instances to multiply (Gray 1960) facilitates its spread from one individual to another. Under certain conditions, however, it may become pathogenic, when the mortality rate may be high. Infection usually occurs in individuals who are immunologically deficient, have a debilitating disease or are undergoing unusual stress. Malignant disease, wintering in animals, pregnancy and extremes of age are examples.

Studies of the epidemiology of the disease have been hampered by the lack of suitable techniques for isolating the organism (Gray & Killinger 1966). Although it is comparatively easy to isolate listerias from clinical material, it is very difficult to recover them from heavily contaminated environmental samples. The ability of the organism to multiply at 4°C has undoubtedly played an important part in the development of suitable isolation methods, but where low temperature is used as a sole source of enrich-

*Present address: Yorkshire Water Authority, Skeldergate Laboratory, Skeldergate, York YO1 1HL, UK.

Isolation and Identification of
Micro-organisms of Medical and
Veterinary Importance

Copyright © 1985 by the Society for Applied Bacteriology
All rights of reproduction in any form reserved
0-12-181460-2

ment, heavy contamination and slow growth of other bacteria together with the sometimes inhibitory nature of specimen, (e.g. septic liquors) may result in failure to isolate listerias by direct plating. Indeed, samples may have to be incubated for up to 12 months before the organisms are isolated. The opportunities for any serious epidemiological studies are therefore clearly restricted. A combination of cold enrichment and selective agents may also be counter productive, as it may minimise the recovery of environmentally damaged cells.

The methods used for the isolation of salmonellas from sewage may usefully be applied to the recovery of other pathogens in heavily contaminated environments. A period of pre-enrichment in a suitable nutrient medium is essential to resuscitate and promote multiplication of environmentally damaged organisms before selective enrichment is applied. Once recovered, organisms will grow rapidly in suitable enrichment media. The technique developed for isolating listerias includes a period of pre-enrichment in a nutrient medium at a low temperature followed by sub-culture to an enrichment medium and incubation at a higher temperature. The two steps allow the organisms to recover and then multiply rapidly, making identification easy when enrichment broths are plated onto a suitable non-selective agar. Even with such techniques, dilution of the enrichment medium before plating is essential to avoid overgrowth and inhibition by competing bacteria.

The method described here is suitable for examining a wide variety of specimens including sewage, sewage sludge, soil, silage and veterinary material. Qualitative results may be obtained by direct inoculation into pre-enrichment broth; quantitative results are obtained by using a most probable number (MPN) system.

Sample Preparation

Solid samples

Solid samples (e.g. soil) are inoculated directly into a suitable volume of sterile Nutrient Broth no. 2 (Oxoid). Quantitative results are obtained by inoculating 50 g, five lots of 10 g and five lots of 1 g into suitable volumes of nutrient broth, e.g. 200 ml, five volumes of 50 ml and five volumes of 10 ml. These are incubated at 4°C and, after enrichment, isolation and identification, are compared with MPN tables.

Liquid samples

Sewage is inoculated directly into double-strength nutrient broth for qualitative results or quantitatively into double- and single-strength broth using an MPN system as above and 100 ml of sewage. Tenfold dilution of sewage in sterile quarter-strength Ringer solution may be necessary if counts are expected to be in excess of 1800 organisms/litre. This is preferable to the inclusion of five lots of 0.1 ml in the MPN system, as it will reduce the workload when a large number of samples is being examined. River or drinking-water samples are filtered through a suitable cellulose acetate or cellulose nitrate filter of pore size 0.45 μm or less. The membranes are placed directly into sterile nutrient broth. Quantitative results are obtained by filtering 500 ml and five 100-ml volumes and placing each membrane into nutrient broth. Where counts are expected to be in excess of 10 organisms/litre, the addition of 100 ml of the sample to double- and single-strength nutrient broth should be used. All pre-enrichment broths are incubated at 4°C.

Isolation

As shown in Fig. 1, at the beginning of each week, 1-ml amounts of each pre-enrichment broth are added to 10 ml of Nutrient Broth no. 2 (Oxoid) containing 3.75% w/v potassium thiocyanate AR (BDH) and 100 μg/ml of naladixic acid/sodium salt (Sigma Chemical Company). The potassium thio-

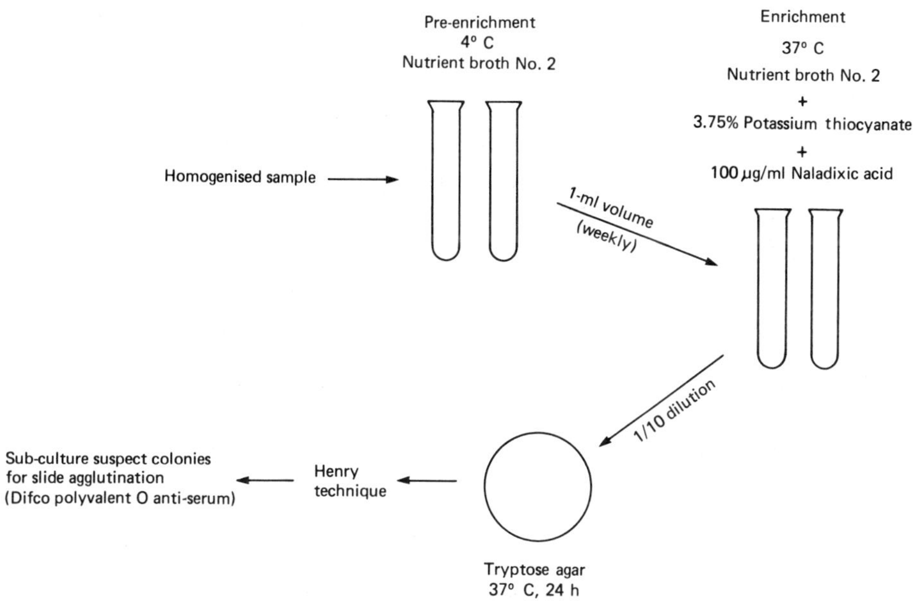

FIG. 1. Scheme for the isolation of *Listeria monocytogenes*.

cyanate can be sterilised in the nutrient broth by autoclaving at 121°C for 15 min, with the naladixic acid added separately as a solution prepared in sterile distilled water. For convenience, the thiocyanate–naladixic acid (TN) broth can be prepared in large volumes (250 ml), the antibiotic added, and the medium dispensed aseptically into sterile 25-ml screw-capped (Universal) bottles. After sub-culture, cold enrichment broths are returned to the refrigerator and TN broths are incubated at 37°C.

After 24 hours incubation at 37°C, 0.1-ml amounts of TN broth are removed, diluted to 1.0 ml with sterile Ringer solution, and 0.1 ml is plated on Tryptose Agar (Difco) using a glass spreader which has been sterilised with alcohol. Dilution to prevent confluent growth was found to be necessary at this stage. The plates are incubated at 37°C for 24 h. Listerias can be recognised by examining plates by the technique of Henri (1933). Obliquely transmitted light is obtained by using a suitable microscope with low-power magnification that will permit removal of the sub-stage condenser so that light passes through the centre of the stage. An ordinary microscope light provides suitable illumination (Fig. 2). Colonies of *L. monocytogenes* can be recognised by characteristic lilac to blue-green colouration and then sub-cultured onto tryptose agar with a straight wire. Colonies which are white or pink in colour should not be sub-cultured. The periphery of the plate is the best area for examination as colonies are larger, becoming suppressed by other microbial growth towards the centre of the plate.

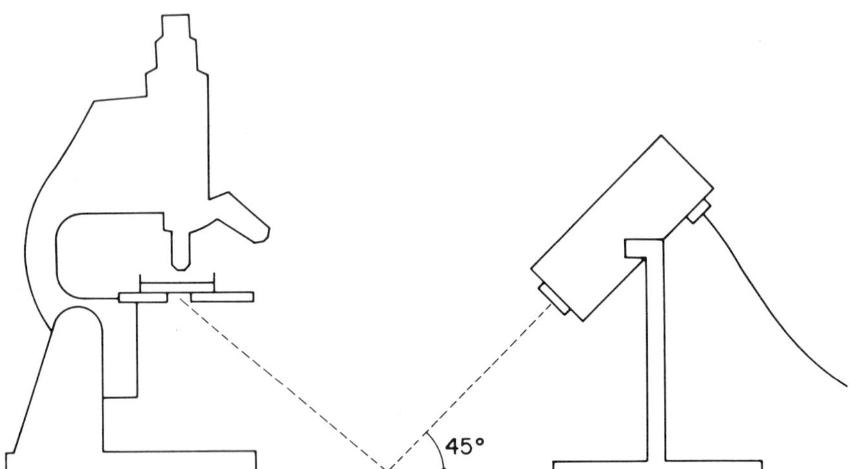

FIG. 2. The Henry technique for examining colonies of *Listeria monocytogenes*.

Identification

Listerias are identified by slide agglutination using polyvalent Listeria O anti-serum (Difco). Colonies which do not suspend easily or which autoagglutinate are not *L. monocytogenes*. The purity of isolates and the characteristic colour can be used as a double check where serology results are doubtful. All negative TN broths should be re-plated after incubation for 48 h. All positive TN broths and corresponding cold enrichment broths are discarded. Negative cold enrichment broths can be sub-cultured at the beginning of subsequent weeks for as long as is considered necessary.

Rapid identification of colonies on tryptose agar may prove difficult at first, and it is recommended that control cultures be made with a known strain of *L. monocytogenes*. With experience, identification becomes easy, even where colonies are heavily overgrown with contaminants. The MPN values can be computed from published tables.

This technique has permitted the isolation of *L. monocytogenes* from a wide range of sewages, sewage sludges, river waters and animal byproduct effluents. The organism was not isolated from samples free from faecal contamination as determined by the presence or absence of faecal indicators (Watkins & Sleath 1981). The method allows *L. monocytogenes* to recover and grow rapidly in enrichment media, thus aiding its recognition on tryptose agar by the presence of large numbers of colonies. A pure culture is often obtained from what would be considered heavily contaminated environmental samples.

This method permits the rapid quantification of *L. monocytogenes* often within 2–3 weeks, and also enables differing and dominant serotypes to be found within a single sample. It can therefore be used as a useful tool for epidemiological studies in both veterinary and human medicine.

References

GITTER, M. 1976 *Listeria monocytogenes* in 'oven ready' poultry. *Veterinary Record* **99**, 336.

GITTER, M., BRADLEY, R. & BLAMPIED, P. H. 1980 *Listeria monocytogenes* infection in bovine mastitis. *Veterinary Record* **107**, 390–393.

GRAY, M. L. 1960 Isolation of *Listeria monocytogenes* from oat silage. *Science (Washington, D.C.)* **132**, 1767.

GRAY, M. L. & KILLINGER, A. H. 1966 *Listeria monocytogenes* and listeric infections. *Bacteriological Reviews* **30**, 309–382.

HENRI, B. S. 1933 Dissociation of the genus *Brucella*. *Journal of Infectious Diseases* **52**, 374–402.

HYSLOP, N. ST. G. & OSBORNE, A. D. 1959 Listeriosis: A potential danger to public health. *Veterinary Record* **71**, 1082–1095.

KAMPELMACHER, E. H. & VAN NOORLE JANSEN, L. M. 1969 Isolation of *Listeria monocytogenes* from faeces of clinically healthy humans and animals. *Zentralblatt für Bakteriologie, Parasitenkunde, Infectionskrankheiten und Hygiene, Abteilung 1: Originale* **211,** 353 –359.

KAMPELMACHER, E. H. & VAN NOORLE JANSEN, L. M. 1972 Further studies on the isolation of *L. monocytogenes* in clinically healthy individuals. *Zentralblatt für Bakteriologie, Parasitenkunde, Infectionskrankheiten und Hygiene, Abteilung 1: Originale* **221,** 70–77.

SCHLECH, W. F., LAVIGNE, P. M., BORTOLUSSI, R. A., ALLEN, A. C., HALDANE, E. V., WORT, A. J., HIGHTOWER, A. W., JOHNSON, S. E., KING, S. H., NICHOLLS, E. S. & BROOME, C. V. 1983 Epidemic listeriosis—Evidence for transmission by food. *New England Journal of Medicine* **308,** 203–206.

WATKINS, J. & SLEATH, K. P. 1981 Isolation and enumeration of *Listeria monocytogenes* from sewage, sewage sludge and river water. *Journal of Applied Bacteriology* **50,** 1–9.

WEIS, J. 1975 The incidence of *Listeria monocytogenes* on plants and in soil. In *Problems of Listeriosis,* ed. Woodbine, M., pp. 61–65. Leicester, England: Leicester University Press.

WELSHIMER, H. J. 1960 Survival of *Listeria monocytogenes* in soil. *Journal of Bacteriology* **80,** 316–320.

Bacillus anthracis

J. A. CARMAN, P. HAMBLETON AND J. MELLING

Vaccine Research and Production Laboratory, Public Health Laboratory Service, Centre for Applied Microbiology and Research, Porton Down, Salisbury, Wiltshire, UK

Anthrax is normally a disease of herbivores, but it is occasionally contracted by man through contact with infected animals or products derived from them. The disease occurs in three forms depending on the site of infection, namely, cutaneous, pulmonary and intestinal. The commonest of these, malignant pustule of the skin, normally responds rapidly to antibiotic therapy, although signs and symptoms of the disease may persist for some time because of the effects of residual toxins.

In our experience, two types of specimen are presented for examination. The first is any material that could contain spores of *Bacillus anthracis*. Such specimens, for example animal hair, hides, bone meal and soil, usually contain a large number of spore-bearing organisms, of which very few, if any, will be *B. anthracis*. Because of improvements in the sterilization of imported materials of animal origin in the last 20 or 30 years, problems from such materials are rare, but recently concern over possible anthrax contamination of land has resulted in a number of soil samples requiring examination. The second consists of isolates usually, but not always, from clinical material and consisting of non-motile Gram positive spore-bearing bacteria, which the isolating laboratory requires to be identified to confirm or refute their suspicions that the organism is not *B. anthracis*.

Safety

Although man is remarkably resistant to infection by *B. anthracis* (Christie 1980), suspect material should be handled with care, preferably in a microbiological safety cabinet (class I or III) (Collins & Lyne, 1984).

Isolation and Identification of Micro-organisms of Medical and Veterinary Importance

Copyright © 1985 by the Society for Applied Bacteriology
All rights of reproduction in any form reserved
0-12-181460-2

Isolation

Clinical specimens

The isolation of *B. anthracis* from clinical material is relatively easy, as there is likely to be only one *Bacillus* sp. present in any one specimen, and primary isolation is normally carried out by the original laboratory.

Occasional specimens

Research workers occasionally send isolates they have made, usually from soil, for checking. (To date, we have never received a positive culture in either this or the preceeding category.)

Specimens requiring isolation techniques

Samples of animal origin such as hair or bone meal and of soil are usually heavily contaminated with spores and, in the case of soils with a high organic content, counts on nutrient agar in excess of a million spores per gram are common, which makes finding one or two of *B. anthracis* a daunting task.

Various selective media have been developed, but ultimately these all require the subjective recognition of suspicious colonies for selection for further tests. A simple, and probably (when used by experienced workers) very effective, method involves pouring plates of a not-too-rich nutrient medium mixed with a suspension, suitably diluted, of the test material and searching for colonies that look like dahlia tubers or Chinese artichokes in the depth of the agar (Collins 1964). This method can be made more selective by adding lysozyme or dibromopropamidine isethionate BP to the medium (Collins & Lyne 1984).

Pearce & Powell (1951) devised a medium consisting of a simple peptone agar with added lysozyme and haematin, and they state that this suppressed germination and growth of 95% of the spores present in soil samples. This medium was apparently more successful if the plates were incubated at 40°C rather than 37°C, although at this temperature some anthrax spores failed to germinate. Morris (1955) also used a peptone agar base, but relied on polymyxin and propamidine isethionate to impart selectivity. The medium also contained 1% packed red cells to help distinguish the rather similar colonies of *B. cereus*, which are usually haemolytic, from those of the non-haemolytic *B. anthracis*. Recent attempts to use this medium were, however, very disappointing (M. Broster, personal communication), and it was thought that newer production methods may have eliminated an impurity in the propamidine, which was really responsible for the selective action.

The PLET medium of Knisley (1966) is that routinely used in this laboratory. It consists of Difco heart infusion agar made according to the manufacturers instructions, to which the following (per ml of medium) are added: polymyxin, 30 units; lysozyme, 40 μg; EDTA, 300 μg and thallous acetate, 40 μg.

The EDTA and the thallous acetate may be added to the agar before autoclaving, but the polymyxin and the lysozyme are sterilised by membrane filtration, and care has to be taken to add them to the agar only after it has cooled to 50°C. The use of this medium proved very successful on material from Gruinard Island (Manchee *et al.* 1981, 1983).

Samples are weighed and suspended in sterile water and, after allowing a few seconds for the larger particles to settle, the supernatant is removed and heated at 60°C for 30 min to kill vegetative micro-organisms. The amount of water required varies considerably from sample to sample and should be enough to allow a reasonable quantity to be withdrawn for heating to inactivate vegetative organisms. As a guide, ordinary moist soil normally requires two to three times its weight of water. If samples are difficult to wet, a drop of the non-bacteriostatic detergent Tween 80 can be added.

Portions (usually 0.25 ml) of this suspension are pipetted on to each of four nutrient agar and four PLET agar plates, and the liquid, which contains varying amounts of suspended matter, is spread over the agar using a large disposable plastic loop. The plates are incubated at 37°C, and the nutrient agar plates are examined for colony growth after 18 h. The latter serve mainly to indicate the background count, but there is little chance of finding *B. anthracis* colonies unless they are numerous. The PLET plates are incubated for 2 days, although sometimes a preliminary examination after 1 day may be useful. The suspension of the sample often needs to be diluted before plating on the nutrient agar (as much as 10^{-4} for soils containing a lot of organic matter) but dilution is rarely required for the PLET plates; soils very rich in organic matter may need only a tenfold dilution.

Colonies of *B. anthracis* on PLET are a little smaller than on nutrient agar, but otherwise they are similar and there is as much inter-strain variation as there is on ordinary media, although this variation seems to be much less with fresh isolates than with laboratory strains from our collection. It is difficult to describe the difference between colonies of *B. anthracis* and the very similar ones of *B. cereus*, but the former tend to be rougher and greyish white in colour. Possible *B. anthracis* colonies are subcultured on to nutrient agar for further examination, and in our experience the majority subsequently prove to be *B. cereus*. With experience, it is possible to lower the level of *B. cereus* isolates to 10 to 1 of *B. anthracis*. As nearly all samples prove to be negative, however, there is a natural tendency to select more colonies than would seem necessary, in the hope of finding one of *B. anthracis*.

When organisms selected from PLET agar were grown in blood at 37°C for carrying out M'Fadyean's test, it was noticed that though *B. anthracis* grew well in blood, most soil organisms either failed to grow or took an appreciable time to commence growth; fresh blood was therefore tried as a selective medium. Heated soil suspensions were incubated in five volumes of fresh sheep blood for 5 h and then plated on nutrient agar. This proved to be as selective as PLET, but unfortunately was no better, so this idea was not pursued further.

The last, and most effective, method of isolation uses guinea pigs, and for this reason, it is difficult to justify except where enhanced sensitivity is essential. This method was developed at the Ministry of Agriculture, Fisheries and Food, Central Veterinary Laboratory, Weybridge. It involves injecting the animals subcutaneously with 1.5 ml of mixed gas–gangrene antisera followed 24 h later by intra-muscular injection of 1.0 ml of a heated extract of soil as used for plating on PLET agar; with some soil samples it is advisable to give protection against tetanus as well, using an appropriate antiserum. Samples of heart blood are taken from any animals which die, and any isolates from these are tested further. It is difficult to be precise, but this technique is at least four times as sensitive as that using PLET agar.

Identification

Over the years a large number of tests have been proposed for the identification of *B. anthracis*, and a comprehensive review of these is given by Parry *et al.* (1983). Only the more important tests are described here.

Motility and haemolysis

The characteristics which most commonly arouse suspicions that an organism is *B. anthracis* are the lack of motility, which can be shown by a hanging drop preparation or preferably by a Craigie tube in a 1-oz bottle of sloppy agar, and the lack of haemolysis on blood agar. Neither of these is conclusive, but they are useful pointers.

Penicillin sensitivity

Bacillus anthracis is normally sensitive to penicillin, and this can be checked by using an ordinary sensitivity disc. A refinement of this test is the "string of pearls" reaction (Jensen & Kleemeyer 1953), which involves growing the suspect organism on nutrient agar containing 0.5 unit penicillin/ml. After incubation at 37°C for 3–6 h, the growth observed under a coverslip using an oil immersion lens can be seen to consist of strings of spherical bodies, hence the name. Leise *et al.* (1959) state that penicillin-resistant strains also give a positive reaction, but this is not true for the only such strain in our collection when the test is carried out on nutrient agar. On brain heart agar, however, although there is no typical "string of pearls" reaction, the cells have a slightly beaded look when compared with those grown on the same medium without penicillin.

Capsule formation

Virulent strains of *B. anthracis* produce capsular material and therefore mucoid colonies when they are grown on media containing serum in an atmosphere containing 10% carbon dioxide. Non-capsulated strains of low virulence do not form such colonies. Prolonged growth under these conditions was used to produce a veterinary vaccine strain (Sterne 1937).

M'Fadyean's reaction

This is usually done by veterinarians on the blood of dead or dying animals, but it gives quite good results on cultures containing serum.

In this laboratory, suspect organisms are first plated on brain heart agar to obtain a pure culture. After incubation at 37°C overnight, a loopful of pure culture is inoculated into 2 or 3 ml of defribrinated horse blood (the species is not critical) which is incubated for about 7 h. A thick film is made and stained for 30 sec with polychrome methylene blue. (Gurr has produced an improved M'Fadyean's stain, which is available from BDH Chemicals Ltd., Poole, England.) If the organism is virulent *B. anthracis*, most, but not all, of the dark-blue-stained bacterial cells will be surrounded by pink capsular material as though the blood had been taken from an animal freshly dead of anthrax. The bacteria will usually be in long chains and the ends of most of the cells will be square. *Bacillus cereus* normally occurs singly or in short chains, and its cells have rounded ends. The appearance of the blood is also a useful guide: some *Bacillus* species leave it unchanged because their growth is suppressed, *B. cereus* darkens the blood and lakes it, and *B. anthracis* darkens it without lysis.

Bacteriophage sensitivity

Although Brown & Cherry (1955) and McCloy (1958) discovered anthrax phages independently and both called them gamma phages, they are not the same. The one we use here is probably that of McCloy.

A "lawn" of the organisms is made on a segment of a brain heart agar plate and a loopful of the gamma phage is placed in the middle of it. All strains of *B. anthracis* that we have examined have been sensitive to the phage, although a few strains may show slight growth within the plaque. The test is not conclusive, however, as an occasional isolate which is not *B. anthracis* is sensitive.

A strain which gives a positive reaction with both the M'Fadyean and phage tests is considered to be virulent, but those that give negative M'Fadyean and positive phage reactions may be either avirulent *B. anthracis*, phage-sensitive *B. cereus*, or some other organisms. An immunodiffusion test is used to distinguish between such strains.

Immunodiffusion test—protective antigen production

The organisms are grown on the medium used to produce human anthrax vaccine (Belton & Strange 1954; modified by Thorne & Belton 1957) and are tested by immunodiffusion against an anti-anthrax serum prepared in horses against the Sterne strain of *B. anthracis*.

Culture media

Base medium

This contains the following ingredients, dissolved in water and made up to 900 ml: casamino acids (Difco), 5.5 g; L-cystine, 18 mg (dissolved in a little water, using a few drops of concentrated HCl); serine, 46.8 mg; thiamine HCl, 0.15 mg; $CaCl_2 \cdot 2H_2O$, 14.8 mg; $MgSO_4 \cdot 7H_2O$, 22.5 mg; KOH, 465 mg; charcoal (amount varies with batch), 250 mg. The pH is adjusted to 6.9 with concentrated HCl before adding the charcoal. The medium is dispensed in 45-ml amounts and sterilised at 117°C for 20 min.

Supplement

This contains (g/100 ml distilled water): $NaHCO_3$, 6.0; glucose, 2.0; $MgSO_4H_2O$, 0.001. It is dispensed in 5-ml amounts and sterilised by filtration.

Immunodiffusion agar

This consists of (g/l in distilled water): agar (Oxoid, purified), 6.75; NaCl, 9.0; Na_2HPO_4, 8.22; KH_2PO_4, 2.33. It is heated in a steamer to dissolve the agar, adjusted to pH 7.3 and dispensed.

Supplement (5 ml) is added to the base medium (45 ml) in a 100-ml Erlenmeyer flask to give the correct depth of medium. This is inoculated with the test strain and incubated at 37°C for 24 h. A culture of the Sterne strain is prepared in the same way. Both strains are then tested in alternate wells in the immunodiffusion agar against the anti-anthrax serum.

Cultures of *B. anthracis* give strong lines of precipitation due to the protective antigen; these are continuous with those given by the Sterne strain.

Virulence Tests

Mice are used in groups of 10. Each group is given an intraperitoneal dose of spores ranging from 10^2 to 10^6. A virulent strain will usually kill most of the animals receiving 10 spores. Infection by *B. anthracis* can be confirmed by the M'Fadyean test on blood obtained by cutting off the tip of the tail. Doses of 10^5 and 10^6 spores of avirulent strains of *B. anthracis*, such as the Sterne strain, and some of the other *Bacillus* species usually kill some animals, but media inoculated with the animals' blood remain sterile, although the organisms can be recovered from the spleen.

Observations

The methods described permit the recovery of *Bacillus anthracis* from samples containing as few as two or three spores per gram, but there is a need either for a better selective medium or for an enrichment broth. Development of either of these is made difficult, however, by the similarity between *B. anthracis*, *B. cereus* and their close relatives.

References

BELTON, F. C. & STRANGE, R. E. 1954 Studies on a protective antigen produced *in vitro* by *Bacillus anthracis*: medium and methods of production. *British Journal of Experimental Pathology* **35**, 144–152.

BROWN, E. R. & CHERRY, W. B. 1955 Specific identification of *Bacillus anthracis* by means of a variant bacteriophage. *Journal of Bacteriology* **96**, 34–39.

CHRISTIE, A. B. 1980 *Infectious Diseases: Epidemiology and Clinical Practice* 3rd edn, p. 713. Edinburgh: Churchill-Livingstone.

COLLINS, C. H. 1964 *Microbiological Methods*. London: Butterworths.

COLLINS, C. H. & LYNE, P. M. 1984 *Microbiological Methods* 5th edn. London: Butterworths.

JENSEN, J. & KLEEMEYER, H. 1953 Die bakterielle differentialdiagnose des anthrax mittels eines neuen spezifischen tests ('perlschnurtest'). *Zentralblatt für Bakteriologie, Parasitenkunde, Infektionskrankheiten und Hygiene, Abteilung 1: Originale* **159**, 494–500.

KNISLEY, R. F. 1966 Selective medium for *Bacillus anthracis*. *Journal of Bacteriology* **92**, 784–786.

LEISE, J. M., CARTER, C. H., FRIEDLANDER, H. & FREED, S. W. 1959 Criteria for the identification of *Bacillus anthracis*. *Journal of Bacteriology* **77**, 655–660.

McCLOY, E. W. 1956 Lysogenicity and immunity to bacillus phage W. *Journal of General Microbiology* **18**, 198–220.

MANCHEE, R. J., BROSTER, M. G., MELLING, J., HENSTRIDGE, R. M. & STAGG, A. G. 1981 *Bacillus anthracis* on Gruinard Island. *Nature (London)* **294**, 254–255.

MANCHEE, R. J., BROSTER, M. G., ANDERSON, I., HENSTRIDGE, R. M. & MELLING, J. 1983 Decontamination of *Bacillus anthracis* on Gruinard Island. *Nature (London)* **303**, 239–240.

MORRIS, E. J. 1955 A selective medium for *Bacillus anthracis*. *Journal of General Microbiology* **13**, 456–460.

PARRY, J. M., TURNBULL, P. C. B. & GIBSON, J. R. 1983 *Wolfe Medical Atlases—19. A colour atlas of Bacillus species* 1st edn, pp. 60–65. London: Wolfe Medical Publications Ltd.

PEARCE, T. W. & POWELL, E. O. 1951 A selective medium for *Bacillus anthracis*. *Journal of General Microbiology* **5**, 387–390.

STERNE, M. 1937 Variation in *Bacillus anthracis*. *Onderstepoort Journal of Veterinary Science and Animal Industry* **8**, 271–349.

THORNE, C. B. & BELTON, F. C. 1957 An agar diffusion method for titrating *Bacillus anthracis* immunizing antigen and its application to a study of antigen production. *Journal of General Microbiology* **17**, 505–516.

Clostridia

K. D. Phillips, J. S. Brazier, P. N. Levett and A. T. Willis

Public Health Laboratory, Luton and Dunstable Hospital, Luton, UK

The clostridia are anaerobic or "facultatively aerobic" bacilli, usually Gram positive, which produce spores that usually distend the cell. Some species decompose protein, some ferment carbohydrates, others have both properties. Some species produce exotoxins and are pathogenic for man. Useful information on the systematics of the clostridia is to be found in the publications of Willis (1969, 1977), Buchanan & Gibbons (1974), Smith (1975) and Holdeman *et al.* (1977).

In this chapter some areas that seem to cause confusion are highlighted and others in which new developments have occurred are emphasized.

Classification

In Table 1 the main species of clostridia are listed and are classified according to their proteolytic and saccharolytic properties, as shown by their action on milk agar (or casein agar) and on glucose, maltose, lactose and sucrose.

The clostridia listed next are those most likely to be encountered in clinical material, some of them much more frequently than others. With the general exceptions of *Clostridium botulinum* and *C. difficile,* which cause botulism and pseudomembranous colitis, respectively, in man and animals, they are all associated with wound infections. The most important of these are *C. tetani,* the causal organism of tetanus; and *C. perfringens, C. novyi, C. septicum, C. histolyticum, C. sordellii, C. bifermentans* and *C. sporogenes,* which are most commonly associated with gas gangrene. The other species are less frequently encountered and many of them, like *C. bifermentans* and *C. sporogenes,* are non-pathogenic.

Isolation and Identification of
Micro-organisms of Medical and
Veterinary Importance

Copyright © 1985 by the Society for Applied Bacteriology
All rights of reproduction in any form reserved
0-12-181460-2

TABLE 1

Division of some clostridia according to their saccharolytic and proteolytic properties

Saccharolytic and proteolytic	Proteolytic but non-saccharolytic	Saccharolytic but non-proteolytic	Non-saccharolytic and non-proteolytic
Clostridium sporogenes	*Clostridium histolyticum*	*Clostridium perfringens*	*Clostridium cochlearium*
C. bifermentans	*C. botulinum* type G	*C. barati*	*C. tetani*
C. sordellii		*C. butyricum*	
C. botulinum		*C. tertium*	
type A		*C. fallax*	
type B		*C. chauvoei*	
type F		*C. septicum*	
		C. sphenoides	
		C. novyi (A–D)	
		C. botulinum	
		type B	
		type C	
		type D	
		type E	
		type F	
		C. cadaveris	
		C. difficile	

Some characteristics of the commonly encountered clostridia are shown in Table 2, from which the following useful items of information emerge:

(1) Gelatinase is not a particularly discriminating feature.
(2) Clostridia that are lactose fermenters are never strongly proteolytic.
(3) Apart from *C. cochlearium*, which is entirely inactive and is rarely encountered, *C. tetani* is the only non-saccharolytic, non-proteolytic species.
(4) *Clostridium histolyticum* and the rarely encountered *C. botulinum* type G are the only species that are non-saccharolytic but strongly proteolytic.
(5) Lecithinase C-producing species are *C. perfringens* (all types), *C. novyi* (types A and B), *C. haemolyticum* (*C. novyi* type D), *C. bifermentans*, *C. sordellii* and *C. barati*. (Some non-toxigenic strains of *C. perfringens* fail to produce this enzyme.)
(6) Lipolytic species are *C. botulinum* (all types except G), *C. sporogenes* and *C. novyi* type A.
(7) *Clostridium novyi* type A is the only species that produces both a lecithinase C and a lipase.
(8) It is useful to note here that all the clostridia except *C. perfringens* are motile and that all except *C. perfringens* and *C. butyricum* are non-capsulated.

TABLE 2

*Reactions of some commonly encountered clostridia**

Organism	Acid from				Gelatinase activity	Indole	Milk agar digestion	Egg yolk agar		Pathogenicity for laboratory animals
	Glucose	Maltose	Lactose	Sucrose				Lecithinase C activity	Lipase activity	
Clostridium perfringens (A–E)	+	+	+	+	+	–	–	+	–	±
C. barati	+	+	+	+	–	–	–	+	–	–
C. tertium	+	+	+	+	–	–	–	–	–	–
C. fallax	+	+	+	+	–	–	–	–	–	±
C. carnis	+	+	+	+	–	–	–	–	–	+
C. chauvoei	+	+	+	+	+	–	–	–	–	+
C. septicum	+	+	+	–	+	–	–	–	–	+
C. paraputrificum	+	+	+	+	–	–	–	–	–	–
C. sphenoides	+	+	+	–	–	+	–	–	–	–
C. botulinum										
(A,B,F)	+	+	–	+	+	–	+	–	+	+
(C,D,E)	+	+	–	±	+	–	–	–	+	+
C. novyi										
(A)	+	+	–	–	+	–	–	+	+	+
(B)	+	+	–	–	+	–	–	+	–	+
(C)	+	+	–	–	+	–	–	–	–	–
C. haemolyticum	+	–	–	–	±	+	–	+	–	+
C. bifermentans	+	+	–	–	+	+	+	+	–	–
C. sordellii	+	+	–	–	+	+	+	+	–	±
C. sporogenes	+	+	–	–	+	–	+	–	+	±
C. cadaveris	+	–	–	–	+	±	–	–	–	–
C. difficile	+	–	–	–	–	–	–	–	–	+
C. cochlearium	–	–	–	–	–	–	–	–	–	–
C. tetani	–	–	–	–	+	+	–	–	–	±
C. histolyticum	–	–	–	–	+	–	+	–	–	+
C. lentoputrescens	–	–	–	–	+	–	+	–	–	–

* After Willis (1977).

Isolation

The three important methods for the surface culture of obligate anaerobes that have been developed are: (1) anaerobic jar methods, (2) anaerobic cabinets and (3) pre-reduced anaerobically sterilised (PRAS) roll tube techniques.

Anaerobic jars

Evacuation–replacement methods

A partial vacuum, drawn in the jar by a pump, is replaced from a cylinder by a gas mixture of either 90% hydrogen and 10% carbon dioxide or 10% hydrogen, 10% carbon dioxide and 80% nitrogen (Watt *et al.* 1976; Willis 1977). The latter mixture contains a concentration of hydrogen which is non-explosive in any proportion with air. Evacuation should be such that, on replacement, the initial proportions of hydrogen and oxygen are approximately 2 : 1 by volume so that the maximum secondary vacuum will develop in the presence of active palladium catalyst. This is important because a substantial secondary vacuum is a reliable indication, both that the jar is airtight and that an adequate quantity (at least 1 g/litre of jar volume) of active catalyst is present. The importance of manometry of pressure changes occurring within the jar cannot be over-emphasised, as jar integrity and catalyst activity are two criteria that must be fulfilled if development and maintainance of anaerobiosis is to be achieved.

Failure of the catalyst can occur between consecutive cycles of operation; this emphasises the importance of checking the catalyst activity by manometry each time the jar is used. The catalyst must be changed if the anticipated degree of secondary vacuum is not achieved. The inclusion of chemical and biological indicators of anaerobiosis is good practice, but these should be considered as merely adjuncts to the prospective control afforded by pre-incubation manometry.

Recent developments in modern anaerobic jar design have contributed greatly to the efficiency and safety of anaerobic jar operations. These include the use of Schraeder valves in place of needle valves for evacuation and replacement operations and the employment of catalyst sachets in which the palladium pellets are protected from damage by an aluminium foil heat sink.

Carbon dioxide is invariably included as a component of gas mixtures for use in anaerobic jars usually at a concentration of 10% v/v in the cylinder. It is worth noting that unless there is virtually complete initial evacuation, the actual concentration of carbon dioxide achieved in the anaerobic jar will fall short of this value, a fact which is not uncommonly overlooked in many published reports. For example, a jar evacuated to an internal pressure of 46 cm mercury will, after replacement of the primary and secondary vacuums with a gas mixture of 90% H_2 and 10% CO_2, contain an atmosphere of only 7.5% v/v CO_2.

Carbon dioxide is essential or stimulatory for many obligate anaerobes and is a well-established supplement to anaerobic atmospheres. A precise concentration is probably not critical for the growth of clostridia but may assume great importance when well-controlled atmospheres are required, as for example in antibiotic sensitivity testing.

Hydrogen/carbon dioxide generators

Commercially available generators create anaerobiosis in sealed jars equipped with active palladium catalyst. They have become very popular, particularly in clinical laboratories. The unit cost of internal generators is high, however, compared with cylinders of compressed gases. As with evacuation–replacement methods, checks for airtightness and catalyst activity should be performed as a routine before incubation. If these tests are neglected there will inevitably be failures in the development and maintainance of anaerobiosis. For this reason it is recommended that internal generators be used in jars equipped with Schraeder valves or vent systems. It should be noted that control over CO_2 concentrations is much less precise with internal generators than with a cylinder supply.

Anaerobic cabinets

A number of cabinets are now available commercially for the isolation and manipulation of obligate anaerobes in conditions of strict and continuous anaerobiosis. This continuity of anaerobiosis is intrinsically more efficient than the cyclic operation of multiple anaerobic jars, and cultures can be inspected frequently without the disadvantages of exposure to air or the temperature fall associated with anaerobic jar cultures. A technically superior anaerobic cabinet is offered by Don Whitley Scientific (14 Otley Road, Shipley, W. Yorks., BD17 7SE England). This model includes "bare hand" facilities for manipulation of cultures, advanced humidity control, the use of "Anotox" for removal of noxious volatile anaerobic metabolites (Brazier 1982) and maximum economy in consumption of expensive anaerobic gas mixtures.

PRAS techniques

This "classical" anaerobic technique provides a profound degree of anaerobiosis suitable for the isolation and cultivation of the most demanding anaerobic bacteria found in the normal flora of man and animals. Its value in this area of investigation is beyond dispute. This method is not widely used in the UK, largely because of the special apparatus and techniques required.

Direct plate culture

Each specimen is cultured on three horse blood agar plates for incubation in the three different atmospheres: aerobic, aerobic with 10% carbon dioxide, and anaerobic with 10% carbon dioxide. It is convenient to include a fourth culture containing 70 μg/ml of neomycin sulphate for anaerobic incubation; this facilitates the subsequent isolation of clostridia in pure culture by suppressing the growth of most aerobic species.

For the differentiation of the growth of facultative anaerobes from that of obligate anaerobes, metronidazole and gentamicin discs may be placed on the non-selective blood agar plate culture for anaerobic incubation. After incubation, a zone of inhibition about the metronidazole disc indicates the presence of obligate anaerobes; inhibition of growth about the gentamicin disc indicates the presence of facultatively anaerobic bacteria.

All cultures are examined after incubation for 24 and 48 h. Recognition of the growth of obligate anaerobes is facilitated by direct comparison of the aerobic and anaerobic cultures, and by noting the effects of gentamicin and metronidazole on the anaerobic plate. Carbon-dioxide-dependent/facultative bacteria (sometimes incorrectly referred to as "microaerophilic" organisms) may be isolated concurrently with clostridia; these show growth on plates incubated both anaerobically and in air plus carbon dioxide, but not on plates incubated without added carbon dioxide.

It is relevant to note that three species of clostridia, *C. histolyticum*, *C. tertium* and *C. carnis* behave as "facultative aerobes" showing (diminished) growth on aerobic plates.

Cultural appearances

Colonial morphology

Any medium that supports good surface growth of anaerobes is satisfactory for the study of colonial morphology of the clostridia. Media used for this purpose should contain not more than 1.5% New Zealand agar, as higher concentrations appreciably alter some of the colonial characters. Fresh horse blood agar and heated blood agar are suitable media. For the full development of characteristic colonies, some organisms require longer periods of incubation than others.

Most clostridia produce semitranslucent or clear surface growths. Characteristically they produce colonies of irregular shape, some species (e.g., *C. sporogenes*) forming frankly rhizoidal colonies. Others such as *C. tetani* and *C.*

septicum rarely produce discrete colonies, except in young culture; their characteristic habit of growth is a swarming film. *Clostridium perfringens* is one of the few commonly occurring clostridia which produces a circular, entire colony. Swarming growth is by no means limited to *C. tetani* and *C. septicum*; many different organisms may show confluent or coalescing growth, especially under conditions of excessive moisture. Aberrant swarming may be reduced by using well-dried culture media, and conditions of "low humidity" may be encouraged by the inclusion in the lid of the culture plate of a filter disk impregnated with glycerol. Natural swarming may be largely prevented by increasing the agar concentration of the medium, although, as noted above, this does interfere with the colonial appearances of other organisms. If indicator effects are also being sought (e.g. haemolysis, egg yolk reactions, proteolysis, anti-microbial inhibition patterns), it should be remembered that higher concentrations of agar will also profoundly alter these appearances.

Finally, it is useful to remember that the size and shape of colonies are not static features; with continued incubation, convex colonies may become umbonate, circular colonies may become irregular, and clear or semi-translucent colonies may become opaque or pigmented.

Excellent descriptions and illustrations of the types of colonies produced by some commonly encountered species of clostridia were published by Batty & Walker (1966).

Microscopic morphology

The shape, size, arrangement, Gram reaction, presence or absence of spores, and presence or absence of branching help determine the genus to which a particular anaerobe belongs. Although the clostridia are defined as Gram positive spore-forming rods, many individual isolates may elude this definition by showing variability with the Gram stain, failure to sporulate, or the presence of cocco-bacillary forms. It is thus important to recognize that, except in its broadest sense, microscopic morphology is of little taxonomic value.

The difficulties that are imposed by a morphological classification of the clostridia include the following: The presence of vacuoles in some strains of *Bacteroides* may be mistaken for spores, as it is well recognized that many clostridia lose their Gram positive reaction in cultures more than a few hours old. Some clostridia, on the other hand, may fail to produce discernable spores and might, therefore, be regarded as *Eubacterium* or *Propionibacterium*, or even as *Bacteroides* in older cultures. In the same way, strains of *Bifidobacterium* that do not show the classic bifid form may be confused with *Clostridium* or *Eubacterium*. Coccoid or cocco-bacillary forms of *C. perfringens*, which frequently occur in rapidly growing cultures, may be confused with *Peptococ-*

cus. Despite these comments, many anaerobic isolates fit neatly into the morphological scheme of separation; indeed, it is on the basis of the *typical* that such a scheme is devised. In common with many aerobic organisms, the typical morphology of the clostridia is often best developed in pathological material.

Gram reaction. The conventional methods of Gram staining are quite satisfactory, but overdecolorization is more likely to occur if acetone is used instead of alcohol. Gram positive anaerobes not infrequently stain Gram negatively, especially in old cultures. This is especially the case with exacting species such as *C. tetani* and *C. novyi*, and is probably due, directly or indirectly, to toxic effects of oxidation on these organisms. In the case of the clostridia, the presence of spores will often resolve the generic identity of such "Gram negative" anaerobic bacilli.

Spores. In Gram stained preparations a spore shows as a circular or oval unstained area in the body of the bacillus. The size, shape and position of the spore should be noted, because sporulating cultures of some clostridia, for example *C. tetani*, present a characteristic microscopic appearance on which a presumptive diagnosis may be made. In Gram stained films, free spores show as circular or oval unstained bodies, each delineated by a faintly Gram negative "wall." Spores may be stained by any of the common methods. Generally, however, little is to be gained by staining the spores of clostridia as they are readily seen in the Gram stained preparations. A complete absence of spores in culture should also be noted, for with some clostridia, notably *C. perfringens*, this is a characteristic feature.

When spores cannot be detected microscopically their presence may sometimes be inferred by demonstrating resistance of the culture to heat (pasteurization) or to treatment with alcohol.

Capsules. Capsule staining does not find much application, as only two of the commonly occurring clostridia (*C. perfringens* and *C. butyricum*) are capsulated.

Flagella. Flagella staining is not recommended in the routine examination of clostridia, as the techniques are difficult and time consuming and the information gained is not highly definitive. With the important exceptions of *C. perfringens* and *C. barati*, most of the clostridia are motile and flagellated.

Motility. Absence of motility is of no significance in identifying any organism; not only do non-motile variants of motile species occur, but motility may be lost because of death of the organism or interference with its flagellar mechanism. Hence the routine examination of anaerobic cultures for motility is not an essential procedure.

Moreover, most anaerobes are adversely affected by the presence of free oxygen, a result of which may be loss of motility. If cultures are examined for motility, it is inadvisable to examine wet-mount preparations. The best method is to fill a capillary tube with an actively growing fluid culture (4–10 h), seal it at both ends, and examine it under the high-power objective with reduced illumination.

Haemolytic activity

The haemolytic activity of clostridia is conveniently observed in cultures on a nutrient agar base medium containing 7.5% defibrinated blood which is a general purpose medium for the isolation and propagation of most clinically significant clostridia and upon which certain patterns of haemolysis may be recognized, depending on the source of the erythrocytes used. It also forms the basis of many selective and fermentation media with the addition of antimicrobial agents or fermentable substrates. As with many plating media, the addition of reductants or enrichments may be helpful or essential for the satisfactory growth of some species.

On horse blood agar, which is the most frequently used blood agar medium in the UK, some clostridia cause partial or complete haemolysis, but the latter is more usual. Different strains of a single species vary considerably in this respect, however; for example, some strains of *C. perfringens* are non-haemolytic on horse blood agar, others produce extensive zones of haemolysis, and yet others produce only limited partial lysis. In this setting, complete haemolysis is due to *C. perfringens* theta-toxin (oxygen-labile haemolysin); partial haemolysis is frequently seen among type A strains, however, and absence of haemolysis is characteristic of some heat-resistant food poisoning type A strains, and of some type C strains, as these do not produce any theta-toxin. Partial haemolysis on horse blood agar is due to *C. perfringens* alpha-toxin (lecithinase C), which is relatively inactive against the red cells of the horse. The addition of 0.5% $CaCl_2$ to horse blood agar renders the erythrocytes more susceptible to the action of alpha-toxin, so that colonies are surrounded by a wide zone of partial haemolysis; strains producing theta-toxin in addition to alpha-toxin produce two zones of haemolysis ("target" haemolysis): a narrow zone of complete lysis due to theta-toxin, and a much wider peripheral zone of partial haemolysis.

It is noteworthy that the haemolysins elaborated by individual species of clostridia are not equally active against the erythrocytes of all animals. For example, the haemolysis produced by *C. perfringens* on horse blood agar is due, as we have noted, to its theta-toxin, which is relatively inactive against the erythrocytes of the mouse, while *C. perfringens* alpha-toxin haemolyses the erythrocytes of the mouse but not those of the horse; both alpha- and theta-toxins, however, are active against the red cells of the sheep. In the same way, the growth of *C. septicum* on horse blood agar is usually associated with frank beta-haemolysis, although the closely related species *C. chauvoei* is commonly non-haemolytic; haemolysis due to *C. chauvoei* is often overt on sheep blood agar. Apart from these species specificities that some clostridial haemolysins exhibit, haemolysis as an indicator effect is also influenced by such factors as the degree of anaerobiosis (e.g. oxygen-labile haemolysins), the hot–cold haemolysin effect (e.g. *C. perfringens* alpha-toxin), the concentration of substrate (erythrocytes), the concentration of toxin (haemolysin), the agar concentration (which affects the diffusibility of exotoxins) and the presence or absence of toxin activators (calcium ions) and inhibitors (fluoride ions).

With some organisms, such as *C. sporogenes*, zones of pseudo-haemolysis similar in appearance to true partial haemolysis are sometimes seen as narrow bands around colonies on horse blood agar. This is probably the result of proteinase activity.

Thus, considered alone, haemolysis is not of much direct diagnostic value in the identification of clostridia. It is, however, of the utmost value in alerting the observer to the possible presence of particular species in mixed cultures, and of recognizing the presence of single species variants.

The CAMP test

Synergistic haemolysis by alpha-toxin producing *C. perfringens* and beta-haemoytic group B streptococci forms the basis of the so-called CAMP test for the confirmation of *C. perfringens*. The test derives its name from the surnames of the three authors (Christie *et al.* 1944) who described haemolytic synergy between group B streptococci and beta-lysin-producing *Staphylococcus aureus*.

Procedure. A horse blood agar plate is inoculated with a group B streptococcus as a single streak. At right angles to this streptococcus inoculum a culture of the test strain of *C. perfringens* is streaked, placed so that the inocula of the two organisms are not touching (about 1–2 mm apart). The plate is then incubated overnight anaerobically at 37°C. A positive result shows a crescent of beta-haemolysis capping the streak of growth of *C. perfringens* adjacent to that of the streptococcus (Fig. 1) (Gubash 1980; Hansen & Elliott 1980).

Lecithinase C and lipase activity

Lecithinase C and lipase activities are two very distinctive reactions shown by some species of clostridia. They are demonstrated in cultures on a suitable nutrient agar base containing 7% egg yolk suspension.

Some clostridia produce an intense diffuse opacity in egg yolk agar media which is due to specific lecithinases C, and which is inhibited by specific antitoxic sera. Some other clostridia produce an intense restricted opacity which is accompanied by a superficial pearly layer covering the colonies and

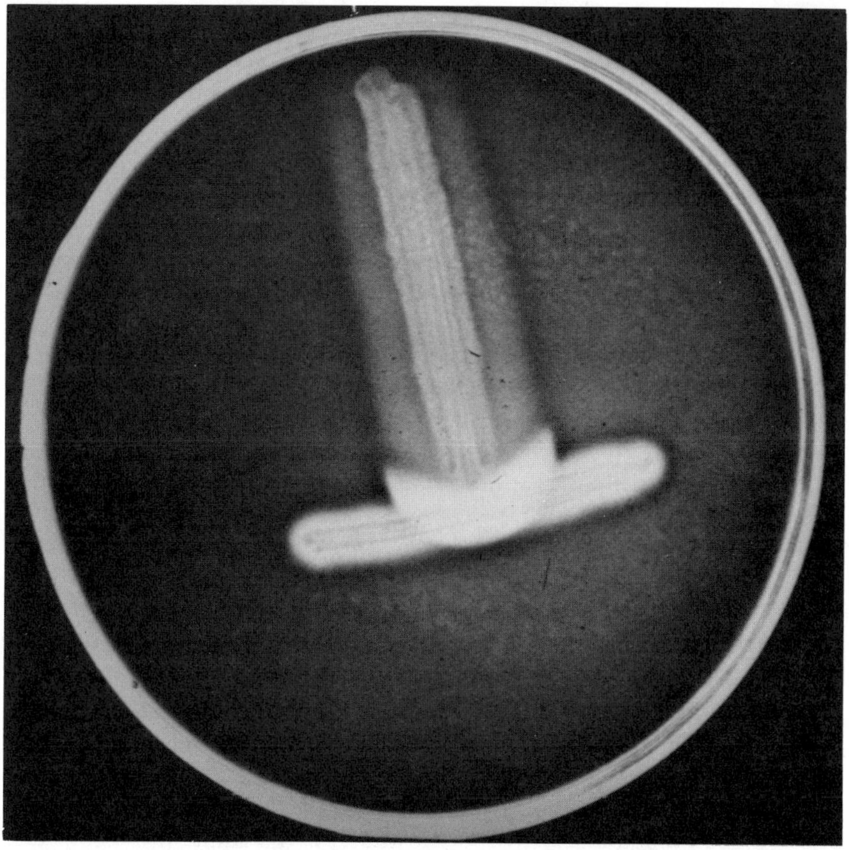

FIG. 1. The CAMP test: a positive result shows a crescent of beta-haemolysis capping the streak of growth of *Clostridium perfringens* adjacent to that of the streptococcus.

sometimes extending beyond their edge onto the surface of the medium; this reaction is due to lipolysis. The restricted opacity is easily seen by scraping away the surface growth.

These reactions, which are easily distinguished from one another, are of great value in identifying clostridia of clinical importance (Table 3). Lecithinase C opacities are partly due to fats derived from lecithin breakdown, while lipase opacities are composed of fatty acids produced by hydrolysis of triglycerides. The pearly layer seen in lipase reactions consists of a thin film of fatty acid which may be floated off intact from the culture surface by gently flooding the plate with water.

It is important to recognize that action on egg yolk is not confined to the

TABLE 3

*Reactions of some anaerobes on egg yolk media**

Organism	Diffuse lecithinase C† opacity	Restricted opacity and pearly layer (lipolysis)
Clostridium perfringens A–E	+	−
C. barati	+	−
C. bifermentans–sordellii	+	−
C. botulinum A–F	−	+
C. sporogenes	−	+
C. novyi		
type A	+	+‡
type B	+	−
type C	−	−
C. haemolyticum	+	−

* After Willis (1977).
† Non-toxigenic strains may fail to produce an egg yolk reaction.
‡ Note that the restricted lipase opacity is obscured by the lecithinase C opacity.

clostridia. *Fusobacterium necrophorum*, for example, is strongly lipolytic, as are many strains of *Staphylococcus aureus* and *Pseudomonas aeruginosa*. The latter organism appears to produce a multiplicity of lecithinases including lecithinase C. *Bacillus cereus* is also characterized by lecithinase C activity.

When egg yolk media are used for direct plating of anaerobic material, they may be rendered selective for clostridia by the addition of neomycin sulphate (70 μg/ml). This is particularly valuable for the presumptive recognition of *C. botulinum* in contaminated foods.

More complex egg yolk media, such as lactose–egg yolk–milk agar, provide additional information about the saccharolytic and proteolytic activities of clostridia (Willis 1977; Willis & Phillips 1983) and enable a presumptive identification of a number of different species in a single culture to be made (Table 4) (see also Hobbs *et al.* 1970).

Half-antitoxin egg yolk plate cultures

Antitoxic sera are now rarely used in the management of anaerobic infections and are virtually unobtainable for use as diagnostic reagents. The loss of half-antitoxin plate tests should not cause too much inconvenience, however, as they are not essential for the accurate or rapid identification of any bacterial species *in vitro*.

TABLE 4

*Reactions of some clostridia on lactose egg yolk milk agar**

Organism	Lactose egg yolk milk agar: Opalescence: Lecithinase	Lipase	Lactose fermentation	Proteolysis
Clostridium perfringens	+	−	+	−
C. bifermentans	+	−	−	+
C. sordellii	+	−	−	+
C. botulinum				
(A)	−	+	−	+
(B)	−	+	−	+
(C)	−	+	−	−
(D)	−	+	−	−
(E)	−	+	−	−
(F)	−	+	−	+
C. sporogenes	−	+	−	+
C. novyi				
(A)	+	+	−	−
(B)	+	−	−	−
(C)	−	−	−	−
C. haemolyticum	+	−	−	−
C. histolyticum	−	−	−	+
C. septicum	−	−	+	−
C. chauvoei	−	−	+	−
C. tetani	−	−	−	−
C. cochlearium	−	−	−	−
C. cadaveris	−	−	−	−
C. sphenoides	−	−	+	−
C. tertium	−	−	+	−
C. tetanomorphum	−	−	−	−
C. butyricum	−	−	+	−
C. fallax	−	−	+	−
C. difficile	−	−	−	−

* After Willis (1977).

Identification

Specific identification of clostridia requires the performance of relatively few biochemical tests. These include determination of the ability of an isolate to hydrolyze a range of fermentable substrates, reactions on egg yolk medium, production of indole, reduction of nitrate, hydrolysis of gelatin, and analysis of end-products of carbohydrate and amino acid metabolism by gas-liquid chromatography. In this laboratory the criteria of Holdeman *et al.* (1977) are generally followed.

Plate method for fermentation reactions

The method used is that of Phillips (1976). Blood agar plates containing anaerobe identification medium (e.g. Lab M Lab 66) and 6% defibrinated horse blood are poured. They are dried and then flooded with 1 ml of 20% w/v carbohydrate solutions as required. After further drying, ditches are made across the plates with a sterile scalpel and they are spot-inoculated with the organisms under test. Following incubation anaerobically at 37°C for 48 h, plugs of agar are withdrawn from each segment and two drops of 0.04% bromothymol blue are added to each plug in order to determine acid production. To avoid false positive reactions in maltose fermentation tests, blood must be omitted from the agar fermentation base; it is conveniently replaced by haemin/menadione solution.

Aesculin hydrolysis is determined in a similar manner, but an aqueous solution of aesculin (2% w/v) and ferric citrate (1% w/v) is used. In the basal medium employed for starch fermentation and hydrolysis the horse blood is replaced by haemin/menadione solution (Willis & Phillips 1983). This basal medium is also used for the determination of gelatin hydrolysis by the method of Frazier (1926).

Miniaturised identification methods

Conventional methods such as those described above become laborious and costly when large numbers of isolates are examined. For this reason miniaturised biochemical methods have become popular (Wilkins & Walker 1975; Jayne-Williams 1975, 1976; Occhionero *et al.* 1982; Levett 1984). A variety of identification kits is also available commercially. These include API 20A (Starr *et al.* 1973), Becton & Dickinson Minitek II (Hanson *et al.* 1979), and Flow Anaerobe-Tek (Lombard *et al.* 1982; Buesching *et al.* 1983). The use of API Zym enzyme strips for anaerobe identification has also been described (Tharagonnet *et al.* 1977; Kilian 1978; Slots 1981).

Miniaturised methods offer several advantages, largely related to the potential saving in time, media and materials. Commercial kits offer even greater economies of time than laboratory-prepared media, but are significantly more expensive. Commercial test systems which employ fermentation tests as their principal component often fail to identify accurately asaccharolytic organisms. Hanson *et al.* (1979) compared API and Minitek systems for the identification of clinical anaerobic isolates and found that both systems produced an identification for only 51% of strains. Moreover, these identifications were found to differ in 16% of cases. In contrast, the Flow Anaerobe-Tek system correctly identified 70% of 204 common anaerobes (Lombard *et al.* 1982) and 85% of 216 strains examined by Buesching *et al.* (1983). Both reports stressed that the value of the Flow system was reduced largely by its poor data base. A miniaturised identification method developed in this laboratory correctly identified all of 121 anaerobic isolates (Levett 1984).

A miniaturised method has been devised for the detection of indole production (Sutter & Carter 1972). The indole spot test utilises a filter paper soaked in a 1% solution of *p*-dimethylaminocinnamaldehyde (DMACA) in 10% (v/v) concentrated hydrochloric acid. A colony from a pure culture is smeared onto the filter paper; the development of a blue or green colour indicates the presence of indole.

Nitrate reduction may be determined by the disc test of Wideman *et al.* (1977). Filter-paper discs are impregnated with a filter-sterilised solution containing sodium nitrate (300 g/litre) and potassium molybdate (1 g/litre). Discs are placed on blood agar plates inoculated with pure cultures and incubated anaerobically at 37°C for 48–72 h. The discs are then removed and one drop of each of the two nitrate reagents placed upon them; the result is read in the usual fashion (Willis & Phillips 1983).

Gas–liquid chromatography

The definitive identification of most common clostridia does not require the application of gas chromatography, although the technique finds an indispensible place in the recognition of many less well defined species.

Antimicrobial Sensitivity Testing

Sensitivity testing of clostridia is carried out by the standard method of Stokes (Stokes & Ridgeway 1980) using a known sensitive strain of *C. perfringens* as the control organism. Because *C. perfringens* is not a highly exacting anaerobe (it will grow in the presence of 10% oxygen), inadequate anaerobiosis, while permitting growth of both test and control organisms, may produce aberrant sensitivity patterns. There are no inherent difficulties in the performance or interpretation of anaerobic sensitivity tests, despite reports to the contrary. It is important to recognize that CO_2 in the anaerobic atmosphere may interfere with the activity of some antimicrobial agents (e.g. lincomycin and erythromycin).

All relevant isolates of clostridia should be tested for their sensitivity to metronidazole (5 μg), chloramphenicol (25 μg), clindamycin (2 μg), ben-

TABLE 5

Sensitivity of some common clostridia to antibiotics

Species	Number of strains tested	Strains sensitive to antibiotics (%)[†]						
		Metro	Pen	Tet	Ery	Clind	Chlor	Aug
Clostridium bifermentans	14	100	100	86	100	100	100	100
C. difficile[*]	22	100	4	95	86	9	91	100
C. perfringens	39	100	95	74	95	79	100	100
C. ramosum	9	100	67	44	100	22	100	100
C. septicum	8	100	100	100	100	100	100	100
C. sordellii	9	100	100	67	100	33	100	100
C. sporogenes	60	100	70	98	97	30	100	100
C. tetani	5	100	100	100	80	80	100	100

[*] All strains sensitive to vancomycin.
[†] Antibiotics: Metro, metronidazole; Pen, penicillin; Tet, tetracycline; Ery, erythromycin; Clind, clindamycin; Chlor, chloromphenicol; Aug, augmentin.

zylpenicillin (1 unit), erythromycin (5 μg), tetracycline (10 μg) and amoxycillin/clavulanic acid (Augmentin) (30 μg). Table 5 records the disc diffusion "antibiograms" of 166 strains of commonly encountered clostridia submitted recently to this laboratory.

All obligate anaerobes are sensitive to metronidazole, but are resistant to the aminoglycosides and to most cephalosporins.

Detection and Isolation of *Clostridium difficile*

The presence of *C. difficile* in faecal specimens may be confirmed by isolation of the organism and/or by detection of its toxins.

Clostridium difficile is known to produce two proteinaceous exotoxins (Banno *et al.* 1981; Taylor *et al.* 1981). Both possess cytotoxic activity in tissue culture assays, but the activity of toxin B is much greater than that of toxin A, and it is toxin B which is detected in the cytotoxin assay commonly employed.

Toxin A induces fluid accumulation in rabbit ileal loops (Bartlett *et al.* 1980; Lyerly *et al.* 1982) and is referred to as an enterotoxin. Its detection is discussed by Redmond *et al.* (Ch. 17, this volume).

Cytotoxin detection (toxin B)

The following method is used in this laboratory: a portion of the stool is emulsified in an approximately equal volume of phosphate-buffered saline containing penicillin (1000 μg/ml), streptomycin (1000 μg/ml) and metronidazole (100 μg/ml). After centrifugation at 2000 **g** for 20 min the supernatant is removed and four drops (100 μl) added to each of two tubes containing a tissue culture monolayer (MRC V cells are preferred, but a number of cell lines are equally sensitive to the cytotoxin). *Clostridium sordellii* antitoxin (100 μl) is added to one tube, and a characteristic cytopathic effect neutralised by the antitoxin after 24–48 h of incubation is regarded as a positive result. Miniaturised methods have been developed which enable the screening of large numbers of stools for cytotoxin (Borriello 1979; George 1979).

The rapid detection of cytotoxin by counterimmunoelectrophoresis has been reported (Welch *et al.* 1980; Ryan *et al.* 1980), but the use of impure antitoxin preparations may lead to errors in diagnosis (Poxton & Byrne 1981). Wu & Fung (1983) reported that counterimmunoelectrophoresis was positive in only 36% of specimens in which *C. difficile* or its cytotoxin were present.

Isolation of Clostridium difficile

Clostridium difficile may be isolated by a variety of methods, most of which use antibiotics as selective agents. A medium containing egg yolk, fructose, neutral red, cycloserine (500 μg/ml) and cefoxitin (16 μg/ml) (CCFA) was devised by George *et al.* (1979). As some strains of *C. difficile* are inhibited by these concentrations of cycloserine/cefoxitin, Willey & Bartlett (1979) employed a blood agar–based selective medium containing 250 μg/ml and 10 μg/ml, respectively. An alternative egg yolk selective medium containing 250 μg/ml cycloserine and 8 μg/ml cefoxitin facilitates the detection of *p*-cresol production by *C. difficile* by the incorporation of *p*-hydroxyphenylacetic acid as a precursor in the medium (Phillips & Rogers 1981). A minimal selective medium employing streptomycin (500 μg/ml) and cefoxitin (16 μg/ml) was described by Hubert *et al.* (1981).

The use of ethanol as a selective agent for spores of *C. difficile* has been reported (Borriello & Honour 1981). Incubation of a 50% stool/alcohol mixture at room temperature for 1 h is followed by culture on both selective and non-selective media.

Several workers have recommended the incorporation in media of agents that stimulate spore germination. Lysozyme (10 μg/ml) was added to egg yolk CCFA by Nakamura *et al.* (1981). Wilson *et al.* (1982) replaced the egg yolk with sodium taurocholate (0.1% w/v), and reported enhanced recovery of *C. difficile* from both spore suspensions and faecal specimens.

In this laboratory both direct plating on selective media and alcohol treatment are employed. Care should be exercised in examining primary isolates on CCFA medium, as *C. difficile* does not form spores in the presence of these selective agents. Suspect colonies must be subcultured promptly, for the organism is rapidly killed by exposure to air. Once sporing cultures on unselective media are obtained, however, the organism may be handled with confidence under aerobic conditions.

Colonies of *C. difficile* on CCFA medium fluoresce yellow-green (chartreuse) under long-wave u.v. light (George *et al.* 1979). This property results from the reduction of the neutral red indicator, and is not restricted to *C. difficile* (Willis 1977). Fluorescence also occurs on media not containing neutral red (Borriello & Honour 1981), but in our experience this characteristic varies between strains and between batches of media. Despite the use of efficient selective media, clostridia other than *C. difficile* are not uncommonly isolated from faecal specimens.

This is more common when using alcohol treatment followed by plating onto unselective media. Species such as *C. perfringens*, *C. sordellii* and *C. sporogenes* are readily distinguished from *C. difficile* by their reactions on egg yolk-based media, but egg yolk-negative, fructose-fermenting clostridia, including *C. butyricum*, *C. glycolicum*, *C. innocuum* and *C. paraputrificum*, may also be isolated, and may exhibit colonial morphology similar to *C. difficile*. These species, however, do not produce *p*-cresol on primary isolation on CCFA medium.

It is nevertheless necessary to identify fully all presumptive isolates following subculture and purification on blood agar.

Identification of Clostridium difficile

In our laboratory the criteria of Holdeman *et al.* (1977) are followed. Most isolates of *C. difficile* hydrolyse aesculin and produce acid from fructose, glucose, mannitol and mannose. Maltose, lactose and sucrose are not fermented. Gelatin and starch are not hydrolysed; indole is not produced and nitrate is not reduced. Acetic, propionic, iso-butyric, butyric, iso-valeric, valeric and iso-caproic acids are produced in cooked meat medium containing 0.5% w/v glucose. Para-cresol is produced from tyrosine via *p*-hydroxyphenyl acetic acid.

Typing of Clostridium difficile

The possible occurrence of cross-infection with *C. difficile* (Mulligan *et al.* 1979; Larson *et al.* 1980; Kim *et al.* 1981; Greenfield *et al.* 1981; Rogers *et al.* 1981) has led to the development of typing schemes. These fall into three main categories: antibiotic resistance patterns, biochemical types and interactions with other organisms.

Antibiogram typing seems to be of little value. In our experience *C. difficile* strains from widely divergent sources exhibit a similar pattern, being resis tant to penicillin and clindamycin. Biochemical variation between strains, however, is common. Poxton (1982) described several "biotypes" based upon a limited range of tests performed upon 43 isolates.

The inhibition of *C. difficile* by other organisms was reported by Rolfe *et al.* (1981) and Barclay & Borriello (1982), as was inhibition by *C. difficile* of other faecal anaerobes (Rolfe *et al.* 1981). These biological interactions appear to be constant for all strains of *C. difficile* so far examined. Malamou-Ladas & Tabaqchali (1982) described the inhibition of *C. difficile* by faecal streptococci, and speculated upon the development of a bacteriocin typing scheme. More recently, Sell *et al.* (1983) reported the construction of a phage typing scheme, and noted the production of bacteriocins. Typing isolates of *C. difficile* by bacteriophage and bacteriocins holds promise for a better understanding of the epidemiology of *C. difficile* infections, particularly in view of the success of these methods when applied to *C. perfringens* (Mahony 1979; Watson *et al.* 1982).

Our thanks are due to Messrs Butterworth and Co. (Publishers) Ltd. for permission to reproduce copyright material.

References

BANNO, Y., KOBAYASHI, T., WATANABE, K., UENO, K. & NOZAWA, Y. 1981 Two toxins (D1 and D2) of *Clostridium difficile* causing antibiotic-associated colitis: purification and some characterization. *Biochemistry International* **2**, 629–635.

BARCLAY, F. E. & BORRIELLO, S. P. 1982 *In vitro* inhibition of *Cl. difficile*. *European Journal of Chemotherapy and Antibiotics* **2**, 155–156.

BARTLETT, J. G., TAYLOR, N. S., CHANG, T. W. & DZINK, J. A. 1980 Clinical and laboratory observations in *Clostridium difficile* colitis. *American Journal of Clinical Nutrition* **33**, 2521–2526.

BATTY, I. & WALKER, P. D. 1966 Colonial morphology and fluorescent labelled antibody staining in the identification of species of the genus *Clostridium*. In *Identification Methods for Microbiologists: Part A*, ed. Gibbs, B. M. & Skinner, F. A., pp. 81–96. London & New York: Academic Press.

BORRIELLO, S. P. 1979 An evaluated micromethod for the detection of *Clostridium difficile* enterotoxin. *Microbios Letters* **7**, 25–28.

BORRIELLO, S. P. & HONOUR, P. 1981 Simplified procedure for the routine isolation of *Clostridium difficile* from faeces. *Journal of Clinical Pathology* **34**, 1124–1127.

BRAZIER, J. S. 1982 Appraisal of Anotox, a new anaerobic atmospheric detoxifying agent for use in anaerobic cabinets. *Journal of Clinical Pathology* **35**, 233–238.

BUCHANAN, R. E. & GIBBONS, N. E. 1974 *Bergey's Manual of Determinative Bacteriology*. 8th edn. Baltimore, Maryland: Williams & Wilkins.

BUESCHING, W. J., SVIRBELY, J. R. & AYERS, L. W. 1983 Evaluation of the Anaerobe–Tek system for the identification of anaerobic bacteria. *Journal of Clinical Microbiology* **17**, 824–829.

CHRISTIE, R., ATKINS, N. E. & MUNCH-PETERSEN, E. 1944 A note on the lytic phenomenon shown by group B streptococci. *Australian Journal of Experimental Biology and Medical Science* **22**, 197–200.

FRAZIER, W. C. 1926 A method for the detection of changes in gelatin due to bacteria. *Journal of Infectious Diseases* **39**, 302–309.

GEORGE, R. H. 1979 A micro-method for detecting toxins in pseudomembranous colitis. *Journal of Clinical Pathology* **32**, 303–304.

GEORGE, W. L., SUTTER, V. L., CITRON, D. M. & FINEGOLD, S. M. 1979 Selective and differential medium for isolation of *Clostridium difficile*. *Journal of Clinical Microbiology* **9**, 214–219.

GREENFIELD, C., BURROUGHS, A., SZAWATHOWSKI, M., BASS, N., NOONE, P. & POUNDER, R. 1981 Is pseudomembranous colitis infectious? *Lancet* **1**, 371–372.

GUBASH, S. M. 1980 Synergistic haemolysis test for presumptive identification and differentiation of *Cl. pefringens*, *Cl. bifermentans*, *Cl. sordellii*, and *Cl. paraperfringens*. *Journal of Clinical Pathology* **33**, 395–399.

HANSEN, M. V. & ELLIOTT, L. P. 1980 New presumptive identification test for *Clostridium perfringens*. Reverse CAMP test. *Journal of Clinical Microbiology* **12**, 617–619.

HANSON, C. W., CASSORLA, R. & MARTIN, W. J. 1979 API and Minitek systems in identification of clinical isolates of anaerobic Gram-negative bacilli and *Clostridium* species. *Journal of Clinical Microbiology* **10**, 14–18.

HOBBS, G., WILLIAMS, K. & WILLIS, A. T. 1970 Basic methods for the isolation of clostridia. In *Isolation of Anaerobes*, ed. Shapton, D. A. & Board, R. G., p. 1–23. London & New York: Academic Press.

HOLDEMAN, L. V., CATO, E. P. & MOORE, W. E. C. 1977 *Anaerobe Laboratory Manual* 4th edn. Blacksburg, Virginia: Virginia Polytechnic Institute and State University.

HUBERT, J., IONESCO, H. & SEBALD, M. 1981 Detection of *Clostridium difficile* on minimal and selective medium and by immunofluorescence. *Annales de Microbiologie (Paris)* **132A**, 149–157.

JAYNE-WILLIAMS, D. J. 1975 Miniaturized methods for the characterization of bacterial isolates. *Journal of Applied Bacteriology* **38**, 305–309.

JAYNE-WILLIAMS, D. J. 1976 The application of miniaturized methods for the characterization of various organisms isolated from the animal gut. *Journal of Applied Bacteriology* **40**, 189–200.

KILIAN, M. 1978 Rapid identification of *Actinomycetaceae* and related bacteria. *Journal of Clinical Microbiology* **8**, 127–133.

KIM, K.-H., FEKETY, R., BATTS, D. H., BROWN, D., CUDMORE, M., SILVA, J. & WATERS, D. 1981 Isolation of *Clostridium difficile* from the environment and contacts of patients with antibiotic associated colitis. *Journal of Infectious Diseases* **143**, 42–50.

LARSON, E. H., PRICE, A. B. & BORRIELLO, S. P. 1980 Epidemiology of experimental enterocecitis due to *Cl. difficile*. *Journal of Infectious Diseases* **142**, 408–413.

LEVETT, P. N. 1984 Development and evaluation of a miniaturised method as an aid to the identification of clinically important anaerobic bacteria. *Journal of Clinical Pathology*. **37**, 70–73.

LOMBARD, G. L., WHALEY, D. N. & DOWELL, V. R. 1982 Comparison of media in the Anaerobe–Tek and Presumpto Plate systems and evaluation of the Anaerobe–Tek system for identification of commonly encountered anaerobes. *Journal of Clinical Microbiology* **16**, 1066–1072.

LYERLY, D. M., LOCKWOOD, D. E., RICHARDSON, S. H. & WILKINS, T. D. 1982 Biological activities of toxins A and B of *Clostridium difficile*. *Infection and Immunity* **35**, 1147–1150.

MAHONY, D. E. 1979 Bacteriocin, bacteriophage and other epidemiological typing methods for the genus *Clostridium*. In *Methods in Microbiology*, ed. Bergan, T. & Norris, J. R., Vol. 13, pp. 1–30. London & New York: Academic Press.

MALAMOU-LADAS, H. & TABAQCHALI, S. 1982 Inhihition of *Clostridium difficile* by faecal streptococci. *Journal of Medical Microbiology* **15**, 569–574.

MULLIGAN, M. E., ROLFE, R. D., FINEGOLD, S. M. & GEORGE, W. L. 1979 Contamination of a hospital environment by *Clostridium difficile. Current Microbiology* **3**, 173–175.

NAKAMURA, S., MIKAMA, M., NAKASHIO, S., TAKABATAKE, M., OKADO, I., YAMAKAWA, K., SERIKAWA, T., OKOMURA, S. & NISHIDA, S. 1981 Isolation of *Clostridium difficile* from the feces and the antibody in sera of young and elderly adults. *Microbiology and Immunology* **25**, 345–381.

OCCHIONERO, M., LUZZI, I., MASTRANTONIO, P., PANICHI, G. & PANTOSTI, A. L. 1982 A note on fermentation reactions of anaerobic bacteria on a solid medium. *Journal of Applied Bacteriology* **52**, 449–551.

PHILLIPS, K. D. 1976 A simple and sensitive technique for determining the fermentation reactions of non-sporing anaerobes. *Journal of Applied Bacteriology* **41**, 325–328.

PHILLIPS, K. D. & ROGERS, P. A. 1981 Rapid detection and presumptive identification of *Clostridium difficile* by *p*-cresol production on a selective medium. *Journal of Clinical Pathology* **34**, 642–644.

POXTON, I. R. 1982 Detection and isolation of *Clostridium difficile. European Journal of Chemotherapy and Antibiotics* **2**, 123–128.

POXTON, I. R. & BYRNE, M. D. 1981 Detection of *Clostridium difficile* toxin by counterimmunoelectrophoresis: a note of caution. *Journal of Clinical Microbiology* **14**, 349.

ROGERS, T. R., PETROU, M., LUCAS, C., CHUNG, J. T. N., BARRETT, A. J., BORRIELLO, S. P. & HONOUR, P. 1981 Spread of *Clostridium difficile* among patients receiving non-absorbable antibiotics for gut decontamination. *British Medical Journal* **283**, 408–409.

ROLFE, R. D., HELEBIAN, S. & FINEGOLD, S. M. 1981 Bacterial interference between *Clostridium difficile* and normal fecal flora. *Journal of Infectious Diseases* **143**, 470–475.

RYAN, R. W., KWASNIK, I. & TILTON, R. C. 1980 Rapid detection of *Clostridium difficile* toxin in human feces. *Journal of Clinical Microbiology* **12**, 776–779.

SELL, T. L., SCHABERG, D. R. & FEKETY, F. R. 1983 Bacteriophage and bacteriocin typing scheme for *Clostridium difficile. Journal of Clinical Microbiology* **17**, 1148–1152.

SLOTS, J. 1981 Enzymatic characterization of some oral and non-oral Gram-negative bacteria with the API Zym system. *Journal of Clinical Microbiology* **14**, 288–294.

SMITH, L. DS. 1975 *The Pathogenic Anaerobic Bacteria* 2nd edn. Springfield, Illinois: Charles C Thomas.

STARR, S. E., THOMPSON, F. S., DOWELL, V. R. & BALOWS, A. 1973 Micromethod system for identification of anaerobic bacteria. *Applied Microbiology* **25**, 713–717.

STOKES, E. J. & RIDGEWAY, G. L. 1980 *Clinical Bacteriology* 5th edn, pp. 215–219. London: Edward Arnold.

SUTTER, V. L. & CARTER, W. T. 1972 Evaluation of media and reagents for indole-spot tests in anaerobic bacteriology. *American Journal of Clinical Pathology* **58**, 335–338.

TAYLOR, N. S., THORNE, G. M. & BARTLETT, J. G. 1981 Comparison of two toxins produced by *Clostridium difficile. Infection and Immunity* **34**, 1036–1043.

THARAGONNET, D., SISSON, P. R., ROXBY, C. M., INGHAM, H. R. & SELKON, J. B. 1977 The API Zym system in the identification of Gram-negative anaerobes. *Journal of Clinical Pathology* **30**, 505–509.

WATSON, G. N., STRINGER, M. F., GILBERT, R. J. & MAHONY, D. E. 1982 The potential of bacteriocin typing in the study of *Clostridium perfringens* food poisoning. *Journal of Clinical Pathology* **35**, 1361–1365.

WATT, B., COLLEE, J. G. & BROWN, R. 1976 Tests of performance of anaerobic jars. *Journal of Clinical Pathology* **29**, 534–536.

WELCH, D. F., MENGE, S. K. & MATSEN, J. M. 1980 Identification of toxigenic *Clostridium difficile* by counterimmunoelectrophoresis. *Journal of Clinical Microbiology* **11**, 470–473.

WIDEMAN, P. A., CITRONBAUM, D. M. & SUTTER, V. L. 1977 Simple disk test for detection of nitrate reduction by anaerobic bacteria. *Journal of Clinical Microbiology* **5**, 315–319.

WILKINS, T. D. & WALKER, C. B. 1975 Development of a micromethod for identification of anaerobic bacteria. *Applied Microbiology* **30**, 825–830.

WILLEY, S. H. & BARTLETT, J. G. 1979 Cultures for *Clostridium difficile* in stools containing a cytotoxin neutralized by *Clostridium sordellii* antitoxin. *Journal of Clinical Microbiology* **10**, 880–884.

WILLIS, A. T. 1969 *Clostridia of Wound Infection*. London: Butterworths.

WILLIS, A. T. 1977 *Anaerobic Bacteriology: Clinical and Laboratory Practice* 3rd edn. London: Butterworths.

WILLIS, A. T. & HOBBS, G. 1959 Some new media for the isolation and identification of clostridia. *Journal of Pathology and Bacteriology* **77**, 511–521.

WILLIS, A. T. & PHILLIPS, K. D. 1983 *Anaerobic Infections* 2nd edn, Public Health Laboratory Service Monograph Series No. 3. London: HMSO.

WILSON, K. H., KENNEDY, M. J. & FEKETY, F. R. 1982 Use of sodium taurocholate to enhance spore recovery on a medium selective for *Clostridium difficile*. *Journal of Clinical Microbiology* **15**, 443–446.

WU, T. C. & FUNG, J. C. 1983 Evaluation of the usefulness of counter-immunoelectrophoresis for diagnosis of *Clostridium difficile*-associated colitis in clinical specimens. *Journal of Clinical Microbiology* **17**, 610–613.

Detection of *Clostridium difficile* Enterotoxin

SUSAN C. REDMOND, J. M. KETLEY, T. J. MITCHELL, AND J. STEPHEN
Department of Microbiology, University of Birmingham, Birmingham, UK

D. W. BURDON
The General Hospital, Birmingham, UK

D. C. A. CANDY
Institute of Child Health, Birmingham, UK

A cytotoxin was found in the stools of patients with antibiotic-associated pseudomembranous colitis (PMC) by Larson *et al.* (1977) and *Clostridium difficile* was later found to be the organism responsible for its production. (Bartlett *et al.* 1978; George *et al.* 1978). Tremendous interest was generated in the possibility that this cytotoxin was the bacterial determinant responsible for the PMC syndrome. PMC could be diagnosed, however, in patients who did not have a detectable cytotoxin in their stools. The three diagnostic criteria of PMC—the presence of rectal pseudomembranes, faecal cytotoxin and faecal *C. difficile*—were present in less than half of the 57 patients studied by Burdon *et al.* (1981*b*) (Table 1). Furthermore, cytotoxin titres did not correlate with the severity of the disease (Burdon *et al.* 1981*a*).

Following the discovery of an enterotoxin which was produced by *C. difficile* and which could be separated from the cytotoxin by ion exchange chromatography (Taylor *et al.* 1980), interest has grown in the role of this enterotoxin in disease, either on its own or in conjunction with the cytotoxin. For this to be investigated, it is necessary to assay the enterotoxin specifically and independently of the cytotoxin. An enzyme-linked immunosorbent assay (ELISA) would be a rapid sensitive method by which this could be achieved. One such assay has been reported. Lyerly *et al.* (1983) described a double, indirect sandwich ELISA using rabbit anti-serum to crude toxins,

Isolation and Identification of
Micro-organisms of Medical and
Veterinary Importance

Copyright © 1985 by the Society for Applied Bacteriology
All rights of reproduction in any form reserved
0-12-181460-2

TABLE 1

*Diagnosis of pseudomembranous colitis**

Group	Number	Rectal pseudomembrane	Faecal cytotoxin	Faecal *C. difficile*
1	25	+	+	+
2	11	−	+	+
3	7	+	−	+
4	1	+	−	−
5	13	−	−	+

* From Burdon *et al.* (1981b).

affinity-purified goat anti-enterotoxin and a commercially available source of rabbit antigoat IgG–alkaline phosphatase conjugate; it was shown to be sensitive and positive in clinical samples which were positive for cytotoxin. These authors, however, recognized the need for a simpler, direct assay. We have developed such an assay and, as the necessary immunoreactants are not readily available, this chapter includes a detailed description of the purification of the enterotoxin and the raising of the corresponding anti-enterotoxin required for the assay. Our initial attempts to purify toxin from ammonium sulphate precipitates of crude toxins followed the published methods of molecular exclusion followed by ion exchange chromatography (Taylor *et al.* 1980), and ion exchange chromatography followed by precipitation of toxin A at pH 5.5 (Sullivan *et al.* 1982). These methods, however, were found to give preparations which were either impure or low-yielding, or were found not to be routinely reproducible in our laboratory.

Other methods by which the enterotoxin can be detected are also discussed. They include whole-animal tests, tissue culture assays and immunoelectrophoretic techniques.

As a result of studies into the biological activities of enterotoxin and cytotoxin, these toxins have now been designated toxin A and toxin B, respectively (Lyerly *et al.* 1982).

Toxin Purification

Preparation of crude culture filtrate

Clostridium difficile strain 23289 (originally isolated from a patient with PMC and found to be a good toxin producer *in vitro*) is grown in 2% (w/v) Bacto Tryptose broth (Difco) supplemented with 0.1% (w/v) glucose, 0.2% (w/v) Na_2HPO_4, 0.1% (w/v) KNO_3, 0.03% (w/v) sodium formaldehyde sulphoxylate (BDH) (included as a reductant), and 5 μg/ml of clindamycin (Upjohn), which is known to enhance toxin production by this strain (George *et al.* 1979). The only attempts made to enhance anaerobic conditions are the addition of the reductant and keeping the flask firmly stoppered from the time the broth had cooled to approximately 37°C following autoclaving until it is inoculated. The inoculum consists of 50 ml/litre of actively growing culture. The culture (2- to 7-litre batches) is grown at 37°C for 4 days, clarified by centrifugation (25,000 **g** for 30 min at 4°C), precipitated with an equal volume of saturated ammonium sulphate solution, and stirred for 30 min at 4°C. The precipitate is collected by centrifugation (25,000 **g** for 30 min at 4°C) and resuspended in water to give a final concentration of not less than 50 times by volume of the original culture. The crude culture concentrate is dialysed against several changes of phosphate-buffered saline (PBS) containing (g/litre) NaCl, 4.25; KH_2PO_4, 2.44; Na_2HPO_4, 8.09; pH 7.0, sterilised by filtration and stored at 4°C. The protein concentration at this stage is typically 20 mg/ml.

Preparative electrophoresis

Preparative electrophoresis is carried out in Ultrodex (LKB) using the discontinuous buffer system described by Davis (1964). The apparatus used is that described by Winter *et al.* (1980) for preparative isoelectrofocussing. It consists of a glass plate 25 × 12.5 × 0.1 cm with a silicone rim (0.5 cm deep × 0.7 cm wide). Plastic dividers are placed at 2.5 and 5.5 cm from the cathode end of the tray, and the compartments are filled with 4% (w/v) Ultrodex in buffers as shown in Fig. 1. The gel is dried to its evaporation limit; before this is reached, the plastic dividers are removed while the gel is still sufficiently liquid to expand into the gaps with minimal indiscriminate mixing of the buffers. The gel is then transferred to the cooling plate of a 2117 Multiphor electrophoresis tank (LKB). Electrical contact between gels and reservoir buffers is ensured using filter-paper wicks (10.5 × 12.5 cm, Whatman no. 1; three at each end) and the gel is run at 1 kV (15 mA), 10°C, for 5–7 h. During this time a bromophenol blue marker dye which is included in the sample forms a sharp zone (<0.5 cm width) and migrates approximately 18 cm. Zones are collected using the plastic funnels supplied with the isoelectrofocussing kit (Winter *et al.* 1980), and toxin A is detected by its cytotoxity towards L cells and by fused rocket immunoelectrophoresis. Toxin A migrates very slowly in this system and remains near the cathode end of the gel.

Toxin B has a high electrophoretic mobility in this system and separates from toxin A. It can be purified further by ion exchange using the system of Sullivan *et al.* (1982).

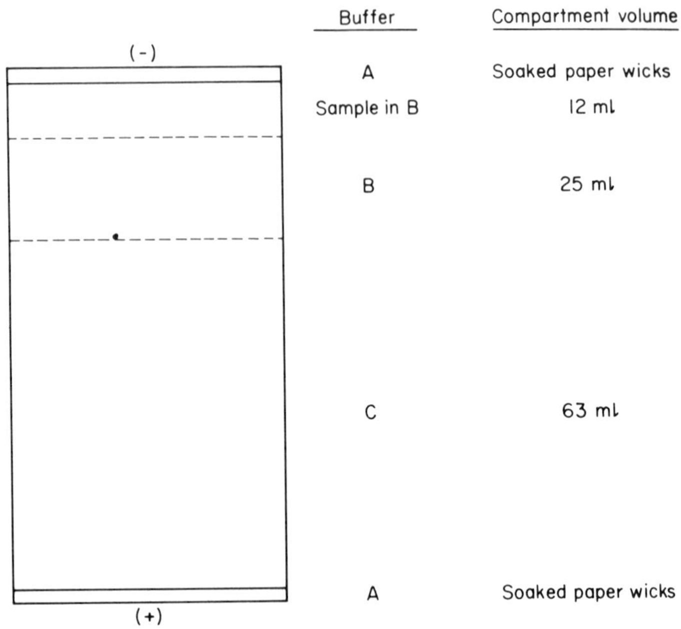

FIG. 1. Arrangement and volumes of buffers for preparative electrophoresis: A, 0.01 *M* Tris–glycine, pH 8.3 (this buffer is also placed in the reservoirs); B, 0.06 *M* Tris–HCl, pH 6.7; C, 0.4 *M* Tris–HCl, pH 8.9. The buffer concentrations quoted are those reached at the evaporation limit of the gel (i.e. those effective during electrophoresis).

Ion exchange chromatography

Fractions from the electrophoresis step which contain toxin A are dialysed against 0.5 *M* Tris HCl buffer at pH 7.5. The sample (typically 3 mg protein in 10 ml buffer) is applied to a column (2.5 × 10 cm) of DEAE sepharose CL6B (Pharmacia) and eluted using the conditions and salt gradient described by Sullivan *et al.* (1982). Fractions (10 ml) are collected and analysed for toxin A by cytotoxic activity and by fused rocket immunoelectrophoresis. Toxin A elutes in the 0.1–0.2 *M* NaCl fractions which are concentrated to 10 ml using polyethylene glycol 6000 and dialysed against PBS. The sample typically contains 150 μg protein/ml.

At each stage in the purification, the purity of samples is checked by crossed immunoelectrophoresis against the polyspecific anti-serum.

Sodium dodecylsulphate polyacrylamide gel electrophoresis (SDS-PAGE)

SDS–PAGE is carried out in 8% acrylamide gels according to the standard method of Laemmli (1970), and electrophoresis is carried out at 4°C overnight (7 mA per gel). The apparatus used was purchased from Raven Scientific Ltd. Protein bands are demonstrated by silver staining.

Silver staining

The silver staining technique used is that of Merril *et al.* (1981). The procedure consists of fixing the proteins in methanol : acetic acid : water (50 : 12 : 38) for 30 min, then washing the gel in ethanol : acetic acid : water (10 : 5 : 85) three times for 10 min each time. The gel is oxidised for 5 min in 200 ml of potassium dichromate solution (0.0034 mole/litre in 0.0032 mole/litre nitric acid). After rinsing (four times, 30 s each) in deionised water, the gel is placed in 200 ml 0.12 mole/litre silver nitrate for 30 min. The gel is rinsed twice in sodium carbonate solution (0.28 mole/litre in 0.05% (v/v) commercial formalin) before being left in 300 ml of fresh sodium carbonate solution until the edges of the gel begin to turn yellow. At this point the developing solution is replaced by 5% (v/v) acetic acid.

Protein concentration

Protein concentrations are estimated spectrophotometrically by E_{280}/E_{260} measurement.

Anti-Sera

Antisera to crude toxins

These are described by Burdon *et al.* (1981*b*).

Antitoxin A

Monoprecipitin anti-sera to toxin A were raised as follows. Male Californian rabbits (2.5 kg; Hylyne) were injected subcutaneously with 10 μg of pure toxin A (0.2 ml) mixed 1:1 with Freund's complete adjuvant. At weekly intervals thereafter a total of three similar subcutaneous injections in Freund's incomplete adjuvant were given. Six weeks later, one rabbit received an intravenous boost of 20 μg toxin A (0.6 ml) without adjuvant; this rabbit became moribund after 4 days and was bled out. Its serum contained low levels of antitoxin A and was used to neutralise toxin A before intravenous boosting of other rabbits with 20 μg of toxin A. Three days later the rabbits received 8 μg of neutralised toxin A, were bled from the ear vein 10 days later, and then rested.

Purification of IgG

The globulin fraction of the rabbit serum is precipitated with an equal volume of 32% (w/v) sodium sulphate, and the precipitate is collected by centrifugation (14,000 **g** for 30 min at 20°C). The precipitate is resuspended to give one-tenth of the original volume of serum, dialysed extensively against PBS and clarified by centrifugation (3500 **g** for 30 min at 20°C). These concentrates are suitable for immunoelectrophoresis, but further purification is necessary to carry out the ELISA test. The IgG fraction is therefore further purified by ion exchange chromatography on DEAE Sephadex A50 in 0.071 mole/litre acetate buffer pH 5.0, (Axelsen *et al.* 1973). All immunoglobulin fractions are stored at −20°C.

Conjugation of antitoxin A to horseradish peroxidase

The method used is based on that of Nakane & Kawaoi (1974). Sodium periodate solution (0.2 ml of 0.1 mole/litre) is added to 4 mg horseradish peroxidase dissolved in 1 ml of water and the mixture is stirred for 20 min at room temperature. Following overnight dialysis (4°C) against 1 mmol/litre sodium acetate buffer (pH 4.4), the pH of the mixture is raised to 9.5 by the addition of 20 μl of 0.2 mole/litre sodium carbonate buffer (pH 9.5). Antitoxin A immunoglobulin, 8 mg in 1 ml of 0.01 mole/litre sodium carbonate buffer (pH 9.5), is added immediately. This preparation is stirred for 2 h at room temperature before any unconjugated enzyme is reduced by the addition of 0.1 ml sodium borohydride solution (4 mg/ml). The conjugate is allowed to stand at 4°C for 2 h before dialysis against 0.1 mole/litre borate buffer (pH 7.4) and stored at 4°C diluted 1 : 1 in 60% (v/v) glycerol in borate buffer. The A_{403}/A_{280} ratio of the conjugate is 0.42.

Assays

Rabbit ileal loop test (RILT)

RILTs are carried out by the standard procedure described by Evans *et al.* (1973). Loops (4 cm) were used and the results read after 12 h. A positive result is taken as a V (ml)/L (cm) ratio ≥1.

Cytotoxic activity

Mouse L929 cells are cultured in minimal essential medium (MEM; Flow Laboratories), supplemented with 10% (v/v) of newborn calf serum (Gibco) and 1% (v/v) of foetal calf serum (Gibco), 0.03% (w/v) of L-glutamine,

0.008% (w/v) of crystamycin (Glaxo), 1% (v/v) of non-essential amino acids (Gibco), and buffered with 0.2% w/v sodium bicarbonate; this is termed complete MEM. Microtitre plates (Corning) are seeded to give 2×10^4 cells per well (containing 100 μl of complete MEM) and the cells are allowed to settle and spread to confluence overnight at 37°C in 5% CO_2. Test samples (10 μl) are added to the cells in the presence of 100 μl of complete MEM. The plate is incubated at 37°C in 5% CO_2 for 6 h. The end point of a titration is taken as the dilution of toxin which causes 50% of the L929 cells to round up and is arbitrarily assigned a value of 1 unit.

Immunoelectrophoresis

All immunoelectrophoresis is carried out with Shandon electrophoresis apparatus based on methods described by Axelsen *et al.* (1973).

Crossed immunoelectrophoresis. This is carried out in 1% (w/v) agarose (Litex type HSA; Park Scientific Ltd.) on 75 × 50 × 1 mm glass slide supports. The buffer system throughout is 0.06 mole/litre barbital at pH 8.6. A 10-μl sample is placed in a well (2 mm diameter) and electrophoresed along the 75 mm length of gel at 2 mA/cm width for 2 h. For the second dimension, a 1% agarose gel containing antiserum equivalent to 0.5 ml of original serum is cast alongside the separated antigens, which are then electrophoresed into the antibody containing gel at 0.45 mA/cm width over night. Temperature regulation is unnecessary.

Rocket immunoelectrophoresis. Wells, 2 mm diam and 5 cm apart, are filled with 10-μl samples and electrophoresed into a gel containing anti-serum carried out as in the second dimension of crossed immunoelectrophoresis.

Fused rocket immunoelectrophoresis. A series of wells (2.5 mm diameter) arranged in two rows such that the centre of each well is 5 mm from the one next to it and diagonally above or below it are cut into an agarose gel. A gel containing anti-serum is cast alongside these wells. Samples (15 μl) of consecutive fractions from a preparative electrophoresis run or a column run are placed in the wells and after 30 min diffusion at 20°C in a humid chamber, electrophoresis is carried out as in the second dimension of crossed immunoelectrophoresis.

Staining. Gels are pressed, washed in 2% NaCl, pressed, dried, stained in Coomassie blue and destained by the methods described by Axelsen *et al.* (1973).

ELISA

A direct sandwich ELISA has been developed for the quantitative detection of toxin A. Each well of a flat bottomed microtitre plate (Nunc type II) receives 100 μl of rabbit antitoxin A (100 μg/ml purified IgG) in coating buffer (0.05 mole/litre carbonate at pH 9.6) at 37°C for 2 h. Without emptying the wells, 20 μl of 5% (w/v) bovine serum albumin (BSA) (Sigma A 7030 for ELISA) in coating buffer are added to each well and the plate is incubated for a further 30 min at 37°C. The plate is then washed four times in PBS-T containing (g/litre) NaCl, 8.0; KH_2PO_4, 0.2; Na_2HPO_4, 1.1; KCl, 0.2; pH 7.4. To this is added 0.05% (v/v) of Tween 20 (polyoxyethylene sorbitan monolaurate, Sigma), with 3 min incubation at room temperature between each wash. The plate can be stored at this stage in PBS-T at 4°C for at least 5 days; stored plates are washed immediately before use. Test antigen solution is added in 100 μl PBS-T and incubated at 37°C for 2 h. The plate is washed twice before addition of 100 μl of 0.1% (v/v) normal rabbit serum in PBS-T to each well and the plate incubated at 37°C for 30 min. The plate is washed again. Rabbit antitoxin A conjugated to horseradish peroxidase (as described above) is diluted 1/1000 in PBS-T and 100 μl is placed in each well. The plate is incubated at 37°C for 2 h and washed four times in PBS-T. Substrate solution, 100 μl [containing *o*-phenylenediamine, 34 mg and hydrogen peroxide (30%), 40 μl, in 100 ml of 0.1 mole/litre citrate phosphate buffer, pH 5.0], is added to each well, and after 30 min incubation at 37°C in the dark, the reaction is stopped with 50 μl of 12% (v/v) sulphuric acid. The colour reaction is read at 492 nm on a Beckman DU 8 spectrophotometer (fitted with a microplate module) against a "no conjugate" blank.

All incubations were at 37°C in a humid chamber.

Observations

Purification of toxin A

The resuspended ammonium sulphate precipitate of the cell-free broth contains a complex mixture of antigens including toxin A (causes fluid accumulation in the rabbit ileal loop test (RILT) and toxin B ($>10^6$ cytotoxic units). Preparative electrophoresis facilitates the isolation of one major antigen (stage yield at least 50%) with some minor contaminants which can be removed completely by ion exchange on DEAE Sepharose CL6B (stage yield at least 85%) (see Fig. 2). The remaining antigen is toxin A, since it is mildly cytotoxic (10 cytotoxic units) and causes fluid accumulation in the RILT (V/L = 2.2). The preparation is considered to be pure as it contained only one peptide as judged by silver-stained SDS-PAGE and because on challenging rabbits with toxin A prepared by this method, a monoprecipitating anti-serum is raised (Fig. 2*d*).

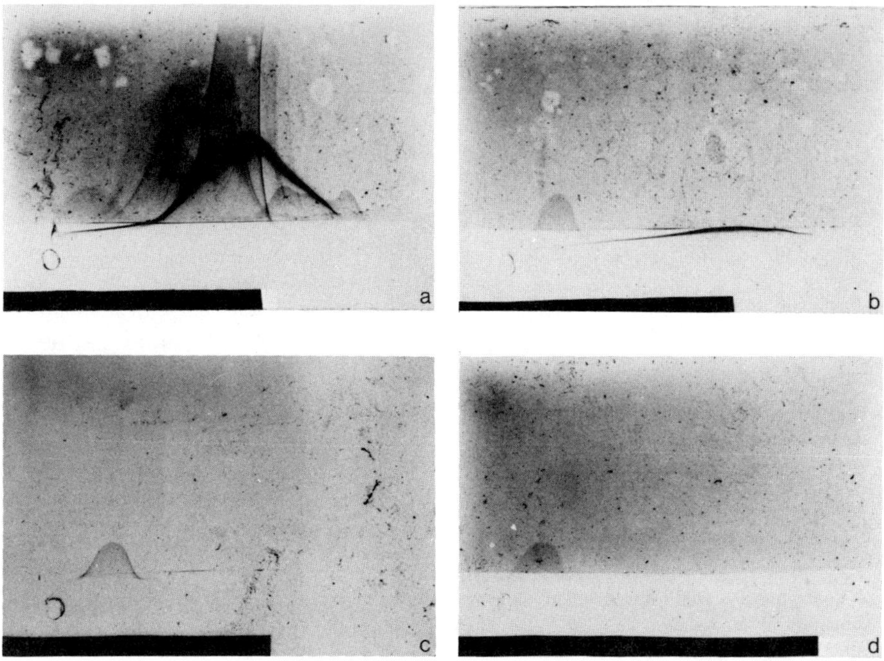

FIG. 2. Crossed immunoelectrophoresis showing purification of toxin A. (*a*) Crude toxins (against polyspecific anti-sera, also used in *b*, *c*) applied to preparative electrophoresis; (*b*) toxin A–containing fractions from preparative electrophoresis applied to ion exchange column; (*c*) toxin A recovered from ion exchange column; (*d*) crude toxins (as in *a*) run against antitoxin A.

Assays

The results of assays used to detect toxin A are summarised in Table 2.

RILT

Samples containing more than 6 μg toxin A/ml gave $V/L > 1.0$. At concentrations only slightly greater than this, loops became extremely distended, haemorrhagic and necrotic. The assay is limited to 20 loops per rabbit, but the severe damage caused by toxin A further limits the number of suspected positive samples which can be assayed per rabbit, as the rabbit may not survive the course of the assay.

Toxin B caused no fluid accumulation in RILT. Rabbit ileum did not become haemorrhagic or show any signs of tissue damage in the presence of partially purified toxin B.

Cytotoxin assay

The cytotoxin assay was found to be rather insensitive, 1 unit of cytotoxic activity representing 15 μg toxin A/ml. This was not due to an inherent

TABLE 2

Comparison of methods which can detect toxin A

Assay	Limit of detection (μg/ml)	Volume required for assay (μl)	Comments
RILT	6	500	Expensive, time consuming, requirement for surgical skills
Cytotoxin assay	15	10	Tissue culture facilities required; assay can be carried out in working day (although sensitivity doubled with 24 h incubation)
Immunoelectrophoresis			
Crossed	30	10	Requires an overnight run
Rocket	15	10	Requires an overnight run
ELISA	0.15*	100	Assay easily run in a working day

* Following repeated immunisation, an extremely high titre of antiserum to toxin A has been obtained. When used in the ELISA, this antiserum improved the sensitivity to 5 ng toxin A/ml (0.005 μg/ml).

insensitivity of mouse L929 cells to toxin A, as HeLa cells and BHK cells showed similar sensitivity. Mouse L929 cells were considered to have an advantage over HeLa and BHK cells as the range of concentrations over which partial (10–90%) rounding up of the cells occurred was less in mouse L929 cells.

Immunoelectrophoresis

The sensitivity of these assays can be maximised by reducing the concentration of antibody in the gel to the limit of visualisation of the immunoprecipitate. This limit was found to be 10 μg of the DEAE Sephadex A50 purified immunoglobulins/cm^2. At this concentration and with the limit of detection being taken as the concentration at which a toxin A–antitoxin A precipitin rocket could be recognised and measured, rocket immunoelectrophoresis was slightly more sensitive than crossed immunoelectrophoresis. A quantitative estimate of toxin A concentration can be made as rocket height is proportional to toxin A concentration (Fig. 3).

ELISA

The method described includes two washing steps which are not usually included in ELISA methods. After coating the microtitre plates with antitox-

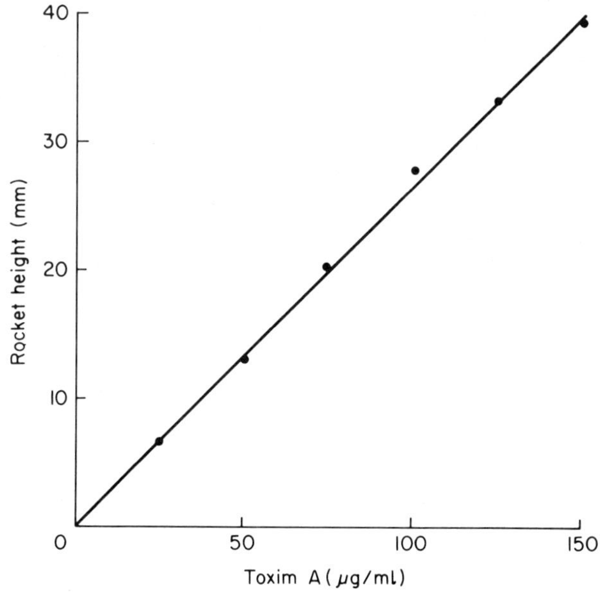

FIG. 3. Quantitation of toxin A using rocket immunoelectrophoresis.

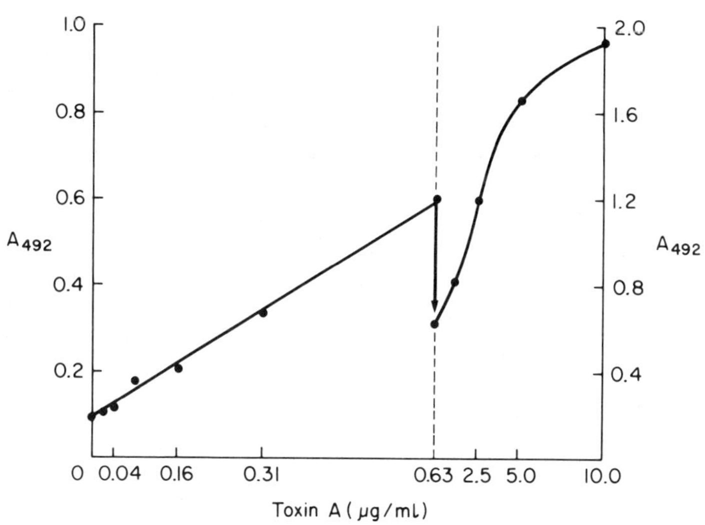

FIG. 4. Quantitation of toxin A using ELISA. All A_{492} values > 0.20 are statistically significant ($p < 0.003$). See footnote to Table 2.

in A, the plates are incubated in the presence of 1% BSA; this additional step was found to be essential to block sites on the plate to which both toxin A and the antitoxin A–HRP conjugate would have bound non-specifically during the latter steps of the assay. Similarly, incubation in the presence of normal rabbit serum blocked any further non-specific binding of the conjugate in the final step. Each blocking step was ineffective without the other, and both steps reduced the background to acceptable levels, improving the sensitivity of the assay.

The assay described in text gives a linear response over the range 0.1–0.5 μg toxin A/ml; above 0.5 μg toxin A, the response was non-linear (Fig. 4). Wells containing no anchor antibody had an A_{492} of approximately 0.10, whilst wells containing no antigen had an A_{492} in the range 0.10–0.15. A positive result could be recorded with 99.7% certainty (three standard deviations from the mean blank) if a sample had an A_{492} greater than 0.20, that is, if the toxin A concentration was greater than 0.15 μg/ml.

Discussion

A two-step method for the purification of toxin A from a 50% saturated ammonium sulphate-insoluble fraction of cell-free medium is described. The method involves preparative electrophoresis followed by ion exchange chromatography. We found that if ion exchange chromatography was carried out before electrophoresis, toxin A was obtained impure. This observation would suggest interactions between components in the crude preparations, which may affect the purification steps. Our inability to purify toxin A by the previously mentioned published methods may be explained in terms of such interactions, as the strains of *C. difficile* and the media used for toxin preparation are different. Also, our purification procedure does not use an initial ultrafiltration step. All of these differences may result in alterations in the relative concentrations of the components of the crude toxin preparations.

The toxin A purified by the method described here has been used to raise a monoprecipitating anti-serum to toxin A. Such an anti-serum is essential to assay toxin A by any of the methods described in this communication.

RILT was previously thought to be an extremely insensitive method of detecting toxin; our results show that it is in fact sensitive. The technique requires surgical skills, however, and becomes expensive if the assays are run (as they should be) with replicated samples, in more than one rabbit, and with the neutralising antibody controls included. The technique has been used successfully to show the presence of toxin A in the stools of one patient with PMC (Aronsson *et al.* 1982).

The cytotoxin assay is subject to interference by other cytotoxic substances which are often present in the stools of patients (notably toxin B). The neutralising antibody controls are essential to this assay, but even then, another cytotoxic substance which is present at a higher titre than toxin A would totally mask the presence of toxin A in the sample.

Immunoelectrophoresis is a simple technique and as sensitive as the cytotoxin assay. Crossed immunoelectrophoresis would not be considered useful in clinical diagnosis, as it is very demanding on space and antibody supply. It is a useful technique in the laboratory though, as toxin A can be readily recognised even in a crude mixture of antigens electrophoresed into a polyspecific anti-serum. Rocket electrophoresis is much less demanding on space and antibody supplies. As the starting materials required for the assay are the same as those required for ELISA, however, it is considered that these materials are used more efficiently in the ELISA, which is at least 100 times more sensitive and yields results more quickly than any of the other methods described here. ELISA also has the advantage that it requires no specialised skills or equipment.

Preliminary experiments have indicated that the ELISA yields negative results with diarrhoeas known not to be of *C. difficile* origin and that toxin A added to such samples is recovered quantitatively. Toxin A was present in 19 of 84 human faecal supernatants from which *C. difficile* had been isolated. None of these samples contained toxin A in the absence of toxin B, whilst 37 contained neither toxin. Only 4 of the patients in this study had undergone sigmoidoscopy, and none had PMC. One other patient was found to have had PMC post mortem; faecal samples had been positive for both toxins. More patients with authenticated PMC are required to evaluate the usefulness of the ELISA test in the diagnosis of PMC.

S. C. Redmond and J. M. Ketley are supported by the Medical Research Council (MRC); T. J. Mitchell is supported by Science and Engineering Research Council (SERC)/Pfizer.

References

ARONSSON, B., GRANSTRÖM, M., MÖLLBY, R. & NORD, C. E. 1982 Toxin A (enterotoxin) from *Clostridium difficile* in antibiotic associated colitis. *Lancet* **ii**, 1279.

AXELSEN, N. H., KRØLL, J. & WEEKE, B. 1973 *A Manual of Quantitative Immunoelectrophoresis: Methods and Applications*. Oslo: Universitetsforlaget.

BARTLETT, J. G., MOON, N., CHANG, T. W., TAYLOR, N. & ONDERDONK, A. B. 1978 Role of *Clostridium difficile* in antibiotic-associated pseudomembranous colitis. *Gastroenterology* **75**, 778–782.

BURDON, D. W., GEORGE, R. H., MOGG, G. A. G., ARABI, Y., THOMPSON, H., JOHNSON, M., ALEXANDER-WILLIAMS, J. & KEIGHLEY, M. R. B. 1981*a* Faecal toxin and severity of antibiotic-associated pseudomembranous colitis. *Journal of Clinical Pathology* **34**, 548–551.

BURDON, D. W., THOMPSON, H., CANDY, D. C. A., KEARNS, M., LEES, D. & STEPHEN, J. 1981*b* Enterotoxin(s) of *Clostridium difficile*. *Lancet* **ii**, 258–259.

DAVIS, B. J. 1964 Disc electrophoresis—II. Method and application to human serum proteins. *Annals of the New York Academy of Sciences* **121**, 404–427.

EVANS, D. G., EVANS, D. J. & PIERCE, N. F. 1973 Differences in the response of rabbit small intestine to heat-labile and heat stable enterotoxins of *Escherichia coli*. *Infection and Immunity* **7**, 873–880.

GEORGE, R. H., SYMONDS, J. M., DIMOCK, F., BROWN, J. D., ARABI, Y., SHINAGAWA, N., KEIGHLEY, M. R. B., ALEXANDER-WILLIAMS, J. & BURDON, D. W. 1978 Identification of *Clostridium difficile* as a cause of pseudomembranous colitis. *British Medical Journal* **i**, 695.

GEORGE, R. H., JOHNSON, M., YOUNGS, D. & BURDON, D. W. 1980 Induction of *Clostridium difficile* toxin by antibiotics. In *Current Chemotherapy and Infectious Diseases*, ed. Nelson, J. D. & Grassi, C., pp. 80–82. Washington DC: American Society for Microbiology.

LAEMMLI, U. K. 1970 Cleavage of structural proteins during the assembly of the head of bacteriophage T4. *Nature (London)* **227**, 680–685.

LARSON, H. E., PARRY, J. V., PRICE, A. B., DAVIES, D. R., DOLBY, J. & TYRRELL, D. A. J. 1977 Undescribed toxin in pseudomembranous colitis. *British Medical Journal* **i**, 1246–1248.

LYERLY, D. M., LOCKWOOD, D. E., RICHARDSON, S. H. & WILKINS, T. D. 1982 Biological activities of toxins A and B of *Clostridium difficile*. *Infection and Immunity* **35**, 1147–1150.

LYERLY, D. M., SULLIVAN, N. M. & WILKINS, T. D. 1983 Enzyme-linked immunosorbent assay for *Clostridium difficile* toxin A. *Journal of Clinical Microbiology* **17**, 72–78.

MERRIL, C. R., GOLDMAN, D., SEDMAN, S. A. & EBERT, M. H. 1981 Ultrasensitive stain for proteins in polyacrylamide gels shows regional variation in cerebrospinal fluid proteins. *Science (Washington, D.C.)* **211**, 1437–1438.

NAKANE, P. K. & KAWAOI, A. 1974 Peroxidase-labelled antibody: a new method of conjugation. *Journal of Histochemistry and Cytochemistry* **22**, 1084–1091.

SULLIVAN, N. M., PELLETT, S. & WILKINS, T. D. 1982 Purification and characterization of toxins A and B of *Clostridium difficile*. *Infection and Immunity* **35**, 1032–1040.

TAYLOR, N. S., THORNE, G. M. & BARTLETT, J. G. 1980 Separation of an enterotoxin from the cytotoxin of *Clostridium difficile*. *Clinical Research* **28**, 285A.

WINTER, A., PERLMUTTER, H. & DAVIES, H. 1980 *Preparative Flat-bed Electrofocusing in a Granulated Gel with the LKB 2117 Multiphor*. LKB Application Note No. 198 (Revised). Bromma: Sweden.

Leptospires and Leptospirosis

SHEENA A. WAITKINS

Public Health Laboratory Service, Leptospira Reference Unit, County Hospital, Hereford, UK

Leptospirosis is a zoonosis caused by the spirochaetal organism *Leptospira interrogans*. It is not synonymous with Weil's disease or necessarily associated with rats. It may present with a variety of symptoms (Table 1), ranging from a mild, "flu-like" illness to severe hepato–renal failure (Turner 1968*a*, 1969; Waitkins 1983*a*).

Three main serogroups of leptospires are responsible for almost all infections occurring in the United Kingdom. They are *L. interogans* var *Icterohaemorrhagiae* (Weil's disease), *L. interrogans* var. *Canicola* (leptospirosis in dogs and occasionally in humans), and *L. interrogans* var. *Hebdomadis* serotype hardjo (cattle-associated leptospirosis, CAL, which often infects cows and man) (Little *et al.* 1980; Mackintosh *et al.* 1980; Ellis *et al.* 1982; Hathaway 1982; Waitkins 1983*b*).

As leptospirosis is a zoonosis, those occupations that bring man into direct or indirect contact with infected animals offer a major risk. There is no person-to-person transmission. Examples of direct contact include milking cows, inspecting animals for slaughter and examining sick animals. Farmers, abattoir workers and veterinarians are therefore at greatest risk (Blackmore & Schoolum 1982). It is often forgotten, however, that indirect contact with infected animals may be a risk factor in the transmission of leptospirosis. Water contaminated with the urine of rats is particularly hazardous, especially to those who pursue water sports such as fishing, canoeing and raft racing (Coghlan *et al.* 1981; Waitkins 1983*a*) (Table 2).

After gaining entry to the susceptible host, leptospires multiply rapidly in the bloodstream and may therefore reach any tissue or organ in the body. In the leptospiraemic phase, which is the acute stage of the disease, the incubation period varies between 3 and 21 days. During this stage the organisms

Isolation and Identification of Micro-organisms of Medical and Veterinary Importance

Copyright © 1985 by the Society for Applied Bacteriology
All rights of reproduction in any form reserved

0-12-181460-2

TABLE 1

Symptoms associated with leptospirosis for serotypes of Leptospira interrogans

Icterohaemorrhagiae	*Canicola*	*Hebdomadis* (Hardjo)
Severe hepato–renal failure	Jaundice	Flu-like illness with *severe* headache
Conjunctivitis	Conjunctivitis	Prolonged recovery
Meningitis	Meningitis	Meningitis occasionally
Death (5–10%)	Occasionally death (0–2%)	Rarely death

may be isolated with ordinary good-quality blood culturing techniques. The convalescent period is associated with leptospiruria, when the organisms may still be recovered, and the development of immunity (Turner 1970). Antibodies may be detected 5 to 8 days after the onset of clinical symptoms, but their appearance may be delayed for up to 21 days. The normal antibody response results in detectable antibodies in the acute sera within 5 days, reaching a peak level 10–15 days later. Immunity to the infecting leptospira group may be transient and the antibodies may disappear gradually during the months of convalescence (Turner 1968*b*).

If leptospirosis is to be diagnosed successfully, the laboratory worker must not only be aware of the technical methods that are available, but also be familiar with the epidemiology of the disease, the clinical symptoms and the course of infection (WHO 1982). Table 3 is a helpful checklist of relevant

TABLE 2

Human leptospirosis, British Isles, 1978–1982

Occupation	*Icterohaemorrhagiae*	*Hebdomadis*	*Canicola*	Total
Farmers	41	73	1	115
Abattoir workers	2	4	—	6
Fish farmers	3	—	—	3
Veterinarians	1	1		2
Miners	1	—	—	1
Sewer workers	2	—	—	2
Other risks				
Water/sewage	29	2	—	31
Rats	21	1	—	22
Dogs	—	—	7	7
Misc./unknown	38	14	15	67
Totals	138	95	23	256

TABLE 3

Checklist of relevant clinical and epidemiological information that should be known about the patient

1. PATIENT NAME, Surname . . . . Given name . . . .
2. DATE OF BIRTH/AGE . . . . 3. SEX, Male ☐ Female ☐
4. OCCUPATION . . . .
5. CONTACT WITH ANIMALS Yes ☐ No ☐ 6. IF YES, When D D M M Y Y ☐☐☐☐☐☐
7. WHICH ANIMALS . . . .
8. CONTACT WITH POTENTIALLY CONTAMINATED WATER Yes ☐ No ☐
9. IF YES, When D D M M Y Y ☐☐☐☐☐☐ 10. WHERE . . . .
11. TYPE OF SPECIMEN SUBMITTED . . . . 12. DATE COLLECTED D D M M Y Y ☐☐☐☐☐☐
13. WAS THERE A PREVIOUS SPECIMEN Yes ☐ No ☐ 14. IF YES, give ref. no. . . . .
15. CLINICAL DETAILS (e.g. flu-like illness; meningism; jaundice; renal failure)

Please specify . . . .

. . . .

16. DATE OF ONSET D D M M Y Y ☐☐☐☐☐☐ 17. IS THE PATIENT HEPTATITIS B POSITIVE Yes ☐ No ☐

18. WERE ANTIBIOTICS GIVEN Yes ☐ No ☐ 19. IF YES, give details . . . .

. . . .

20. REQUESTING LABORATORY . . . .

. . . .

21. TO WHOM SHOULD THE REPORT BE SENT . . . .

22. REQUESTING DOCTOR . . . .

information that the clinician who suspects leptospirosis should submit to the laboratory.

This chapter is not intended to be a compendium on leptospirosis. The object is to describe the diagnostic methods used in Great Britain to confirm the diagnosis of that disease. There are two main laboratory investigations: the isolation of leptospira, and serology. Although the latter is most often attempted because routine methods are available and suspected positives may be confirmed by the Leptospira Reference Laboratory, isolation methods will be described first to conform with the title of this book.

Species and Serotypes

There are two species of *Leptospira*: the pathogenic *L. interrogans*, of which there are 19 serogroups, and the saprophytic *L. biflexa* (Faine & Stallman 1982). The organisms are very delicate, fine spirals with a characteristic microscopic morphology (Fig. 1). They may be seen only by dark ground, immunofluorescence, phase contrast or electron microscopy. Individual organisms have hooked ends (Fig. 1) and are 6–20 μm in length and 0.1 μm in diameter.

Leptospira interrogans is an obligate aerobe and requires various growth factors. The major source of energy is long-chained fatty acids. Carbohydrates are not a suitable source of carbon, and amino acids are not catabolised. The sources of nitrogen are ammonium salts. With such demanding requirements it is not surprising that leptospira are difficult to grow.

Culture media

Blood culture media may be used to maintain leptospira during transit to a reference laboratory. Palmer *et al.* (1984) found that some commercial blood culture media supported the survival of the organisms better than others. Those that provided good nutritive sources and adequate aerobic conditions and were incubated at 30°C enabled *L. interrogans* to survive for up to 9

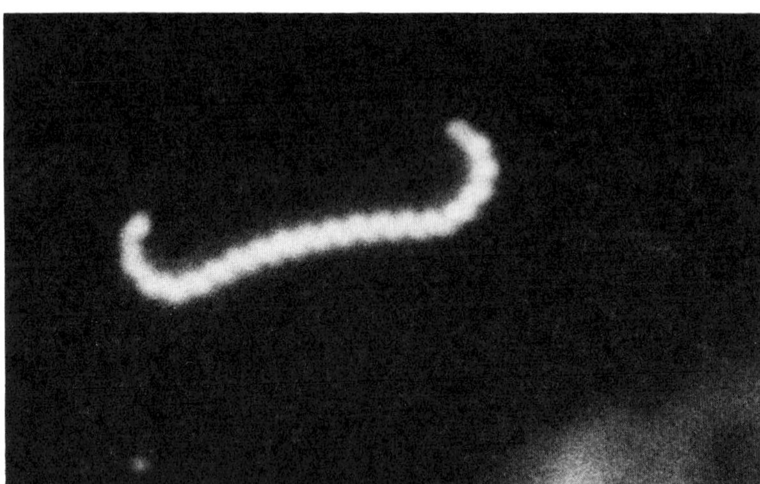

FIG. 1. Characteristic microscopical morphology of leptospira. Note the hooked ends and typical spirochaetal appearance of the organism (× 1000).

weeks. There is, however, a serious need for the development of media that can be used routinely for the isolation of the organisms from cases of human leptospirosis.

Media used for cultivation are usually enriched with fresh rabbit serum or with bovine serum albumin (Turner 1970). Various media, such as those of Fletcher (1928), Korthof (1932) and Stuart (1946), have been used successfully, but they are all difficult to make, require fresh animal products and have limited shelf lives. The semisynthetic medium of Ellinghausen & McCullough (1965), containing bovine serum albumin and Tween 80 (EM medium), and its modification by Johnson & Harris (1967) (EMJH medium) are used in modern isolation techniques. They are available commercially (Difco). They may be made selective by the addition of 5-fluorouracil (5FU).

EMJH medium

The following method is used in this laboratory.

25× Phosphate buffer solution. This contains (g/litre): Na_2HPO_4, 16.6; KH_2PO_4, 2.172.

20× Salt solution. This contains (g/litre): NaCl, 38.5; NH_4Cl, 5.35; $MgCl_2{\cdot}6H_2O$, 3.81.

Tween 80 solution. The stock bottle of Tween 80 and a flask containing 480 ml of distilled water are heated to 56°C in a water bath and 4.8 ml of the Tween is added to the warm water. The pipette is washed out in the solution and the mixture is shaken gently to dissolve. A further 236 ml of distilled water is then added.

Bovine albumin solution. A mixture of 40 ml of the 25× phosphate buffer and 960 ml of distilled water is boiled and allowed to cool. Then 50 g of bovine serum albumin, fraction V (Pentax), is added and dissolved.

Base medium. This contains distilled water, 2800 ml; 25× phosphate buffer solution, 160 ml; 20X salt solution, 200 ml; $CuSO_4{\cdot}5H_2O$ solution (30 mg/100 ml), 4 ml; $ZnSO_4{\cdot}7H_2O$ (40 mg/100 ml), 40 ml. At this point the medium should be clear. $FeSO_4{\cdot}7H_2O$ (250 mg/100 ml), 80 ml, is added and the mixture is shaken for 5 min. A haze will develop. L-Cystine, 0.8 g, is then added and the mixture is shaken for a further 1 min; this will not all go into solution.

The medium is then filtered through three thicknesses of Whatman no. 1 filter paper, and the filtrate should be clear. Vitamin B12 solution (a 1 : 10 solution of the concentrate in water), 80 ml; vitamin B1 solution (10 mg/100 ml), 0.4 ml; and the whole of the Tween 80 solution are now added and the medium autoclaved at 121°C for 15 min.

Complete medium. After cooling, inactivated rabbit serum, 50 ml, is added and the pH is checked and adjusted to 7.3–7.4. The medium is then filtered through a sandwich of Millipore membranes (pore sizes 0.22, 0.45, and 1.2 μ*m*) into sterile bottles. The bottled medium is incubated at 37°C to check sterility.

Selective 5FU medium

The addition of 5FU to this media allows leptospira to grow but inhibits other organisms.

5-Fluorouracil, 1 g, is dissolved in approximately 50 ml of distilled water containing 1–2 ml of 1 mole/litre NaOH by gentle heat (not above 56°C). The pH is adjusted to 7.4–7.6 with 1 mole/litre HCl and the volume is brought to 100 ml with distilled water. It is sterilized by filtration through a membrane (pore size 0.45 μm). This is the stock solution. It is stored at −4°C. For use it is thawed by warming to not more than 56°C and 1.0 ml is added to 100 ml of the complete medium. This gives a final concentration of 5FU of 100 μg/ml.

Isolation procedure

The organisms are most likely to be present in the blood and cerebro-spinal fluid (CSF) during the first 10 days of illness and in the urine after this period. Culture from urine, however, is not easy; it must be adjusted to an alkaline pH.

Two drops of blood, CSF or urine are added to 5 ml of EMJH medium, and serial dilutions of this are made into a further five bottles of the medium to dilute antibodies and contaminants.

Identification

Blood culture and enrichment media are examined daily for one week and thereafter weekly for several weeks. This may lead to contamination, and media containing 5FU at the concentration given above should be used. Serial dilution of contaminated cultures, or removal of contaminants by filtration through membranes (pore size 0.45 and 0.22 μm) may be successful.

The cultures are examined by dark-ground, phase-contrast, or fluorescence microscopy. If leptospira are seen, they may be identified by three simple tests:

(1) Ability to grow at 13°C in a medium in which they grow well at 30°C.
(2) Ability to grow in medium containing 8-azoguanine, 225 μg/ml.
(3) Hamster inoculation.

TABLE 4

Identification of species of leptospira

Strain	Pathogenicity for man	Growth at 13°C	Growth in 8-azoguanine	Hamster inoculation
Leptospira interrogans	+	–	–	Clinical symptoms; death
Leptospira biflexa	–	+	+	No symptoms

As indicated in Table 4, *L. interrogans* does not grow at 13°C, is inhibited by 8-azoguanine, and is pathogenic for hamsters, but *L. biflexa* grows under both cultural conditions and produces no symptoms in hamsters.

It is therefore not surprising that the routine isolation of leptospira from biological fluids is not often attempted. The serological diagnosis of the disease is much easier, and therefore routine laboratories may attempt their own diagnosis.

Serological Tests

The serological diagnosis of leptospirosis is often difficult because of the problems in interpreting the results of the tests. If the ground rules are followed, however, there is no reason why this work should not be done in routine laboratories.

Three serological tests are currently used in Great Britain: the macroscopic slide agglutination test, a rapid method which indicates the presence of leptospiral antibodies; the microscopic agglutination test (MAT), which is serogroup-specific and is used as a confirmatory test; and the complement fixation test (CFT), which is genus-specific and is used routinely as a screening test. The enzyme-linked immunosorbent assay (ELISA), also described below, is being evalued, and it is hoped that it will be an alternative to the CFT.

Unless otherwise specified in the text, the leptospire cultures, antigens and sera may be obtained from the Leptospira Reference Laboratory (address already given).

The macroscopic (slide) agglutination test

This is a simple slide agglutination test, using a genus-specific antigen. It should detect all leptospiral serogroups. Patient's serum, undiluted and diluted 1 : 10, is used. The dilution helps to overcome the prozoning effect which might be interpreted as a weak positive result. This test is less sensitive than the other leptospiral serological tests but tends to indicate early infection (3–7 days after onset), and positive results persist for several months after the patient has recovered.

The following materials are required for the test: glass slides 150 × 75 × 2 mm, engraved with three rows of squares 25 × 25 mm; dropping pipettes ("50" droppers) delivering 0.02 ml (unit volume for test); rotating apparatus; Patoc slide test antigen (Leptospira Reference Laboratory); phosphate-buffered saline (PBS) (Oxoid); sodium azide, 1% w/v in physiological saline (azide saline).

Preparation of antigen

(1) A flask containing 1 litre of EMJH medium is inoculated with *L. biflexa* Patoc I strain and incubated at 30°C for 7 days on a shaking machine (New Brunswick, G25).
(2) The culture is checked for quality of growth and for contamination by dark-ground microscopy and by culture on blood agar.
(3) The organisms are killed by adding neutralised formalin solution to give a final concentration of 0.2% and then letting it stand for 30 min.
(4) The culture is then centrifuged at 16,000 **g** (10,000 rev/min on an MSE HS 18 machine) at 5°C for 30 min.
(5) The deposit is resuspended in azide saline and the centrifugation is repeated.
(6) The deposit is resuspended in 50 ml of azide saline and boiled on a water bath for 30 min.
(7) It is centrifuged again and the deposit is suspended in 50 ml of azide saline.
(8) After a further centrifugation, the deposit is suspended in 10 ml of azide saline. This is stored at 4°C.

Before use the antigen is diluted with azide saline to give a suspension equivalent in density to the McFarland standard no. 20. This standard may be prepared by mixing 2 ml of 1% w/v barium chloride solution with 8 ml of 1% v/v sulphuric acid.

The quality of the antigen is checked against PBS and known positive and negative sera as described next.

The test proper

(1) With a "50" dropper, nine drops (0.18 ml) of PBS are placed in a small tube and 1 drop (0.02 ml) of test serum is added to give a 1 : 10 dilution.

(2) One drop of undiluted serum and one drop of serum diluted 1 : 10 are placed on an engraved slide.
(3) One drop of well-mixed Patoc slide test antigen is mixed with each drop of serum, using a small applicator stick and spreading the fluid to cover an area approximately 15 mm in diameter.
(4) The slide is rotated mechanically at 125 rev/min for 4 min and then examined immediately using an oblique light source.

The agglutination is graded thus:

None	–
Very weak and fine	Trace
Particulate	+
Small aggregates and turbidity	++
Large aggregates and slight turbidity	+++
Large aggregates and no turbidity	++++

The positive control should give a ++++ reaction, undiluted and diluted 1:10. The results are illustrated in Fig. 2.

Complement fixation test

A compound screening antigen containing *Patoc, Canicola, Hebdomadis* and *Autumnalis* strains is used. It is considered that this provides the best antisen system for testing sera in Great Britain (Turner 1970). There are two important factors in this test. (1) The incubation must be overnight at 4°C. (2) The antigen is slightly anticomplementary; the complement in the test should therefore be at $2HC_{50}$, not $3HC_{50}$ as in virus CF tests.

The following materials are required: CFT diluent (Oxoid, BR16); compound CFT antigen (Leptospira Reference Laboratory); rabbit haemolytic serum (HIB; Wellcome, VD15); sheep red blood cells (Tissue Culture Services); Mictotitre trays, U wells (Sterilin); Microdiluter (Dynatech).

Compound CFT antigen

Leptospira interrogans var. *Patoc*, var. *Canicola*, var. *Hebdomadis*, and var. *Autumnalis* are grown separately in EMJH medium to give concentrations of about 10^8 organisms/ml. They are concentrated by centrifugation and treated with hot acid to reduce their natural anticomplementary activity. Each serogroup is tested against its homologous anti-serum to determine the "in use" dilution. They are then pooled in appropriate concentrations and distributed in 20-ml amounts. The compound antigen has a shelf life of 6 months at 4°C.

Complement

This is prepared weekly by reconstituting the dried material according to the manufacturer's instructions. On the first day of the test, 1 volume of the

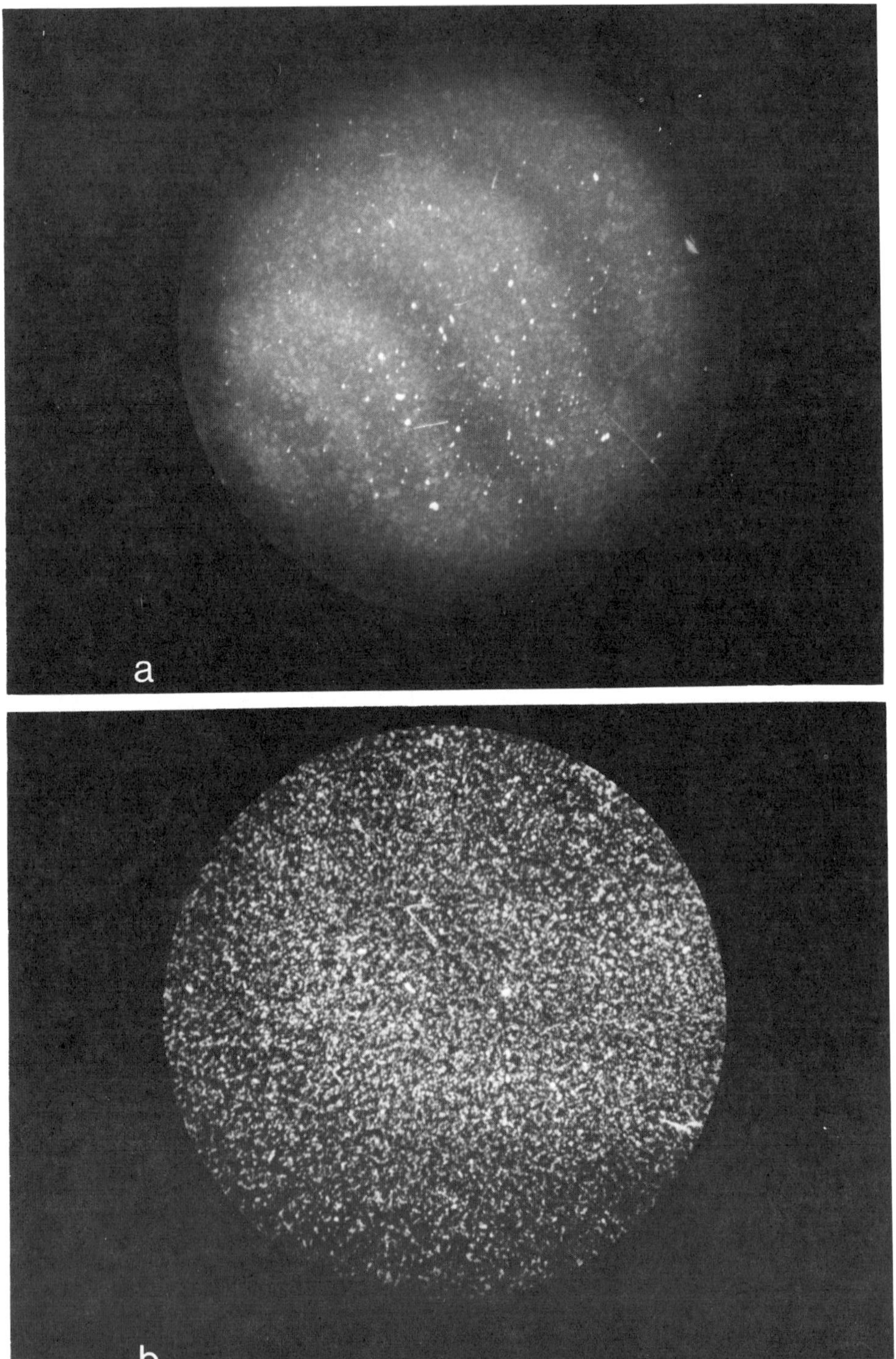

FIG. 2. The macroscopic slide agglutination test: (*a*) no agglutination, negative; (*b*) small aggregates and turbidity (++), negative; (*c*) large aggregates and slight turbidity (+++), weak positive; (*d*) large aggregates and no turbidity (++++), positive.

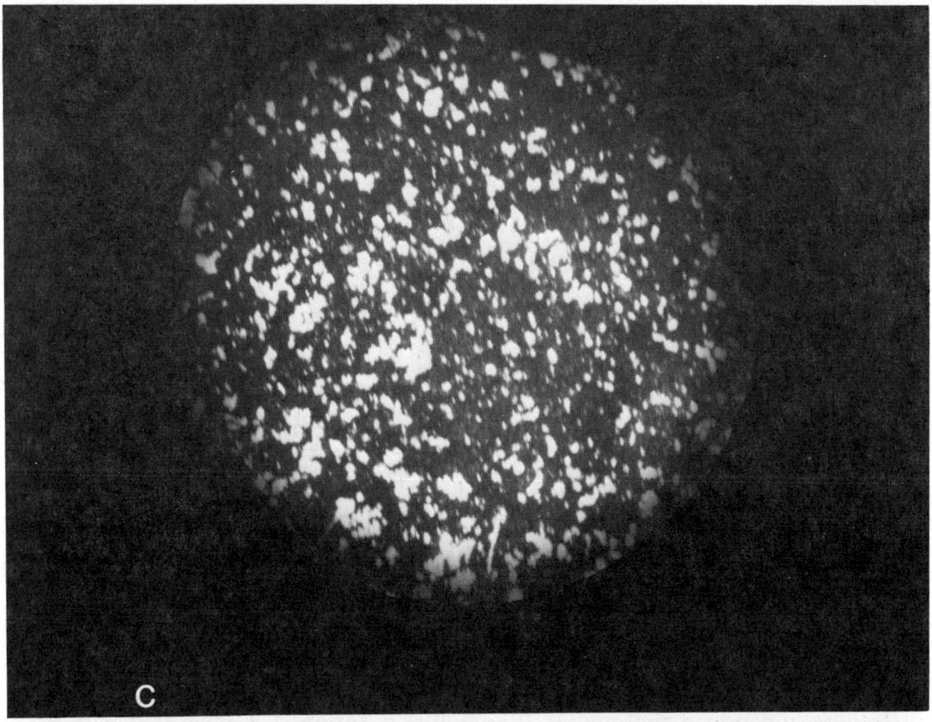
c

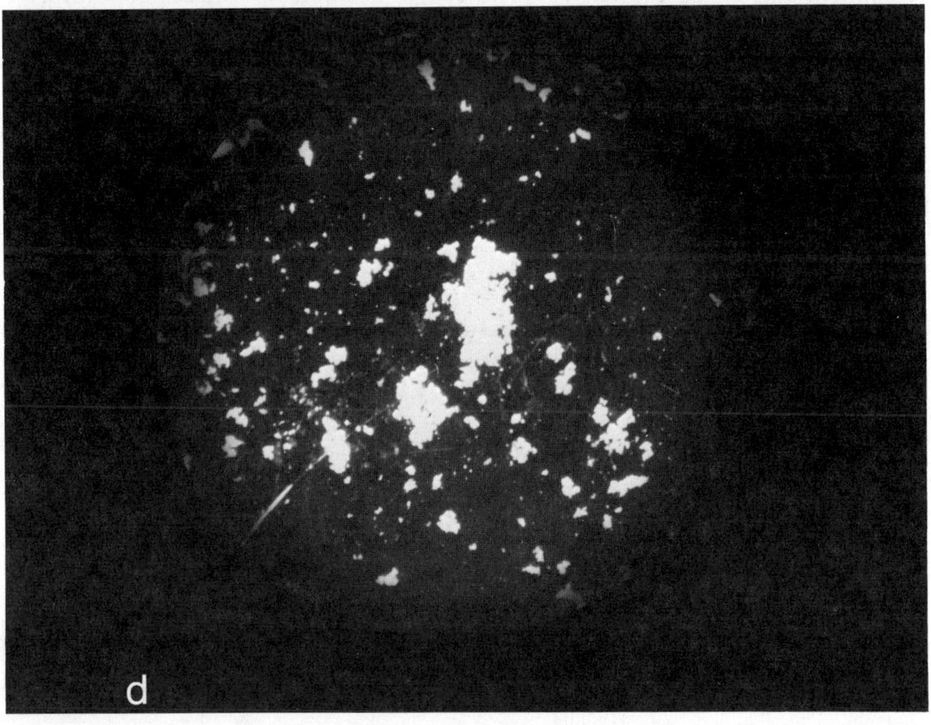
d

reconstituted complement is diluted with 7 volumes of distilled water. Because of the initial dilution before drying, this gives a 1 : 10 dilution. Further ddidilutions to the working strength of $2HC_{50}$ is done in CFT diluent after titrating the complement in the presence of antigen to determine 1 HC_{50}. For example, a working dilution of 1 : 30 is prepared by adding 4.8 ml of diluent to 2.4 ml of the 1 : 10 dilution of the complement.
Complement controls. These are prepared in small tubes as follows:

	$2HC_{50}$	$1HC_{50}$	$0.5HC_{50}$	$0.25HC_{50}$
Diluent (in milliliters)	—	0.5	1.5	3.5
$2HC_{50}$ complement (in milliliters)	0.5	0.5	0.5	0.5

Haemolytic serum

The titre of this is found by a chessboard titration against the complement.

Sensitization of sheep cells

On the first day of the test, the sheep cells are suspended in diluent and centrifuged three times. They are then suspended in twice their volume of diluent and mixed thoroughly. The cell concentration is measured with a spectrophotometer or a haematocrit. It should be about 20%. The cells are stored at 4°C overnight.

On the second day of the test, 10 ml of 4% suspension of the cells in diluent is prepared and mixed well with 10 ml of the haemolytic serum at its working strength by pouring back and forth from one beaker to another.

The test proper

(1) A 1 : 10 dilution of the test serum is made by adding 0.05 ml of serum to 0.45 ml of diluent. A positive control (Patoc) serum is similarly diluted. These are inactivated at 56°C for 30 min.
(2) Starting at row B of a Microtitre plate, 1 volume of diluent is placed in all wells of two columns for each serum to be tested. The odd-numbered (test) column will have the antigen added and the even-numbered (control) column will detect anticomplementary activity.
(3) One volume of the inactivated diluted serum is placed in the first two wells of the test and control columns.
(4) A Microdiluter is wetted with diluent and doubling dilutions are made from row B, mixing 20 times for each dilution. The diluters are sterilized by flaming.

(5) One volume of diluent is placed in all the wells of the even numbered columns.
(6) One volume of diluted complement is added to all wells.
(7) One volume of diluted antigen is added to all wells in the odd-numbered columns.
(8) Complement controls (see above) are set up at 2, 1, 0.5, and $0.25HC_{50}$ with and without antigen. They are prepared in duplicate as follows: (a) One volume of diluent is placed in each well of two rows of four wells for the antigen + complement control. (b) Two volumes of diluent are placed in two rows of four wells for the complement-only control. (c) One volume of each of the complement dilutions is added to the wells in the appropriate columns. (d) One volume of antigen is added to all wells in then appropriate columns.
(9) The plates are shaken mechanically and incubated at 4°C overnight.
(10) One volume of sensitized sheep cells is added to each well and the plates are again shaken.
(11) They are incubated at 37°C for 15 min, shaken again, and returned to the incubator for a further 15 min.
(12) After shaking again, the plates are placed at 4°C to allow the cells to settle.

The results are read as follows: (a) The end point is the highest dilution of serum which shows at least 50% fixation. (b) The control column should show complete lysis. The presence of red cells indicates anticomplementary activity. (c) The complement controls should show complete lysis at $2HC_{50}$ and 50% lysis at $1HC_{50}$.

Microscopic agglutination test (MAT)

This is a specific serogroup-confirmatory test. It uses a battery of antigen suspensions which includes at least one serotype from each of 19 serogroups of *L. interrogans* and one from the *L. biflexa* complex. The following materials are required: micro (syringe) filters with membranes of pore size 0.45 μm; Perspex agglutination trays; glass tiles 124 × 75 × 1 mm (Solmedia).

Antigens

These may be obtained from the Leptospira Reference Laboratory.
Formolised antigens. Leptospiral strains are selected to represent the following serogroups: i, *Icterohaemorrhagiae;* ii, *Javanica;* iii, *Celledoni;* iv, *Canicoli;* v, *Pyrogenes;* vi, *Ballum;* vii, *Autumnalis;* viii, *Australis;* ix, *Grippotyphosa;* x, *Hebdomadis;* xi, *Sejroe;* xii, *Saxkoebing;* xiii, *Wolfii;* xiv, *Mini;* xv, *Bataviae;* xvi, *Tarassovi;* xvii, *Pomona;* xviii, *Cynopteri;* xix, *Semeranga.*

(1) Each selected strain is grown in 20 ml of EMJH medium at 27°C for 7 days.
(2) The cultures are checked for purity by dark-ground microscopy and culture on blood agar.
(3) They are killed by adding neutralized formalin to a final concentration of 0.2%.
(4) Two or three representative serotypes are pooled from each of the 19 serogroups to form the antigen pools required for the test. A pool check is done on new pools and subsequently once each week (method follows).

Live antigens. These are more sensitive than formolised antigens but are easily contaminated and should be sub-cultured weekly.
Pool check. The following reference antisera are used: i, *Icterohaemorrhagiae, RGA;* ii, *Javanica, Veldrat Batavia 46;* iii, *Celledoni, Celledoni;* iv, *Canicola, Vleermuis 90 C;* v, *Pyrogenes, Robinsoni;* vi, *Ballum, Mus 127;* vii, *Autumnalis, Bankinang 1;* viii, *Australis, Lora;* ix, *Gripptyphosa, Moskva V;* x, *Hebdomadis, Hebdomadis;* xi, *Sejroe, M 84;* xii, *Saxkoebing, Mus 24;* xiii, *Wolffi, Hardjoprajitno;* xiv, *Mini, Szwajizak;* xv, *Bataviae, Van Tienen;* xvi, *Tarassovi;* xvii, *Pomona, Pomona;* xviii, *Cynopteri, 3522 C;* xix, *Semeranga, Patoc.*

(1) The reference sera are diluted 1 : 160 by adding 25 μl to 4.0 ml of PBS. These dilutions are discarded after 2 weeks. The final dilution after adding the antigen will be 1 : 320.
(2) One volume of each diluted reference serum is placed in a well of a Perspex agglutination tray.
(3) One volume of PBS is placed in an equal number of parallel wells for the negative antigen controls.
(4) One volume of each antigen pool is placed in the well containing its homologous antiserum and into a well containing PBS.
(5) The trays are tapped to mix the contents of the wells and are placed in a moist chamber overnight.

The results are read as described already for the test proper. An antigen pool is satisfactory if there is (1) at least 50% agglutination with the homologous anti-serum and (2) no clumping with PBS.

The test proper

(1) A 1 : 10 dilution of the test serum is prepared by adding 100 μl to 1.0 ml of PBS in a small test tube. This is filtered through a microfilter.
(2) If the slide agglutination test was positive, doubling dilutions of 1 : 20 to 1 : 2560 are made. The final dilutions after adding the antigen will be 1 : 40 to 1 : 5120
(3) One volume of each dilution is placed in each of 19 wells in a Perspex

plate. It may be necessary to use two adjacent trays to accommodate all the dilutions.

(4) One volume of phosphate-buffered saline (PBS) (Oxoid) is placed in each well of 19 wells for negative antigen controls.

(5) One volume of each antigen is added to each column.

(6) The trays are tapped to mix the contents of the wells. If formolised antigens are used the trays are placed overnight in a moist chamber at room temperature. If live antigens are used the trays are incubated at 30°C for 2–4 h.

(7) A large-bore pipette is used to mix the contents of each well and to transfer one drop to a glass tile for examination by dark-ground microscopy.

If formolised antigens are used, the clumps of organisms are large, loosely packed and have a "fluffy" appearance; live antigens are agglutinated into compact white balls (Fig. 3).

Agglutination is graded as follows: complete, +; 60–90%, (+); 50%, ±; 20–40% (±); 10–20% Traces; none, −.

The final endpoint is defined as the highest serum dilution showing 50% agglutination of leptospira.

After use the trays should be placed in an appropriate disinfectant and left overnight.

Interpretation of results

The microscopic agglutination test (MAT) is both specific and confirmatory. In the early stages of infection the patient's serum may exhibit many cross reactions to several leptospiral groups (Table 5). As the patient recovers, however, a reaction to the infecting serogroup will predominate, although this may take some weeks or even months (Table 6). Several sequential sera may therefore have to be examined before the identity of the infecting serogroup can be confirmed. The MAT antibodies may be detected for several years after the infection, and the technique is therefore a very useful serological tool in epidemiological surveys.

Although the techniques are straightforward, the preparation of the individual pools of antigens is time consuming, and care must be taken not to contaminate the stock cultures. The cross reaction phenomenon is regarded as paradoxical and may lead to misinterpretation, especially by inexperienced operators. For these reasons it is best to regard the MAT as a reference laboratory test.

The ELISA test

The ELISA method used in this laboratory was developed by Terpstra *et al.* (1980). The antigen is prepared from a single serogroup, and the test is

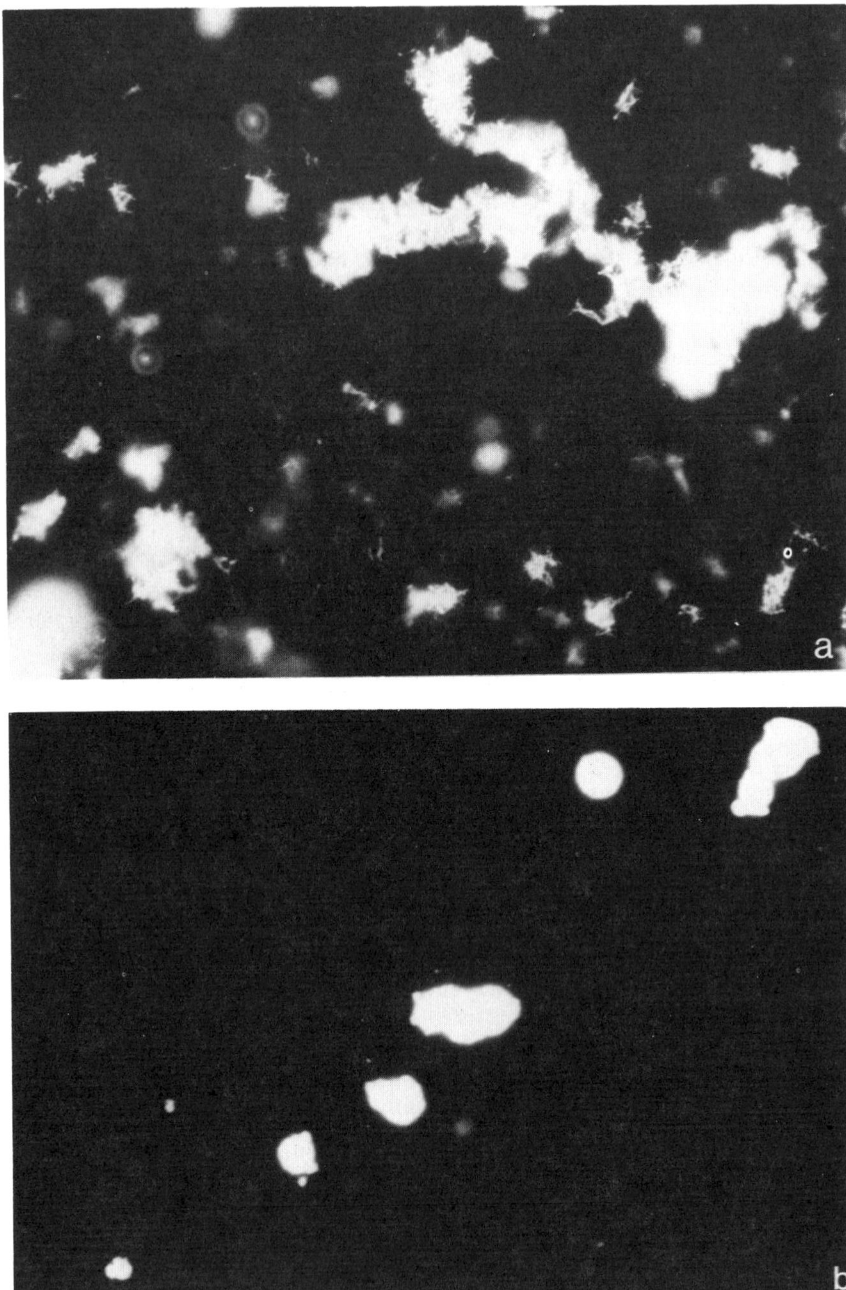

FIG. 3. The microscopic slide agglutination test (MAT). (*a*) With formalised leptospires, 100% or complete agglutination. Note "fluffy" appearance compared with (*b*). (*b*) With live leptospires. The agglutination appears as a bright, dense mass. (*c*) With formalised leptospires, 50% agglutination. (*d*) With formalised leptospires, no agglutination.

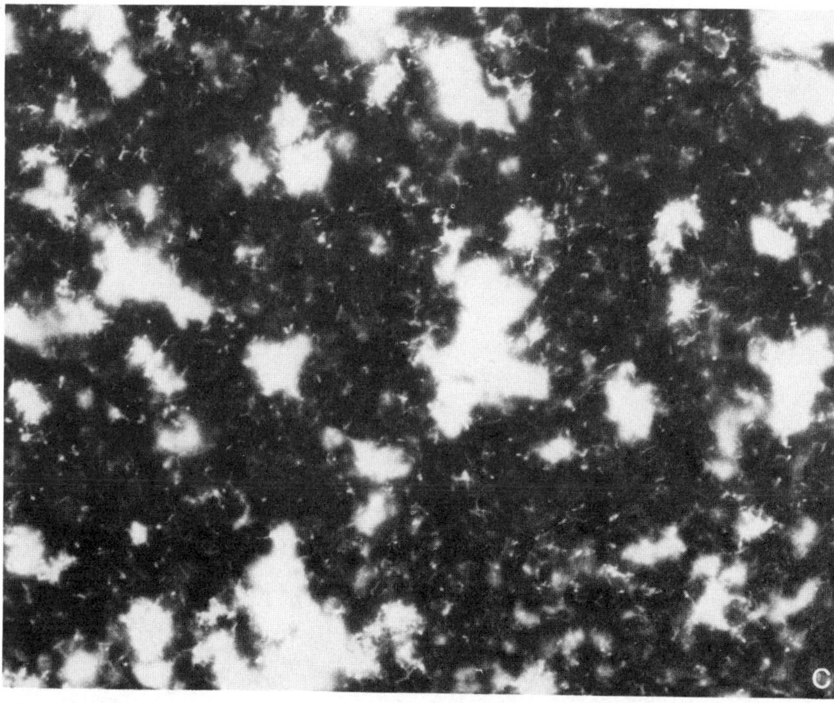
c

d

TABLE 5

Cross reactions seen in the microscopic agglutination test during the early phase of leptospirosis*†

Dilution of sera	*Ict* 1	2	3	*Can* 4	5	6	7	8	9	10	11	12	*Hardjo* 13	14	15	16	17	18	*Patoc* 19
1:40	+	±	±	+	±	±	+	+	+	−	+	±	+	+	+	−	+	±	+
1:80	+		±	+			±	±	+		+		+	±	+		±		+
1:160	±			±					±		+		±		±				+
1:320	Tr			Tr							±		Tr						±
1:640																			
1:1280																			
1:2560																			
1:5120																			

Pools: columns 1–19.

* Ten days after onset.

† Key: +, 100% agglutination; ±, 50% agglutination; −, 0% agglutination; Tr, tracing reaction. *Ict*, *Icterohaemorrhagiae; Can*, *Canicola*.

TABLE 6

The emergence of the predominating serogroup during convalescence[*],[†]

Dilution of sera	*Ict* 1	2	3	*Can* 4	5	6	7	8	9	10	11	12	*Hardjo* 13	14	15	16	17	18	*Patoc* 19
1:40	+	−	±	±	−	−	−	±	+	−	±	±	+	+	+	−	±	−	+
1:80	±		Tr					±	±		±		+	±	+				+
1:160	Tr										Tr		+		±				±
1:320													+						
1:640													+						
1:1280													+						
1:2560													±						
1:5120																			

[*] One month after date of onset.

[†] Key: +, 100% agglutination; ±, 50% agglutination; −, 0% agglutination; Tr, tracing reaction. *Ict*, *Icterohaemorrhagiae; Can*, *Canicola*. Titre 2560, predominant infecting serogroup.

genus-specific but is also capable of being serogroup-specific. It is easier to perform than the CFT, and it indicates whether the antibody is IgM or IgG.

The following materials are required: immunowash apparatus (Nunc); a Multiskan Plate Reader (Titertek); Microtitre plates (Sterilin); and a micro-shaker (Cooke AM69).

Substrate. This is prepared by dissolving 80 mg of 5-aminosalicylic acid (Merck) in 100 ml of distilled water at 75°C. If stored at 4°C in the dark, this solution will remain stable for about 1 month. Before use, the pH is adjusted to 6.0 with 1 mole/litre NaOH. If the pH goes above 6.0, it is brought back with substrate solution, not acid.

Hydrogen peroxide. This is a 1 : 30 dilution of the 30% commercial reagent in distilled water. It is stored at 4°C. Before use, 1.5 ml of this is added to 28.5 ml of distilled water to give a 0.05% solution.

Washing fluid. Tween 20 (BDH, 66368) is warmed to 56°C and 5.0 ml is added to 10 litres of PBS, pH 7.2.

Bovine albumin solution. Bovine albumin fraction V (Armour Pharmaceuticals), 50 g, is dissolved in 1 litre of distilled water. The solution is sterilized by membrane filtration and stored at −20°C in 20-ml amounts.

Control sera. Known positive human sera are tested against all conjugates, and those with titres of 1 : 640 or above are pooled. Known negative human sera are also tested against all conjugates, and those that give titres of 1 : 20 or less are pooled.

Conjugates. IgG (reference no. 75051), IgM (reference no. 75061) and Poly-IgG,M,A (reference no. 75200) are obtained from the Institut Pasteur, Paris.

Optimal dilutions are found by a chessboard titration. The positive control serum is titrated at 1 : 160–1 : 5120 and the conjugates at 1 : 500–1 : 4000. A negative control at 1 : 20 is included. The highest dilutions of serum and conjugate that give an optimal density reading of 1.0 are used.

Antigen. A seed culture (20 ml) of a single serogroup is added to 1 litre of EMJH medium and incubated at 30°C for 6–7 days in a shaking incubator (New Brunswick, G25) at 220 rev/min until there is a density of $9–11 \times 10^8$ leptospira/ml as determined with a Thoma counting chamber. The culture is checked for purity by dark-ground microscopy and culture on blood agar.

The culture is tested for the presence of residual Tween. Neutralised formalin solution, 5 μl, is added to.5 ml of the culture. After standing for 30 min, this is centrifuged at 4500 **g** (5000 rev/min on an MSE Super Minor machine) for 10 min. The supernatant is removed to another tube and boiled. If it remains clear, there is no residual Tween and the culture is ready for harvesting. If it is cloudy the culture is incubated for another 2–3 days and the Tween test is repeated.

Neutralised formalin solution is added to the satisfactory culture to give a final concentration of 0.5%. After standing for 30 min, the culture is boiled on a water bath for a further 30 min, during which time it is shaken every 5 min. After cooling to room temperature, it is centrifuged at 16,000 **g** (10,000 rev/min on an MSE HS18 machine) for 30 min at 5°C. The supernatant (antigen) is stored at 4°C.

Coating the Microtitre plates. Neat antigen, 100 μl, is placed in every well of the plates. These are left at room temperature, protected from sunlight and gases, until evaporation is complete (about 3 days). They are stored in sealed plastic bags at room temperature.

The test proper

(1) The diluting fluid is prepared by mixing 20 ml of stock bovine albumin solution with 180 ml of PBS and adding 100 μl of Tween 20 solution.
(2) The coated Microtitre plates are washed, using the Immunowash apparatus and this routine: (a) The wells in the plate are filled from top to bottom. (b) The fluid is aspirated from top to bottom. (c) The wells are filled from bottom to top. (d) The fluid is aspirated from bottom to top. (e) Both of these operations are repeated, leaving the fluid in the wells for 1 min before removing it. The plates are then dried by laying them on absorbent paper and tapping them. Gloves are worn during these operations.
(3) The test and control sera are diluted, using one row of wells for each conjugate to be tested: (a) 180 μl of diluting fluid is placed in the first well and 100 μl in the remaining wells of each row, and (b) 20 μl of test serum is added to the first well and doubling dilutions are made along the row, discarding the last volume, to give the range 1 : 10–1 : 20480.
(4) The plates are covered and incubated at 30°C for 1 h.
(5) The serum dilutions are aspirated and the plate washed as in (2).
(6) Diluting fluid (100 μl) is placed in the first well (1 : 10) as a control to detect false positives.
(7) Conjugate (100 μl) at its optimal dilution is added to all other wells.
(8) The plates are covered and incubated at 30°C for 30 min.
(9) The conjugate solution is removed and the plates washed as in (2).
(10) Nine parts of substrate and one part of hydrogen peroxide at working strength are mixed (10 ml is required for each plate) and 100 μl of this is added to all wells.
(11) The plates are covered and incubated at 30°C for 2 h and then shaken on the Microshaker.
(12) The Multiskan Plate Reader is switched on during this time to allow it to stabilise for at least 20 min.

(13) A filter giving a wavelength of 492 nm is selected and a blank reading taken on an empty plate.
(14) Readings are taken: an optical density of 0.5 is recorded as the last positive dilution. The controls should give readings not less than one dilution lower than their optimal titre.

Interpretation of results

The ELISA method uses anti-human peroxidase-conjugated immunoglobulin with 5-aminosalicylic acid. The colour detected is a deep brown, which is easily read even without using the Multiskan Plate Reader. Figure 4 illustrates the ease of reading and interpreting the results. The top row (A) is the positive control and the second row (B) the negative control. The next three rows contain test sera. Dilutions of all sera are 1 : 10–1 : 20480 in columns 1–12 across the plate. The figure shows that row C is negative, row D is positive at 1 : 160, and row E is positive at 1 : 640.

Sera are usually screened with Poly-IgG,M,A immunoglobulin, and those that are positive are retested with IgG and IgM. The presence of IgM

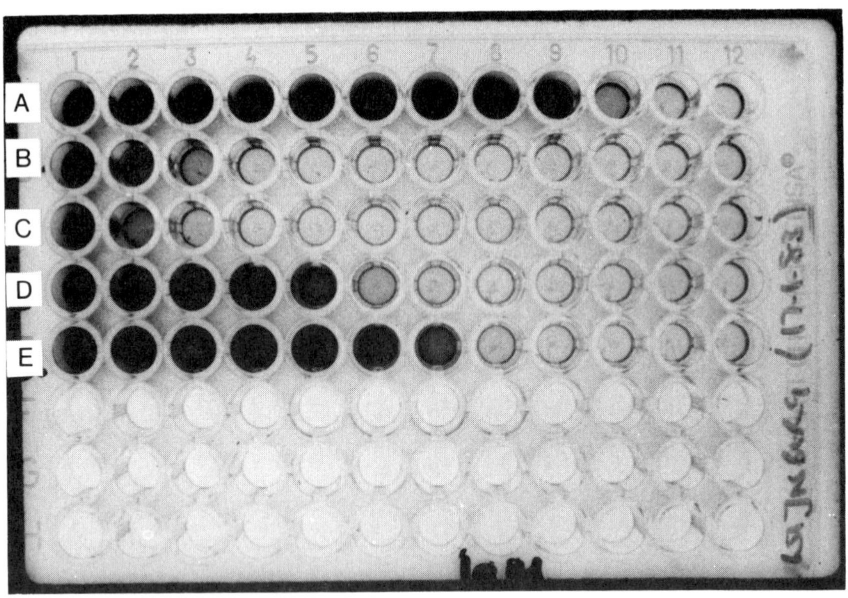

FIG. 4. ELISA results using peroxidase as the enzyme and 5 aminosalicyclic acid as substrate. Note the ease of reading the endpoints. Row A, positive control, titre 1:2560. Row B, negative control, 1:20. Row C, test serum, 1:10, negative. Row D, test serum, 1:160, positive. Row E, test serum, 1:640, positive.

suggests the initial stage of the disease, while IgG alone indicates a past infection.

When ELISA is introduced into routine work it will help in the early detection of leptospirosis, thereby facilitating early treatment of persons with the disease.

References

BLACKMORE, D. K. & SCHOOLUM, L. M. 1982 The occupational hazards of leptospirosis in the meat industry. *New Zealand Medical Journal* **95**, 494–497.

COGHLAN, J. D. & COMMUNICABLE DISEASES SURVEILLANCE CENTRE 1981 Leptospirosis in man, British Isles, 1979–1980. *British Medical Journal* **282**, 2066.

ELLINGHAUSEN, H. C. & MCCULLOUGH, W. G. 1965 Nutrition of *Leptospira pomona* and growth of 13 other serotypes: Fractionation of oleic albumin complex and a medium of bovine albumin and polysorbate 80. *American Journal of Veterinary Research* **26**, 45–51.

ELLIS, W. A., O'BRIEN, J. J., NEILL, S. D., FERGUSON, J. W. & HANNA, J. 1982 Bovine leptospirosis: Microbiological and serological findings in aborted fetuses. *Veterinary Record* **110**, 147–150.

FAINE, S. & STALLMAN, N. D. 1982 Amended description of the genus *Leptospira* Noguchi 1917 and the species *L. interrogans* (Stimson 1907) Wenyon 1926 and *L. biflexa* (Wolbach and Binger 1914) Noguchi 1918. *International Journal of Systematic Bacteriology* **32**, 461–463.

FLETCHER, W. 1928 Recent work on leptospirosis, Tsutsugamushi disease and tropical typhus in the Federated Malay States. *Transactions of the Royal Society of Tropical Medicine and Hygiene* **21**, 265–287.

HATHAWAY, S. C. 1982 Leptospirosis in New Zealand: An ecological view. *New Zealand Veterinary Journal* **29**, 109–112.

JOHNSON, R. C. & HARRIS, V. G. 1967 Differentiation of pathogenic and saprophytic leptospires. 1. Growth at low temperatures. *Journal of Bacteriology* **94**, 27–31.

KORTHOF, G. 1932 Experimentelles Schlammfieber beim Menschen. *Zentralblatt für Bakteriologie, Parasitenkunde, Infektionskrankheiten und Hygiene, Abteilung 1: Originale* **125**, 428–434.

LITTLE, T. W. A., RICHARDS, M. S., HUSSAINI, S. M. & JONES, T. D. 1980 Significance of leptospiral antibodies in calving and aborting cattle in South West England. *Veterinary Record* **106**, 221–224.

MACKINTOSH, C. G., SCHOLLUM, L. M., HARRIS, R. D., BLACKMORE, D. K., WILLIS, A. F., COOK, N. R. & STOKE, J. C. J. 1980 Epidemiology of leptospirosis in dairy farm workers in the Manawatu. Part 1. A cross sectional serological survey and associated occupational factors. *New Zealand Veterinary Journal* **28**, 245–250.

PALMER, M., WAITKINS, S. A. & ZOCHOWSKI, W. 1984 Survival of leptospires in commercial blood culture systems. *Zentralblatt für Bakteriologie, Parasitenkunde, Infektionskrankheiten und Hygiene, Abteilung 1: Originale.* **257**, 480–487.

STUART, R. D. 1946 The preparation and care of a simple culture medium for leptospires. *Journal of Pathology and Bacteriology* **58**, 343–349.

TERPSTRA, W. J., LIGTHART, G. S. & SCHOONE, G. J. 1980 Serodiagnosis of human leptospirosis by enzyme-linked-immunosorbent-assay (ELISA). *Zentralblatt für Bakteriologie, Parasitenkunde, Infektionskrankheiten und Hygiene, Abteilung 1: Originale* **247**, 400–405.

TURNER, L. H. 1968*a* Leptospirosis. In *Some Diseases of Animals Communicable to Man in Britain* pp. 231–245. Oxford & New York: Pergamon Press.

TURNER, L. H. 1968*b* Leptospirosis. Part II. Serology. *Transactions of the Royal Society of Tropical Medicine and Hygiene* **62**, 880–899.

TURNER, L. H. 1969 Leptospirosis. *British Medical Journal* **1**, 231–235.

TURNER, L. H. 1970 Leptospirosis. Part III. Maintenance, isolation and demonstration of leptospires. *Transactions of the Royal Society of Tropical Medicine and Hygiene* **64**, 623–646.

WAITKINS, S. A. 1983*a* Laboratory diagnosis of leptospirosis. *Laboratory Technology* **17**, 178–184.

WAITKINS, S. A. 1983*b* A Survey of Human Leptospirosis in Dairy Farmers. British Cattle Veterinary Proceedings 1982–1983. London.

WHO 1982 *Guidelines for the Control of Leptospirosis*, ed. S. Faine, Publication No. 67. Geneva: World Health Organization.

Mycobacteria

P. A. JENKINS AND L. R. DUDDRIDGE

Mycobacterium Reference Unit, Public Health Laboratory, University Hospital of Wales, Heath Park, Cardiff, UK

C. H. COLLINS AND M. D. YATES

Regional Centre for Tuberculosis Bacteriology, Public Health Laboratory, Dulwich Hospital, London, UK

The bacteriological diagnosis of tuberculosis and most other mycobacterial diseases depends upon the demonstration of acid-fast bacilli in microscopic preparations and the culture of the organisms from pathological material that usually contains many other, and more rapidly growing, micro-organisms. By the 1950s satisfactory technical methods for these procedures had been developed, but problems then arose with the identification of the acid-fast bacilli in the cultures and in testing their susceptibility to chemotherapeutic agents.

It had become apparent that *Mycobacterium tuberculosis* was not the only acid-fast organism that could be seen in and cultured from morbid material. Other species were recognised, variously called 'atypical' or 'anonymous'' mycobacteria, although most of them have now been given specific names. Some of these mycobacteria are now recognised as opportunist pathogens, but others, which gain access to pathological material during collection and processing, are of no clinical significance.

The chemotherapy of tuberculosis, as with other bacterial diseases, resulted in the emergence of drug-resistant strains of tubercle bacilli. It was feared that these strains would spread among the population, and in 1962 a Drug Resistance Register was instituted and held at the Public Health Laboratory Service's Tuberculosis Reference Laboratory in Cardiff. Physicians were requested to send two specimens from each patient thought to be excreting drug-resistant tubercle bacilli to the Cardiff laboratory. A report

Isolation and Identification of
Micro-organisms of Medical and
Veterinary Importance

Copyright © 1985 by the Society for Applied Bacteriology
All rights of reproduction in any form reserved
0-12-181460-2

from this laboratory (Marks 1965) revealed serious discrepancies between its results and those of the laboratories which first isolated and tested the organisms. For example, of 82 patients registered as having organisms "resistant" to streptomycin, 41 were found to be sensitive at Cardiff, and in 11 patients mycobacteria other than tubercle bacilli were isolated. Clearly such errors could seriously affect the health of the patients.

Accurate and reliable identification and drug sensitivity tests are important in the successful treatment of tuberculosis and other mycobacterial diseases. Reasons for discrepancies in the results obtained from different laboratories became apparent when their techniques were examined. They included inadequate identification procedures; varying, and sometimes unsatisfactory ranges of drug concentrations; unequal methods of heating to coagulate the medium; storage of media for periods longer than the half-life of the drugs; inadequate controls; and small numbers of tests, which effectively prevented "in-line" controls.

It was clear that standardised techniques, reliable controls, an adequate number of tests in any one batch, and experienced workers were all necessary to ensure reliability, and in 1966 the Public Health Laboratory Service (PHLS 1966) designated six of its laboratories in England as Regional Centres for Tuberculosis Bacteriology. In Wales, the Tuberculosis Reference Laboratory (now the Mycobacterium Reference Unit) was already acting as a Regional Centre in addition to its function as the Reference Laboratory for England and Wales. Hospital and public health microbiologists were invited to send cultures of mycobacteria, for identification and sensitivity tests, to the Centre of their choice. There was no "regionalisation" in spite of the names given to the centres.

Almost immediate use was made of this service, and the numbers of cultures received have increased considerably since the service was first offered. Figures for 1982 show that approximately 3500 cultures are received annually at Cardiff, 2600 at Dulwich, and that the other centres receive about 200 each. This in addition to specimens received for primary isolation of mycobacteria. It is estimated that in England and Wales 90% or more of all mycobacterial identification and sensitivity tests are done at the PHLS Centres. In a survey of the notification of tuberculosis conducted by the Medical Research Council (MRC) in the last 3 months of 1978 and the first 3 months of 1979 (MRC 1980), 1611 (96%) of 1676 cultures for identification and sensitivity tests were examined at PHLS Centres (MRC, personal communication).

Although techniques for identifying mycobacteria vary slightly in detail, sensitivity tests in all centres are done by the resistance ratio method, using the modal resistance technique of Marks (1961) and Leat & Marks (1970). This has given consistently reliable and reproducible results within and between laboratories over a number of years and has been assured by a quality control system. Each week the Reference Unit sends four cultures of *M. tuberculosis* to the Regional Centres. These are used as controls for the modal resistance method and the results are returned to the Reference Unit for statistical analysis. Local deviations are easily recognised and long-term trends may be observed. In addition, under the Resistance Register system, the Regional Centres send to the Reference Unit all cultures of resistant organisms isolated from new cases of tuberculosis. Finally, other mycobacteria which are difficult to identify can be referred to the Reference Unit for specialist examination.

At the Reference Unit mycobacteria are identified as far as is clinically necessary by cultural, biochemical and chromatographic techniques. Sensitivity tests are done against isoniazid (INAH), ethambutol (Eb), rifampicin (R), streptomycin (Sm), pyrazinamide (Z) and when required, ethionamide (Ea), capreomycin (Ca) and cycloserine (Cy). The Dulwich laboratory uses biochemical tests to identify myocbacteria and does sensitivity tests against INAH, Eb, R, Sm and Z and refers other requests to the Reference Unit. The other centres screen cultures for *M. tuberculosis* (as follows) and test tubercle bacilli against INAH, Eb, R and Sm. Other mycobacteria and cultures for other sensitivity tests are sent to the Reference Unit. In addition to this work, the Reference Unit and the Regional Centres continue to examine locally collected pathological material for mycobacteria.

The methods described in this chapter are those in current use at the Cardiff and Dulwich laboratories (Marks 1961, 1976; Leat & Marks 1970; Jenkins *et al.* 1982; Collins & Lyne 1984).

Safe Handling of Tuberculous Materials

Laboratory-acquired tuberculosis is a constant hazard among laboratory workers who handle tuberculous material (Collins 1982, 1983). Tubercle bacilli (i.e. *M. tuberculosis, M. bovis, M. africanum*) are Risk Group 3 (Category B) organisms, initiating infection mainly by the air-borne route. As most manipulations with material and cultures may produce aerosols and infected air-borne particles, all work with these organisms should be done in microbiological safety cabinets (usually Class I or Class II) in Containment or Category B laboratories under appropriate conditions (DHSS 1978; Collins 1983). Infections can also occur by ingestion and accidental self-inoculation. Mouth pipetting should therefore be banned, the use of hypodermic needles reduced to a minimum, and glass pipettes etc. replaced by plastic varieties. Plastic disposable loops avoid the use of Bunsen burners and possible spattering of infected material, as well as being rigid and therefore better tools for handling sputum and similar material. When fluids are decanted after centrifugation, they should be poured into disinfectant through a funnel which dips into the liquid to avoid aerosol production. The lip of the tube should be wiped with blotting paper which is then discarded into the disinfectant. Saturated mercuric chloride solution should be used instead of water for making films for microscopy. This kills the organisms promptly and reduces the risk of dispersing air-borne particles.

Selection of Specimens

Sputum

Leak-proof, wide-mouth containers should be provided and patients should be instructed to produce sputum, not saliva. A single specimen will miss about 25% of microscopically positive and about 50% of culture positive cases. Efficient laboratory diagnosis requires at least three specimens from ordinary cases, six from "difficult" and an unlimited number from "problem" cases.

When sputum is not available

Gastric contents may be collected, especially from children, early in the morning and before food is taken. The patient should cough and swallow for 10 min before collection. If the gastric contents are scanty the stomach may be washed out with sterile distilled water.

Laryngeal swabs are less efficient than sputum or gastric contents. Some samples of cotton wool kill tubercle bacilli. It is best to extract the wool with alcohol before making the swabs. Tracheal or bronchial lavages are sometimes collected and fibreoptic bronchoscopy is an excellent method of obtaining specimens from specific sites in the lung.

Urine

Early morning urine, the first 50 ml passed after the patient's rising, should be collected on 3 successive days. Midstream specimens are best; 24-h specimens are not recommended.

Pleural and other fluids

These should be collected in two separate containers, one plain bottle to allow the specimen to clot and the other containing 0.2 ml of 20% w/v sodium citrate to prevent clotting.

Faeces

Examination is rarely necessary and should be avoided. Most cases of tuberculous enteritis have lung lesions, and tubercle bacilli in faeces can come from swallowed sputum.

Pus, tissues, blood and marrow

As much pus as possible should be collected. Swabs of pus are unsatisfactory. Tissues should be selected with care by a histopathologist. Examination of blood is of little value; bone marrow is more useful, but in disseminated tuberculosis better results are obtained by examining urine and gastric contents.

Milk

The examination of bulk milk is useless. Samples should be taken from individual udders, or better still quarters.

Direct Microscopic Examination

Sputum

Direct microscopy is much less sensitive than culture. It needs 50,000 acid-fast bacilli/ml of sputum to give a positive direct film, although 10,000/ml will give a 50% chance of finding the bacilli. Careful selection of caseous or purulent particles give the best results. These should be spread carefully on clean slides (one slide for each specimen), using a plastic disposable loop. The films should be dried in air in the safety cabinet.

The patient whose sputum is positive on direct smear is likely to infect his or her close contacts. It is therefore important that smear-positive cases be identified quickly. It is now accepted that approximately 3 weeks after commencement of treatment the infectivity of such patients is greatly diminished, even though their sputum may still be smear positive. Presumably the majority of the bacilli seen in such smears are dead (British Thoracic Society 1983).

Pleural and other fluids and pus

Films should be made from the centrifuged deposits of fluids as well as from clots. A loopful of the deposit from a spinal fluid should be placed in a small area marked with a diamond and allowed to dry. Further loopfuls may then be added and dried. Pus should be spread on slides as for sputum.

Other specimens

There is little to be gained by examining direct films of other materials. False positives, due to the presence of saprophytic mycobacteria and acid-fast artefacts, may be obtained from urines, faeces, tissues, blood and milk. Such material should be cultured.

Staining methods

Ziehl-Neelsen (ZN) method

The staining solution contains (g or ml/litre in distilled water): basic fuchsin, 10; crystalline phenol, 50; 95% ethanol, 100. The fuchsin and phenol are dissolved in the ethanol with a warm water bath and then added to the water.

The decolorising agent is acid alcohol made by mixing 30 ml of concentrated HCl with 970 ml of 95% ethanol. The counterstain may be either 0.5% methylene blue or malachite green in water. Colour-blind workers should counterstain with 0.75% picric acid in water.

Films are stained for about 4 min with the carbol fuchsin solution which has previously been heated in a test tube. They are then washed with water, decolorised with the acid alcohol for about 20 s and counterstained for a few minutes. Alternatively, the slides are flooded with the stain and heated twice with a Bunsen flame until steam rises. They are then washed with water, treated with the acid alcohol until all traces of pink colour have disappeared, and counterstained for about 30 s.

Wide-field microscopy (e.g. with a 3.5-mm oil immersion lens) is advisable, and films should be searched for several minutes before giving a negative report.

Fluorescent method

The auramine–phenol stain contains (g/litre in distilled water): auramine, 3; crystalline phenol, 30. The phenol is dissolved in the water by gentle heat and the dye added, a little at a time, with vigorous shaking. The decolorising agent consists of (g or ml): concentrated HCl, 5; NaCl, 5; distilled water, 250; 95% ethanol, 750. The quenching solution is potassium permanganate, 0.1% in water.

Films are stained with the auramine–phenol for 4 min, washed in water, decolorised for 4 min and washed with the permanganate solution. They are examined by fluorescent microscopy using a low-power lens to locate fluorescent material and then with a high-power (dry) lens to distinguish the bacilli. All positives should be confirmed by the ZN method. It is possible to overstain auramine-stained films by the ZN method.

Culture of Mycobacteria

Culture media

The media used for the isolation of mycobacteria are Lowenstein–Jensen, Marks' buffered egg and Kirchner broth containing an antibiotic mixture.

Lowenstein–Jensen medium (LJ)

This modification contains no starch. The basic mineral salt solution contains (g or ml/litre in distilled water): KH_2PO_4, 4.0; $MgSO_4{\cdot}7H_2O$, 0.4; magnesium citrate, 1.0; asparagine, 6.0; glycerol, 20.0. For culturing *M. bovis*, the glycerol is replaced by sodium pyruvate, 7.0 g. The ingredients are dissolved in this order and the mixture is steamed for 2 h.

Fresh hen eggs are cleaned with soapy water, rinsed in tap water, and then steeped in alcohol for 15 min. The alcohol is allowed to evaporate and the eggs are then broken into a sterile jar containing large glass beads, shaken well and filtered through muslin to remove any shell and chalaza. Egg fluid, 1600 ml, is mixed with 1 litre of mineral salt solution and 40 ml of 1% malachite green. The mixture is then tubed in 5-ml amounts and inspissated at 80°C for 50 min to form slopes.

Acid egg medium

Acid egg medium no. 2 (Zaher & Marks 1977) contains: KH_2PO_4, 2.4 g; $MgSO_4{\cdot}7H_2O$, 0.3 g; and glycerol, 12 ml. These are dissolved in 600 ml of water and autoclaved at 110°C for 15 min. The following are then added, aseptically: whole egg, mixed and strained through gauze, 1000 ml; 1 mole/litre HCl, 40 ml; malachite green (2% w/v), 10 ml; sodium penicillin, 100,000 units. In the pyruvate version the glycerol is replaced by sodium pyruvate, 7.0 g. The medium is distributed and inspissated as already described for LJ medium.

Kirchner medium

This fluid medium (PHLS 1958) contains (g/litre in distilled water): KH_2PO_4, 2; $Na_2HPO_4{\cdot}12H_2O$, 19; $MgSO_4{\cdot}7H_2O$, 0.6; sodium citrate, 2.5; asparagine, 5.0. The ingredients are dissolved by heat; glycerol, 20 ml, and phenol red (0.4% w/v), 3.0 ml, are added. It is distributed in 9-ml amounts in screw-capped bottles and autoclaved at 115°C for 10 min. Sterile horse serum, 1 ml, is added to each tube.

Mitchison *et al.* (1983) recommend the addition of the following antibacterial compounds to suppress the growth of unwanted micro-organisms: polymixin B, 200,000 units/litre; carbenicillin, 50 mg/litre; trimethoprim, 10 mg/litre; ampicillin, 10 mg/litre.

Sputum and gastric contents

Sputum may be transferred from a specimen container to the processing container with sterile wooden swab sticks or pipettes made by cutting glass tubing into 20-cm lengths. One end is smoothed in a flame to accept a rubber teat and the other end left sharp. With the teat compressed, the sharp end is dipped into the sputum, the teat is released, and the pipette is rotated to cut the mucoid material. Enough sputum for culture may be obtained in this way.

Method 1

Sputum or gastric contents, about 2 ml, is mixed with 2 ml of 4% w/v NaOH solution in a screw-capped bottle. The bottle is placed in an incubator at 37°C for 15–30 min (no longer) depending on the speed of liquefaction. The alkali is then neutralised by adding 3 ml of 1 mole/litre K_2HPO_4 (about 14% w/v in distilled water coloured with phenol red, bottled in 3-ml amounts and autoclaved). The bottle is centrifuged (in a sealed centrifuge bucket) at 3000 rev/min for 20 min. The supernatant is decanted into disinfectant, and the deposit is divided between two slopes of LJ medium, one containing glycerol and the other pyruvate (Collins & Lyne 1984).

Method 2

The procedure is the same as in method 1, but the alkali is diluted with sterile distilled water, 16 ml, and after centrifugation the deposit is seeded on two slopes of acid egg medium, one containing glycerol and the other pyruvate (Zaher & Marks 1977).

Laryngeal swabs

The swab is suspended in 3% oxalic acid for 40 min at room temperature and then drained and rubbed on two slopes of LJ medium (one glycerol, one pyruvate).

Pleural and other fluids

The degree of contamination with non-acid-fast bacteria can be assessed by culturing the centrifuged deposit on blood agar and incubating overnight. Meanwhile, the fluid should be refrigerated. If the blood agar shows no growth of other organisms the deposit is seeded directly on LJ medium. Any clot should be broken up (e.g. with sterile scissors) and cultured. As mycobacteria adhere to glass it may be advantageous, especially with spinal fluids, to add Kirchner medium containing the antibiotic mixture to the original container after the fluid has been removed for treatment.

If there is growth on the blood agar plate, 2 ml of 2% sulphuric acid is mixed with the deposit. After standing at room temperature for 10 min the material is divided between several slopes of LJ medium.

If the material is bulky, thick or purulent, or there are large clots, 4% sulphuric acid should be used and the exposure time increased to 20 min.

Pus

The general procedure is the same as with fluids, omitting the centrifugation.

Urine

The degree of contamination may be assessed by examining a wet preparation of the deposit after centrifuging at 3000 rev/min for 20 min. The deposit may also be cultured on blood agar. The remainder of the deposit is refrigerated overnight. It is then mixed with 2 ml of 4% sulphuric acid and transferred to another bottle with a Pasteur pipette. After standing for 15–40 min depending on the number of non-acid-fast organisms present, it is diluted with 16 ml of sterile distilled water, centrifuged and the deposit cultured on several slopes of LJ medium (glycerol and pyruvate).

Tissue

Tissue should be cut into small pieces and homogenised in a suitable (e.g. Griffiths) grinding tube in a safety cabinet. The homogenate is then treated like pus or thick pathological fluids.

Milk

Milk is the only specimen that requires guinea pig inoculation, as it is very difficult to decontaminate. The milk is allowed to stand overnight in the cold. The gravity cream is removed and about 50 ml of the skimmed milk is centrifuged at 3000 rev/min for 20 min. The deposit is combined with about 2 ml of the gravity cream and injected into the thigh muscle of a guinea pig (two milks may be tested in one animal). The animal is observed daily, and killed and autopsied after 6 weeks or earlier if it shows any signs of enlarged lymph nodes, draining local lesion, etc. The autopsy should be done in a safety cabinet and the affected glands and tissue removed and cultured as above.

Storage of specimens

The untreated remains of specimens may be stored in a refrigerator for 7 days. If, at the first examination of the cultures (see next section), the slopes are contaminated, the stored specimens can be given the maximum treatment. This avoids the necessity of requesting further specimens. Surgical specimens are often difficult to repeat, and storage of a portion at −4°C is useful in case the cultures are lost at any time during the incubation.

Incubation and examination

All cultures should be incubated at 37°C. Duplicate cultures from superficial lesions should be incubated at 30–33°C because certain opportunist pathogens (*M. ulcerans*, *M. marinum* and some members of the *M. fortuitum–chelonei* complex) do not grow at 37°C in primary cultures, although the last two organisms will usually adapt to the higher temperature after laboratory subculture.

The cultures should be examined after 3 and 7 days and thereafter weekly for 6–8 weeks. Although tubercle bacilli may take 3–6 weeks to show any growth, the early examination allows prompt recognition of contamination and rapidly growing mycobacteria. Cultures that show growth of white, buff or yellow colonies or effuse growth are removed for microscopy and further tests.

Identification of Tubercle Bacilli

In the United Kingdom tubercle bacilli account for over 90% of all mycobacteria isolated. It is therefore necessary to have a simple screening system that can identify them quickly and accurately. The following tests allow a preliminary report to be given after 3 days stating that the organisms are likely to be tubercle bacilli or are probably "opportunist" mycobacteria. The organisms can be confirmed as tubercle bacilli in 18–21 days (Marks 1976).

Microscopy

ZN-stained films of all "positive" cultures should be examined to determine (1) that the organisms are acid-fast bacilli; (2) that contaminants are not present; and (3) the morphology of the bacilli. The ease with which the organisms emulsify in the mercuric chloride solution when films are made for the ZN stain and the microscopic appearance of the organisms are noted. Tubercle bacilli do not emulsify readily but form solid clumps and usually show evidence of "cording", i.e. the arrangement of the organisms in parallel lines resembling pieces of rope. Other mycobacteria usually emulsify easily and show a different morphology, although this is not enough evidence for identification to species level.

Culture

Suspensions of the organisms are made in small bottles containing a phosphate buffer (pH 7.4), a few small glass beads and a small wire nail. They are placed on a magnetic stirrer for a few minutes and then allowed to stand for the same time.

The following media are inoculated: (1) LJ, for incubation at 37°C in the light; (2) LJ, for incubation at 25°C (not room temperature, which varies); (3) LJ containing 500 mg/litre of *p*-nitrobenzoic acid (PNB) for incubation at 37°C; (4) LJ containing 10 mg/litre of thiacetazone (TSC) for incubation at 37°C.

The primary culture is also re-incubated, this time in continuous light.

Cultures may be exposed to light during incubation by keeping the hot room light on all the time, or by placing a "cold" miniature fluorescent strip lamp in an ordinary incubator (Collins & Lyne 1984).

If a decision is needed quickly, the slopes may be examined after incubation for 3 days. At this time, tubercle bacilli will give no apparent growth. Taken together with the absence of pigment in the reincubated primary culture and an acceptable morphology, an interim report may be issued, stating that "So far the strain has the cultural characteristics of tubercle bacilli".

After incubation for 7 days, a tubercle bacillus will show some growth on the LJ (control) slope incubated at 37°C but no growth on the other slopes. At this stage a more definite report may be issued. After 14–21 days a tubercle bacillus will have grown well on the control slope but not on any of the others.

Opportunist mycobacteria will either have had a distinctive morphology, and/or be pigmented and/or show growth on the other screening slopes in 3–7 days. The report on such an organism will depend on its identity, which may be determined within a few days or weeks of its isolation (to be discussed further).

Typing tubercle bacilli

It is not usually necessary to identify tubercle bacilli as *M. tuberculosis*, *M. bovis* or *M. africanum* and their geographic variants except possibly for epidemiological purposes. Indeed it may cause confusion among physicians and lead to unnecessary contact-tracing investigations, as *M. bovis* infections do not necessarily have their origin in milk and *M. africanum* infections are not confined to those who dwell in Africa (Collins & Grange 1983; Collins *et al.* 1982). A typing system is given here, however, which also permits the identification of Bacille Calmette Guerin (BCG).

A suspension of the organisms, prepared as already described, is used to inoculate the following:

(1) A semi-solid medium made by adding 0.1% (pure) agar and 10% horse serum to Kirchner's medium. This is dispensed in 10-ml amounts in 15-ml screw-capped bottles. The inoculum is carefully mixed with the medium, avoiding air bubbles, and the bottles are incubated undisturbed for 21 days. Growth at the surface of the medium indicates that the organisms are aerobic. A band of growth 1 cm or more below the surface indicates that they are microaerophilic. (This is the oxygen preference test of Marks 1976.)
(2) LJ medium containing thiophene 2-carboxylic acid hydrazide (TCH) at 5 mg/litre. This is examined for growth after incubation for 21 days (Harrington & Karlson 1966).
(3) Middlebrook 7H9 medium (Difco) for the nitratase test of Virtanen (1960).
(4) Pyrazinamide sensitivity media (to be discussed further).
(5) LJ medium containing cycloserine at 20 mg/litre (Rist *et al.* 1967).

The results are read at 21 days. Table 1 shows the reactions of the various species and strains.

There is some doubt about the specific rank of *M. bovis* and *M. africanum*. Some workers believe that these are infraspecific variants of *M. tuberculosis* (Collins *et al.* 1982).

TABLE 1

Variants of Mycobacterium tuberculosis

Variant	Oxygen preference*	TCH†	Nitratase	Pyrazinamide	Cycloserine
Classical human	A	R	+	S	S
Asian human	A	S	+	S	S
African I	M	S	−	S	S
African II	M	S	+	S	S
Bovine	M	S	−	R	S
BCG	A	S	−	R	R

* A, Aerobic; M, microaerophilic; R, resistant; S, sensitive.
† Thiophene-2-carboxylic acid hydrazide, 5 mg/litre.

Identification of "Opportunist" Mycobacteria

It is important that the opportunist mycobacteria which are capable of causing human and animal infections are identified to species. It is less important to identify other mycobacteria, and group or trivial names are sufficient, except possibly when the organisms are recovered from patients who have a natural or induced immunoincompetence. The two approaches to identification in the PHLS are (1) cultural and (2) thin-layer chromatography of lipids.

Cultural identification

Suspensions of the organisms are prepared as already described and used to inoculate the following media, which are incubated at 37°C unless otherwise indicated.

(1) Two LJ slopes; one is incubated in continuous light (see above) and the other is wrapped or incubated in a box so that it is not exposed to light. Pigment production is compared at 18–21 days. Photochromogens produce a yellow pigment only on slopes exposed to light. Scotochromogens produce a yellow or orange pigment on both slopes.
(2) Five slopes of LJ medium; one slope is incubated at each of the temperatures 20, 25,37, 42, and 44 or 45°C in water baths or incubation blocks which have accurate thermostats.
(3) An LJ slope containing thiacetazone (TSC) at 10 mg/litre.
(4) An LJ butt for the catalase test; when good growth is observed, 2 ml of a mixture of equal parts of hydrogen peroxide (100 volumes) and Tween 80 (0.5%) is added and the height of the foam in millimeters recorded after 5 min.
(5) A tube of Middlebrook 7H9 medium containing 6.4 g/litre of potassium phenolphthalein disulphate; when good growth is obtained (3–4 days for rapidly growing organisms, 14–21 days for others), a few drops of ammonia are added. A pink colour of varying intensity indicates the presence of an arylsulphatase.
(6) A tube of Middlebrook 7H9 medium; after incubation for 14–18 days, a few drops of 0.4% potassium tellurite are added. A black precipitate, appearing after a further 5-day incubation, indicates ability to reduce tellurite.
(7) A tube of Tween medium; this is a 0.6 mole/litre phosphate buffer containing 0.5% Tween 80 and 2% of a 0.1% solution of neutral red. It is heavily inoculated, e.g. using a throat swab. A change in colour from yellow to red or cherry red indicates the ability to hydrolyse Tween.
(8) A tube of Middlebrook 7H9 medium; after 18–21 days of incubation, two drops of 0.1% potassium nitrate are added and the culture is re-

TABLE 2

Properties of some mycobacteria: A, opportunist pathogens; B, not likely to be clinically significant

Type	Mycobacteria*	Pigment†	Growth at 20°C	25°C	37°C	42°C	44°C	TSC†	Catalase‡	Sulphatase§	Tellurite reduction	Tween hydrolysis‖	Nitratase
A	*Mycobacterium kansasii*	P	−	+	+	−	−	S	+++	+	NA	+++	+
	M. marinum	P	+	+	V	−	−	R	++	+	NA	++	−
	M. xenopi	Poor or Sc	−	−	+	+	+	R	−	+++	−	+	−
	MAIS	Poor or Sc	V	+	+	+	V	V	−	V	+	−	−
	M. malmoense	None	−	+	+	−	−	S	−	−	+	+ (Late)	+
	M. simiae	P	−	+	+	−	−	V	++	+	−	−	−
	M. szulgai	Sc	−	+	+	−	−	R	+	+++	NA	++	+
	M. fortuitum	None	+	+	+	V	−	R	+++	+++	+	+	+
	M. chelonei	None	+	+	V	−	−	R	+++	+++	+	+	−
B	Scotochromogens	Sc	+	+	V	−	−	R	++	+	NA	+++	−
	Nonchromogens	None	+	+	+	−	−	R	V	+	−	++	V
	M. smegmatis and *M. phlei*	V	+	+	+	+	+	R	++	+	+	+	+

* *Mycobacterium marinum* grows only at 30–33°C on primary culture; MAIS is *M. avium–intracellulare–scrofulaceum* complex; scotochromogens include *M. gordonae* and *M. flavescens*; nonchromogens includes *M. terrae, M. gastri, M. nonchromogenicum.*

† P, Photochromogen; Sc, scotochromogen; S, sensitive; R, resistant; V, variable; NA, test not done on pigmented organisms.

‡ Foam: +++, 10 mm; ++, 5–9 mm; +, 1–4 mm; −, none.

§ +++, Deep pink, ++, pink; +, faint pink; −, no colour.

‖ +++, Cherry red; ++, red; +, pink; −, yellow.

TABLE 3

Some opportunist pathogenic mycobacteria

Species	Other names	Probable source	Pathogenicity
Mycobacterium kansasii	"*M. luciflavum*"	Water	Pulmonary (rarely elsewhere)
M. xenopi	"*M. xenopei*"	Water	Pulmonary (rarely elsewhere)
*M. avium–intracellulare**	*M. avium*, *M. intracellulare*, "*M. marianum*"	Birds, Pigs, Soil, Water	Pulmonary in adults, cervical adenitis in children
*M. scrofulaceum**			
M. simiae	"*M. habana*"	?	Pulmonary (rare)
M. malmoense		?	Pulmonary in adults, cervical adenitis in children
M. szulgai		?	Pulmonary (rare); very rare elsewhere
M. fortuitum	"*M. ranae*", "*M. peregrinum*"	Soil, Water	Superficial trauma; injection abscesses
M. chelonei	"*M. abscessus*"		

* Usually considered together as "*Mycobacterium avium–intracellulare–scrofulaceum*" (MAIS) complex.

incubated for a further 4–18 h. This is used for the nitratase test of Virtanen (1960).

The results of these tests are recorded after 18–21 days and the identity of the organism is deduced from Table 2. Brief information about the species and groups is given in Table 3. It should be noted that a single isolation of an opportunist mycobacterium does not necessarily have any clinical significance, as these organisms are frequently present in the environment and often gain access to clinical material during collection and processing. Only if several consecutive sputum specimens yield the same organism should it be considered as possibly significant. Opportunist mycobacteria in urine and gastric washings are virtually never significant. Isolation from other sites has to be considered in the light of the clinical presentation.

Identification by lipid pattern using thin-layer chromatography

Identification of opportunist mycobacteria by lipid patterns from thin layer chromatography has been described by Marks & Szulga (1965) and Jenkins (1980).

Preparation of extract

Eugonic mycobacteria are grown on LJ medium, and dysgonic organisms are grown in a modified Kirchner medium containing tryptone (Oxoid) instead of asparagine and bovine serum albumin (0.8%) instead of horse serum.

The organisms are harvested by scraping from the egg medium or taking the centrifuged deposit from the Kirchner. The bacilli are transferred to a tared sample tube and dried over phosphorus pentoxide in a vacuum desiccator at 60 cm Hg. Their weight is then found. At this stage they may be stored at 4°C for several months.

The extract is made by adding a freshly prepared mixture of diethyl ether : ethanol : water (17 : 17 : 6 v/v) to the dried bacilli in the proportion 16 μl/mg dry weight within the range 10–25 mg. The tubes are stoppered well and left at room temperature overnight.

Silca gel plates

A slurry is made by shaking 30 g of silica gel with 65 ml of distilled water for 60 s in a 150-ml stoppered flask. There is a variation between different batches of gel, and it may be necessary to vary the volume of water to obtain an optimum consistency so that the slurry spreads evenly and is free from trapped air bubbles. Glass plates 20 × 10 cm, previously cleaned with dichromate solution, are coated with a layer of slurry approximately 0.1 mm thick using Unoplan equipment (Shandon). The plates are left until they are matt and then dried for 30 min in a hot-air oven fitted with a fan. They are washed in acetone, dried at 100°C for 30 min, and activated by heating at 120°C for 30 min. They may be stored over silica gel until needed, but preferably no longer than 7 days.

Application of extract

The extract is applied as four superimposed aliquots of 2.5 μl (10 μl altogether), allowing natural drying between each application. For uni-dimensional chromatograms, the extracts are applied 15 mm apart.

Developing solvents

Uni-dimensional runs using a variety of solvents give satisfactory results when known and unknown strains are run together. Solvents in current use are (1) *n*-propanol : water : ammonia (75 : 22 : 3), (2) *n*-propanol : water : *n*-butanol : ammonia (57 : 20 : 20 : 3), (3) *n*-propanol : water : di-isopropyl ether (45 : 10 : 45), (4) *n*-propanol : water : ammonia : isopropyl ether (45 : 9.5 : 0.5 : 45).

Ammonia (specific gravity 0.880) is used. The di-isopropyl ether must be added slowly to the other ingredients with continuous shaking.

Location of spots

The chromatograms are run for 10 cm, dried at 100°C for 30 min, allowed to cool, and then sprayed with a mixture of two parts of 60% sulphuric acid and one part of freshly prepared 0.1% orcinol in water. The plates are then heated on a hot plate at about 140°C until the spots appear. Over-spraying should be avoided, as it makes all the spots the same brown–black colour. The chromatogram is placed in a glass dish, the bottom of which has been ruled in 5-cm squares with a diamond marker. This is illuminated from below and the position and shape of the spots can be seen clearly and recorded on graph paper in the appropriate colours.

Lipid patterns

As well as specific lipids, all extracts contain contributions from the media on which the cultures are grown and also non-specific or unimportant lipids. These are all easily recognised and rarely cause any confusion. The phenol red in the medium used for dysgonic strains provides a useful marker, as the small amount in the extract appears as a narrow band in the centre of unidimensional chromatograms run in solvents 1 and 2, and in some cases the specific lipids can be related to it.

Antituberculous Drug Sensitivity Tests

In the PHLS, the modal resistance ratio method of Marks (1961) is used. In this, the minimal inhibitory concentration (mic) of the drug for the unknown strain is compared with that for a representative group of recently isolated and known drug-sensitive strains using the same batch of medium and performed at the same time and under the same conditions.

Isoniazid, rifampicin, ethambutol and streptomycin

The same technique is used for all of these. The concentrations of the drugs in the media have to be determined for each laboratory and for each new batch of the chemical. Those in current use are shown in Table 4. The drugs are added to liquid LJ medium and the medium is tubed and inspissated under standard conditions. The final media are stored at 4°C for no more than 21 days before use. Isoniazid, ethambutol and streptomycin are kept as stock solutions in water at −20°C; rifampicin is dissolved in formdimethylamide when required.

Inoculation and incubation

Suspensions of mycobacteria are prepared as described under *Identification* using about 3 mg of bacilli for each test. They are left standing for several

TABLE 4

Sensitivity tests: concentrations of drugs in Lowenstein–Jensen medium

	Final concentrations, μg/ml, in tube number				
Drug	1	2	3	4	5
Isoniazid	0.014	0.028	0.56	0.112	0.224
Ethambutol	0.4	0.8	1.6	3.2	6.4
Rifampicin	4	8	16	32	64
Streptomycin	2.5	5	10	20	40

minutes after mixing to allow any clumps to settle. One drug-free slope and three slopes containing doubling concentrations of each drug (Table 4) are inoculated each with approximately 2.5 μl of the suspension using a Repette (Jencon, Leighton Buzzard, Bedfordshire) and a fresh sterile disposable plastic pipette tip for each suspension. The tips are discarded into disinfectant and subsequently autoclaved. They may be re-used.

The tests are incubated at 37°C for 21 days but are examined after 3, 7 and 14 days to detect contamination and any evidence of resistance which is reported to the physician and which enables other tests to be started if necessary.

Control strains: modal resistance

At least four strains of tubercle bacilli, recently isolated and found to be sensitive to the drugs, are supplied each week by the Mycobacterium Reference Unit to PHLS Regional Centres for Tuberculosis Bacteriology. These are included in each batch of tests, using five doubling dilutions of the drugs determined by initial experiment so that growth appears usually in the two lowest concentrations but not in the next three.

The mic used in calculating the resistance ratio is the modal average of the results for the individual strains, as shown in Table 5.

Recording and reporting

Ideally the drug-free slope should show a confluent growth (CG) or innumerable discrete colonies (IC). Slopes giving 20–100 colonies are reported as positive and those with less than 20 as trace. Only tests where the control is CG or IC are recorded. Any others are repeated.

The mic of the test strain is divided by that of the modal control to give ratios of 1, 2, 4 or 8. Strains giving a resistance ratio of 1 or 2 are reported as *sensitive*, those giving a ratio of 4 as *resistant*, and those giving a ratio of 8 as *highly resistant* (Table 6).

TABLE 5

Example of a modal resistance control set

	Control†	Drug concentration* for tube number 1	2	3	4	5
Strain A	CG	CG	CG	–	–	–
B	IC	IC	+	–	–	–
C	CG	CG	IC	–	–	–
D	CG	CG	IC	+	–	–
E	IC	IC	–	–	–	–
Mode		CG	IC	–	–	–

* Doubling concentrations; see Table 4.
† CG, confluent growth; IC, innumerable discrete colonies; +, 20–100 colonies; –, less than 20 colonies.

Ethionamide, cycloserine and capreomycin

The technique is the same as that already described for the other drugs, but if the resistance ratio appears to be between 2 and 4 resistance should be suspected.

Pyrazinamide

The mic method is unsatisfactory largely because the drug is potentiated at an acid pH. It is thought that it acts on the bacilli that are within mac-

TABLE 6

Sensitivity tests on tubercle bacilli: recording and interpretation

	Control	Drug concentration for tube number* 1	2	3	4	5	Resistance ratio	Interpretation
Mode	CG/IC	CG	IC	–	–	–		
Test strain 1	CG/IC			–	–	–	1	Sensitive
2	CG/IC			+	–	–	1	Sensitive
3	CG/IC			CG/IC	–	–	2	Sensitive
4	CG/IC			CG/IC	+	–	3	Borderline
5	CG/IC			CG/IC	CG/IC	–	4	Resistant
6	CG/IC			CG/IC	CG/IC	CG/IC	8	Highly resistant

* CG, confluent growth; IC, innumerable discrete colonies; +, 20–100 colonies; –, less than 20 colonies.

rophages. Tests therefore have to be made at pH 5.2, but this presents problems because egg medium is reluctant to coagulate when heated at this pH. The problem was overcome by increasing the concentration of egg in the medium, decreasing the concentration of malachite green and by heating for a longer time. Less malachite green is bound at an acid pH and the excess is inhibitory to tubercle bacilli. Control and drug-containing media are made at three different pH levels ranging from 5.0 to 5.4. These are supplemented by a semisolid agar medium for the "stepped pH" technique of Marks (1964). This method is used at the Mycobacterium Reference Unit but is complicated and is not recommended for general use. The following modification of Yates (cited in Collins & Lyne 1984), which is a modification of the "stepped pH" method of Marks (1964), has given satisfactory results at the Dulwich Regional Centre.

LJ medium with half the usual concentration of malachite green is prepared, adjusted to pH 5.2, and divided into two batches. Kirchner medium containing 0.1% agar is also adjusted to pH 5.2 and divided into two batches. After autoclaving, bovine albumin (or OADC supplement, Difco) is added. One batch of LJ and one of Kirchner each receive 65 mg/litre of pyrazinamide. The LJ medium containing the drug is tubed in 1-ml amounts and inspissated in an upright position. When it has set, 2 ml of the Kirchner medium containing the drug is added. The drug-free medium is made up in the same way.

Two tubes, one with and one without pyrazinamide, are each inoculated with 0.02 ml of the suspension. Another two more, paired, are each inoculated with 0.02 ml of a 1 : 10 dilution of the suspension in sterile distilled water. This takes care of possible overgrowth by vigorously growing organisms and poor growth by dysgonic bacilli.

The sets are incubated at 37°C for 18–21 days. Good growth in the drug free tubes and none in the pyrazinamide tubes indicates that the organisms are sensitive, and growth in all four tubes that they are resistant. If the set that received the neat suspension shows resistance but the set inoculated with the diluted suspension shows growth in the drug-free tube but not in the pyrazinamide tube, the strain is reported as sensitive.

Problems and Difficulties

These are usually the result of attempts to undertake identification and, in particular, sensitivity tests when only small numbers of strains are available. The advantages of having all these tests done in laboratories where large numbers of strains are tested are that there are adequate controls, in the form of peer samples, stock cultures, comparisons of results with others in the same and different batches ("line controls"), as well as exchanges of cultures between the specialists concerned.

References

British Thoracic Society 1983 Control and prevention of tuberculosis: a code of practice. *British Medical Journal* iv, 1118–1121.

Collins, C. H. 1982 Laboratory-acquired tuberculosis. *Tubercle* **63,** 151–155.

Collins, C. H. 1983 *Laboratory-acquired Infections*. London: Butterworths.

Collins, C. H. & Grange, J. M. 1983 A review: The bovine tubercle bacillus. *Journal of Applied Bacteriology* **55,** 13–29.

Collins, C. H. & Lyne, P. M. 1984 *Microbiological Methods* 5th edn, pp. 373–395. London: Butterworths.

Collins, C. H., Yates, M. D. & Grange, J. M. 1982 Subdivision of *Mycobacterium tuberculosis* into five variants for epidemiological purposes; methods and nomenclature. *Journal of Hygiene* **89,** 235–242.

DHSS 1978 *A Code of Practice for the Prevention of Infection in Clinical Laboratories*. Department of Health and Social Security. London: HMSO.

Harrington, R. & Karlson, A. G. 1966 Differentiation between *Mycobacterium tuberculosis* and *Mycobacterium bovis* by *in vitro* procedures. *American Journal of Veterinary Research* **27,** 1193–1196.

Jenkins, P. A. 1980 Thin layer chromatography of mycobacterial lipids as an aid to classification. In *Mykobakterien und Mykobakterielle Krankheit,* ed. Meissner, G., Schmiedel, A., Nelles, A. & Pfaffenberg, R., Part 1, pp. 305–318. Jena: Fischer Verlag.

Jenkins, P. A., Pattyn, S. R. & Portaels, F. 1982 Diagnostic bacteriology. In *The Biology of the Mycobacteria,* ed. Ratledge, C. & Stanford, J. R., pp. 441–470. London & New York: Academic Press.

Leat, J. & Marks, J. 1970 Improvements of drug sensitivity tests on tubercle bacilli. *Tubercle* **51,** 68–75.

Marks, J. 1961 The design of sensitivity tests on tubercle bacilli. *Tubercle* **42,** 314–316.

Marks, J. 1964 A stepped pH technique for the estimation of pyrazinamide sensitivity. *Tubercle* **45,** 47–50.

Marks, J. 1965 Bacteriological aspects of the national drug register of drug resistant tuberculosis in its first two years. *Monthly Bulletin of the Ministry of Health and Public Health Laboratory Service* (*Great Britain*) **24,** 2–6.

Marks, J. 1976 A system for the examination of tubercle bacilli and other mycobacteria. *Tubercle* **57,** 207–225.

Marks, J. & Szulga, T. 1965 Thin layer chromatography as an aid to the classification of mycobacteria. *Tubercle* **46,** 400–411.

Mitchison, D. A., Allen, W. B. & Manickavasagar, D. 1983 Selective Kirchner medium for the culture of specimens other than sputum for mycobacteria. *Journal of Clinical Pathology* **36,** 1357–1361.

MRC 1980 National survey of tuberculosis notifications in England and Wales 1978–79 (Medical Research Council). *British Medical Journal* **281,** 895–898.

PHLS 1958 Cultivation of *Mycobacterium tuberculosis*. Part VI. Choice of culture media for the isolation of tubercle bacilli from sputum. *Monthly Bulletin of the Ministry of Health and Public Health Laboratory Service* (*Great Britain*) **17,** 112–118.

PHLS 1966 Regional centres for tuberculosis bacteriology. *Monthly Bulletin of the Ministry of Health and Public Health Laboratory Service* (*Great Britain*) **25,** 36.

RIST, N., CANETTI, G., BOISVERT, H. & LE LIRZIN, M. 1967 L'antibiogramme du BCG. Valeur diagnostique de la resistance a la cycloserine. *Revue de Tuberculose et de Pneumologie* **31**, 1060–1062.

VIRTANEN, S. 1960 A study of nitrate reduction by mycobacteria. *Acta Tuberculosea Scandinavica, Supplementum* No. 48, 1–119.

ZAHER, F. & MARKS, J. 1977 Methods and medium for the culture of tubercle bacilli. *Tubercle* **58**, 143–145.

Primary Isolation of *Chlamydia trachomatis* and *Chlamydia psittaci* in *in Vitro* Cell Culture

SHIRLEY J. RICHMOND, JUDITH M. G. BAILEY, A. S. BAILEY AND GILL MEARNS

North Manchester Regional Virus Laboratory, New Holzel Laboratory, Booth Hall Children's Hospital, Manchester, UK

Chlamydia are Gram negative bacteria which parasitise the cytoplasm of eukaryotic cells. They cannot be grown on synthetic media, as they have no means of generating their own energy-rich phosphates; they rely entirely on the eukaryotic host for their energy requirements, and are therefore obligate intracellular parasites. All chlamydia share a common, lipopolysaccharide antigen, which is located in the outer membrane of the cell envelope, and they undergo a unique developmental cycle within host cell cytoplasm (for a recent review, see Ward 1983).

The genus is of both medical and veterinary importance: *Chlamydia trachomatis* infects man and causes trachoma, lymphogranuloma venereum (LGV) and sexually transmitted infections of the lower genital tract. The latter infections, which are exceedingly common, cause considerable morbidity in both adults and neonates in western societies at present (see reviews in Oriel 1982; Westrom and Mardh 1982; Beem & Saxon 1982). It was the need for feasible isolation techniques for the study of genital chlamydia in human populations that has provided much of the impetus for the development of modern isolation techniques. The other species in the genus, *C. psittaci*, infects many different avian and mammalian hosts and causes a wide spectrum of disease in domestic animals (Stortz 1971). Man is not a natural host, and mammalian *C. psittaci* strains have little or no pathogenicity for man. In contrast, many avian *C. psittaci* strains are capable of invading and damaging the human respiratory tract, causing the zoonosis ornithosis (psittacosis). The main divisions of the genus are shown in Fig. 1.

Until the late 1960s diagnosis of chlamydial infections in both man and

Isolation and Identification of
Micro-organisms of Medical and
Veterinary Importance

Copyright © 1985 by the Society for Applied Bacteriology
All rights of reproduction in any form reserved
0-12-181460-2

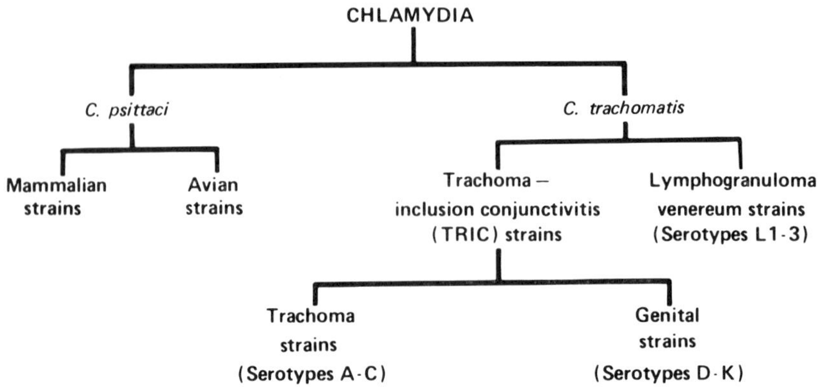

FIG. 1. Classification of the genus *Chlamydia*.

animals depended either on demonstration of the organisms directly in host tissues, or on their isolation in the yolk sac of the embryonated hen's egg; attempts until then to isolate the organisms in cell culture were unsuccessful. In 1969, however, primary isolation of genital *C. trachomatis* strains in *in vitro* cell cultures was achieved, by centrifugation of clinical specimens on to monolayers of irradiated (and therefore non-replicating) McCoy cells, before the cultures were incubated (Gordon *et al.* 1969). The use of centrifugation to initiate productive infection of the eukaryotic host cell, and growth of chlamydia in non-replicating cells are now both recognised as essential components of successful *in vitro* isolation techniques. Isolation of *C. trachomatis* from clinical material has become a routine procedure in many medical microbiology laboratories, and, with minor modifications, similar procedures can be used for isolation of *C. psittaci.*

This chapter concentrates on the factors which are likely to be important for the isolation of all chlamydia from their naturally infected hosts, drawing largely on experience gained in recent years from isolation of the *C. trachomatis* species from man, but also emphasising special factors which must be considered when attempting the isolation of *C. psittaci.* Detailed information on the composition of suitable transport medium, cell growth and maintenance media, staining techniques and a summary of a suitable isolation protocol is listed in later sections of this chapter.

The Chlamydial Developmental Cycle

An understanding of this cycle is essential for successful isolation and manipulation of chlamydia *in vitro*. The main stages in the cycle are shown diagramatically in Fig. 2, and the major morphological features of intracellular

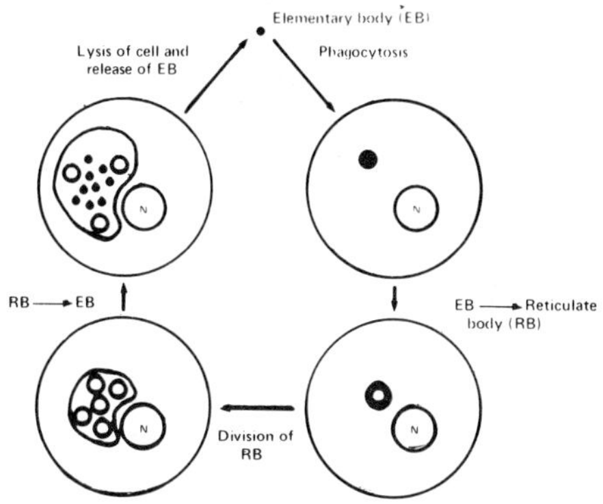

FIG. 2. Diagram of the chlamydial developmental cycle.

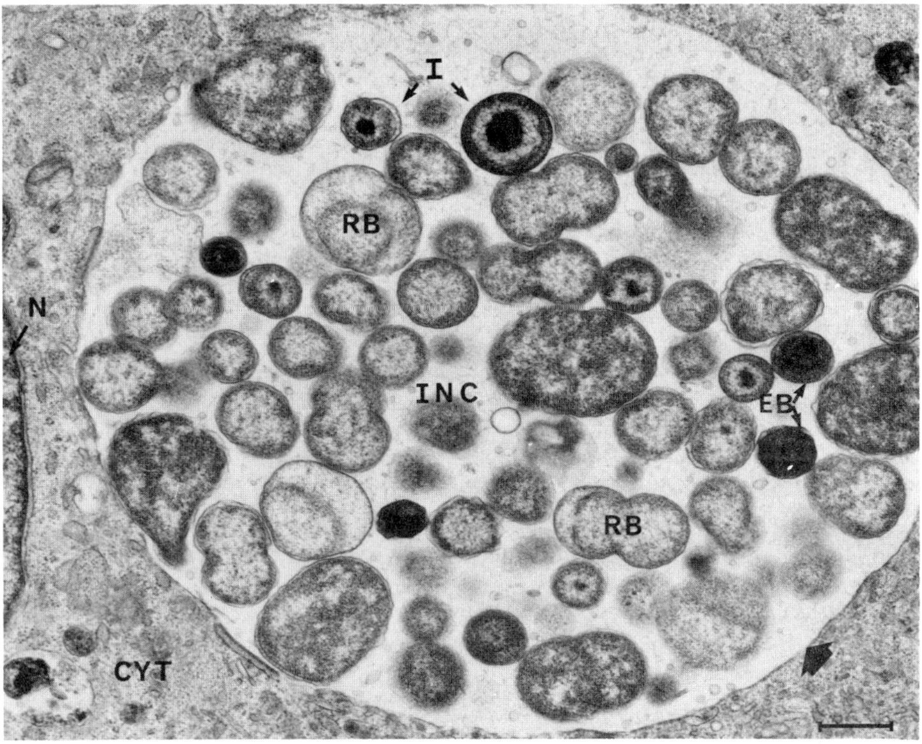

FIG. 3. McCoy cell 30 h after infection with *Chlamydia trachomatis*. The chlamydial inclusion (INC) contains predominantly reticulate bodies (RB), some of them dividing; elementary bodies (EB) and intermediate forms (I), representing RBs in the process of condensation into EBs, are also present. N = McCoy cell nucleus; CYT = McCoy cell cytoplasm; ➧ = inclusion membrane; bar = 500 nm. (Electron micrograph kindly supplied by Penny Stirling, Department of Microbiology, University of Bristol.)

chlamydial development, demonstrated by transmission electron microscopy in cultured cells 30 h after infection, are shown in Fig. 3.

Chlamydia exist in two distinct morphological forms, known respectively as the elementary body (EB) and the reticulate body (RB). The infectious particle, the EB, is a metabolically inert, electron-dense form about 300 nm in diameter, which penetrates the susceptible host cell by phagocytosis. The resulting phagosome evades fusion with host cell lysosomes, and all subsequent chlamydial development takes place within this phagosome, which is known as the chlamydial inclusion. After phagocytosis, the EB is converted into the larger (1000 nm), fragile, metabolically active but non-infectious form, the RB, which divides by fission within the phagosome. For the first 20 h the phagosome contains RBs only, many of them undergoing division. After this time, condensation of RBs back into EBs begins, so that by 2 days in culture the greatly enlarged phagosome contains a microcolony of pleomorphic chlamydial forms, in which infectious EBs predominate. The cycle comes to an end when the host cell lyses to release a new generation of EBs into the extracellular environment.

The mature inclusion is the hallmark of a chlamydial infection; diagnosis depends on identification of such inclusions, either (as in the past) directly in cells obtained from the infected host, or (as more recently) in cells infected *in vitro* with clinical material.

Safety Considerations

As many chlamydia are pathogenic for man, and as the pathogenicity of other strains is uncertain, these organisms should always be handled carefully, with good working practices observed at all times. The major risk from the *C. trachomatis* organisms commonly handled in clinical microbiology laboratories is to the eye, and care must always be taken to avoid accidental inoculation of the conjunctiva. Some *C. psittaci* strains, e.g. the feline pneumonitis strain, may also cause follicular conjunctivitis in man (Schachter *et al.* 1969).

Some chlamydia may also infect the human respiratory tract. Avian *C. psittaci* strains are most likely to do this, and these should always be handled in Class I microbiological safety cabinets.

Principle of Isolation Technique

The principle of *in vitro* isolation of chlamydia is to obtain infectious EBs from the infected host, to inoculate susceptible cell monolayers with these EBs, and then to generate mature inclusions in the cultured cells by incubating them for about 2 days, the time taken for one developmental cycle to take place. Monolayers are then fixed, stained with one of a variety of inclusion-specific stains, and examined by microscopy appropriate for the particular stain that is chosen (light, dark-ground or fluorescence microscopy).

Care must be taken to avoid the use of any antibiotics which act on chlamydia, both in the transport medium and also in cell growth and maintenance media. Penicillin, which is frequently incorporated in cell cultures used for virus isolation, must be avoided; like other β-lactam antibiotics, this profoundly modifies the chlamydial developmental cycle. Aminoglycosides such as streptomycin, vancomycin or gentamicin are suitable alternatives to penicillin, and antifungal agents such as amphotericin B should be included as well if the specimen is likely to be contaminated with yeasts.

Cell culture

Chlamydia have been isolated in a wide range of different cells, for example HeLa 229 (Kuo *et al.* 1972), human thyroid (Hobson *et al.* 1976), human amnion (Harrison & Riggin 1979), buffalo green monkey (Hobson *et al.* 1982), and mouse epithelial cells (Keski-Oja & Paavonen 1982). However, McCoy cells, a continuous line of mouse fibroblasts similar to and probably derived from L cells, have been by far the most frequently used for primary isolation of *C. trachomatis*. We are therefore limiting this chapter to a description of the isolation of chlamydia in these cells.

The original method of Gordon *et al.* (1969) used McCoy cells which had been irradiated with 5000 rads 5–7 days before use. This dose of irradiation produces large, multinucleated, non-replicating cells, in which inclusions develop well and are clearly visible. Irradiation was soon superceded by other, more convenient methods of producing static monolayers. Treatment of cells with 5-iodo-2-deoxyuridine (IUDR) (Wentworth & Alexander 1974) or cytochalasin B (Sompolinsky & Richmond 1974) was introduced as feasible alternatives in 1974, and treatment with cycloheximide (CH) in 1977 (Ripa & Mardh 1977). Although these different regimens affect the eukaryotic cell in different ways, they all ultimately cause inhibition of host cell macro-molecular synthesis, so that a non-replicating monolayer is produced. Irradiation and treatment with IUDR or cytochalasin all require pre-treatment of the cells before inoculation of chlamydia; CH, however, which acts on the eukaryotic ribosomes and therefore selectively inhibits host cell but not chlamydial protein synthesis, can be incorporated in the cell maintenance medium immediately after phagocytosis of chlamydia has occurred, and no pre-treatment of the cells is necessary. So although all these alternative treatments produce cells of comparable sensitivity for the isolation of

C. trachomatis (La Scolea & Keddell 1981), CH treatment is the simplest method to use, and is described here.

It must be emphasised that many chlamydia, including *C. trachomatis*, can be grown in conventional, replicating McCoy cell monolayers (Johnson & Hobson 1976), but nevertheless there are several reasons why non-replicating cells are preferable. Restriction of host cell metabolism prevents the medium from becoming too acid; this is important when a CO_2 incubator is not used, as *C. trachomatis* is inactivated at pH 6 (Johnson & Hobson 1976). Moreover, inclusions are easier to see in monolayers which are not overgrown. In addition to these factors, actual growth of many chlamydial strains is enhanced in CH-treated cells (Spears & Stortz 1979). This is probably because both the eukaryotic host and chlamydia require many of the same essential metabolites, in particular certain amino acids, for macro-molecular synthesis, and chlamydial access to some of these substances is limited under conditions of normal host cell growth (Moulder 1982). Restriction of host-cell requirements for these essential macro-molecular precursors appears to make them more readily accessible for chlamydial metabolism (Hatch 1975). The particular amino acids which may limit chlamydial growth vary from one chlamydial biotype to another (Pearce & Allan 1982); so CH, by reducing the requirements of the eukaryotic cells in the culture for all amino acids, enhances the growth of many chlamydial strains of both species. CH should therefore be incorporated in the maintenance medium of cell cultures when isolation of any chlamydia is attempted from the naturally infected host.

Effect of centrifugation

Chlamydia gain access to host cell cytoplasm by phagocytosis, and for reasons which are not yet fully understood, the growth of chlamydia in cultured cells is greatly enhanced by centrifugation of EBs onto the monolayer after inoculation. This is particularly true for ocular and genital strains of *C. trachomatis*, where centrifugation enhances the uptake of inclusion-forming units about 1000-fold. But *in vitro* growth of both LGV strains of *C. trachomatis* and also many *C. psittaci* strains is also increased by centrifugation, though to a lesser extent. Therefore, this step should always be included when isolation of chlamydia from clinical material is attempted, as the number of infectious particles in the specimen may be low.

There is considerable uncertainty about the optimum centrifugal force that should be used (Darougar & Treharne 1982). It is likely, however, that this is not critical, as comparable *C. trachomatis* isolation rates have been obtained with centrifugal forces which vary from 2000 to 15,000 **g**. The importance of centrifugation probably lies in the fact that it brings the EBs in close contact with the McCoy cell membrane, and thereby facilitates rapid phagocytosis of the parasite by the host cell. In practice, most laboratories find a force of about 2000 **g** for 1 h at 30–34°C is adequate; it is very important, however, that the centrifuge should not overheat, as most chlamydia are rapidly inactivated at temperatures of 40°C and above.

Clinical Specimens and Transport Methods

Clinical specimens containing infectious EBs can be obtained either by swabbing accessible infected mucosal surfaces (for example, conjunctiva, cervix, respiratory tract, rectum, cloaca, etc.) or by releasing EBs from infected tissue (for example, spleen, placenta). As EBs are fragile organisms which will survive only for limited periods outside living hosts, particular care must be taken to ensure that viable specimens reach the laboratory. Sterile cotton-wool swabs should be used to obtain specimens from mucosal surfaces; wood swabs are usually adequate, but wire swabs are preferable for narrow orifices (e.g. male urethra). Swabs should be placed in transport medium (see below), and, provided the specimen can be cultured within 24–48 h, it can be held until then at 4°C. For longer periods of storage it should be deep frozen (−70 or −160°C) as soon as possible after the specimen is obtained. A transport medium in which chlamydia will survive freezing should therefore be used. Two such media utilise 0.2 *M* sucrose and 10% sorbitol respectively as the cryoprotective agent. The sucrose medium has been widely and successfully used; it is, moreover, economical, and therefore particularly useful in laboratories where a large amount of clinical material is examined. The sorbitol medium, which also contains fetal calf serum, is more expensive, but organisms will survive several cycles of freezing and thawing in it, as well as surviving deep frozen for years.

If isolation is attempted from tissue, this should be fresh and sent to the laboratory as soon as possible. Intracellular EBs are released by homogenisation of the cells in transport medium. The supernatant from the tissue is then used to inoculate cell cultures.

Transport media

Sucrose phosphate transport medium

Sucrose buffer (2 *M*) is prepared by dissolving 342.3 g of sucrose, 10.44 g of K_2HPO_4, and 5.44 g of KH_2PO_4 in distilled water and making the volume up to 5 litres. This is autoclaved at 115°C for 15 min and stored at or below 4°C. The following antibiotics are added, per milliliter: streptomycin, 5μg; vancomycin, 100 μg; amphotericin B, 5 μg. The transport medium is then dispensed in 2-ml amounts and stored at −20°C.

Sorbitol transport medium

This contains Earle's saline (e.g. Oxoid), 100 ml; fetal calf serum, 10 ml; NaHCO (4.4% w/v), 8 ml. Antibiotics, as for the sucrose medium, are added, the medium is dispensed in 2-ml amounts, and it is stored at −20°C.

Isolation

McCoy cell media

Growth medium

This consist of Eagles' minimal essential medium (MEM, available commercially), 90 ml; $NaHCO_3$ (4.4% w/v), 5 ml; fetal bovine serum, 7.5 ml; streptomycin, 50 μg/ml; vancomycin, 100 μg/ml.

Maintenance medium

This is similar to growth medium except that it contains 1% rather that 7.5% of fetal bovine serum and additional glucose, 2.5 ml of a 10% w/v solution in MEM, per 100 ml of medium.

Cycloheximide

The stock solution contains 100 μg/ml. It is dispensed in 1-ml amounts and stored at −20°C. The final concentration in maintenance medium should be 1 μg/ml.

Monolayers

Confluent monolayers of McCoy cells are prepared on circular glass coverslips in flat-bottomed tubes. These tubes may be obtained from Gibco Europe and a sterile glass coverslip (Chance Propper) must be added. Alternatively, tubes complete with cover glasses may be obtained from Laboratory Sales, Whitworth Road, Rochdale, Lancs.

Procedure

The cell cultures, one or two per specimen, are each inoculated with 0.25 ml of the material. Two cultures per specimen should be used if serial passage is required. The tubes are centrifuged at 2000 **g** for 1 h at 34°C and then incubated at 35°C for 1 h. The medium in each tube is then replaced by 1 ml of maintenance medium containing cycloheximide. The cultures are incubated at 35°C for 2–3 days.

The medium is then removed, the monolayer fixed in methanol for 10 min, and the cells stained by immunofluorescence, Giemsa, or iodine methods.

Recognition of Inclusions

Although all chlamydia develop within a phagosome derived from the host cell membrane, the actual phagosome morphology varies considerably from one strain to another. All *C. trachomatis* organisms form discrete inclusions, with clearly defined margins; these lie immediately adjacent to the host cell nucleus, which is often indented by the developing phagosome. *Chlamydia psittaci* inclusions, which are often more densely packed with chlamydia than *C. trachomatis* inclusions, vary considerably in their morphology, from round, compact bodies similar to *C. trachomatis* to much more diffuse, lobulated inclusions. Each strain, however, is consistent in its particular inclusion morphology, and a classification of mammalian strains into seven different biotypes, based largely on differences in the appearance of the inclusions, has been described by Spears & Stortz (1979).

Various techniques are available for demonstration of inclusions at different stages of their development. These are summarised in Table 1, and detailed staining procedures described next.

TABLE 1

Recognition of inclusions

Stain	Stage of development detected	Use
Indirect IF (anti-sera to chlamydial genus antigen)	Early and late inclusions	Sensitive diagnostic stain for all chlamydia
Giemsa (DG)	Late inclusion ("autofluorescence" of EB)	Diagnostic stains for *Chlamydia trachomatis*
Iodine (light microscopy)	Late inclusion (glycogen matrix)	
Methylene blue (DG)	Late inclusion ("autofluorescence" of EB)	Diagnostic stain for *C. psittaci*
Giemsa (light microscopy)	Early and late inclusions	Good stains for host cell and inclusion morphology
Modified Ziehl–Neelsen (light microscopy)		
Acridine orange	Early inclusions	Good RB stain

Indirect immunofluorescence (IF) staining

Indirect IF staining of monolayers with anti-serum to the genus antigen is probably the most sensitive technique available for detecting chlamydial inclusions of either species. This is a more complicated procedure than some of the alternative stains, and it requires relatively expensive reagents, such as fluorescein-conjugated immunoglobulins. As the outer membrane of both RBs and EBs contains the genus antigen, however, inclusions of all strains can be detected at all stages of their development once they become large enough to be seen by fluorescence microscopy. Moreover, specificity of the stain can be checked if necessary by incorporating appropriate negative serum controls. Suitable sera with high titres of antibodies to the genus antigen can be obtained from various sources, e.g. human sera from patients with psittacosis or lymphogranuloma venereum, or sheep sera from animals infected with the agent of ovine chlamydial abortion. Alternatively, hyper-immune rabbit serum can easily be prepared by repeated immunisation of rabbits with partially purified chlamydial EBs propagated in yolk sacs (Richmond 1980). In addition, fluorescein-conjugated monoclonal antibodies to the genus antigen are likely to become commercially available in the near future, which would transform the IF stain into a simple, one-stage procedure.

IF staining method

The anti-serum, which has a high titre of antibodies to the chlamydia genus-specific antigen, is diluted for use in phophate-buffered saline (PBS) at pH 7.4 to the highest dilution that gives bright inclusion fluorescence with a known isolate. Appropriate anti-species FITC-conjugated immunoglobulins are diluted for use with PBS.

After fixation the methanol is removed and the cells washed with PBS. Diluted anti-serum, 0.1 ml, is added. The monolayers are shaken to ensure that the cells are covered and are incubated for 30 min. The anti-serum is then removed and the cells are washed twice with PBS. Diluted conjugate, 0.1 ml, is added and the tubes are shaken and incubated at 35°C for 30 min. The conjugate is then removed and the cells washed twice with PBS.

The coverslip is removed from the tube and mounted, cell side down, on a glass slide in a mixture containing equal parts of glycerol and PBS. The preparation is scanned by fluorescent microscopy at ×100 magnification; if inclusions are seen they are confirmed using the high-power lens. Inclusions appear as discrete, bright green granular intracytoplasmic bodies.

Iodine, Giemsa and Ziehl–Neelsen staining

The two classical methods of recognising *C. trachomatis* inclusions are (1) staining the glycogen matrix of the inclusion with iodine (Fig. 4*a*) or some

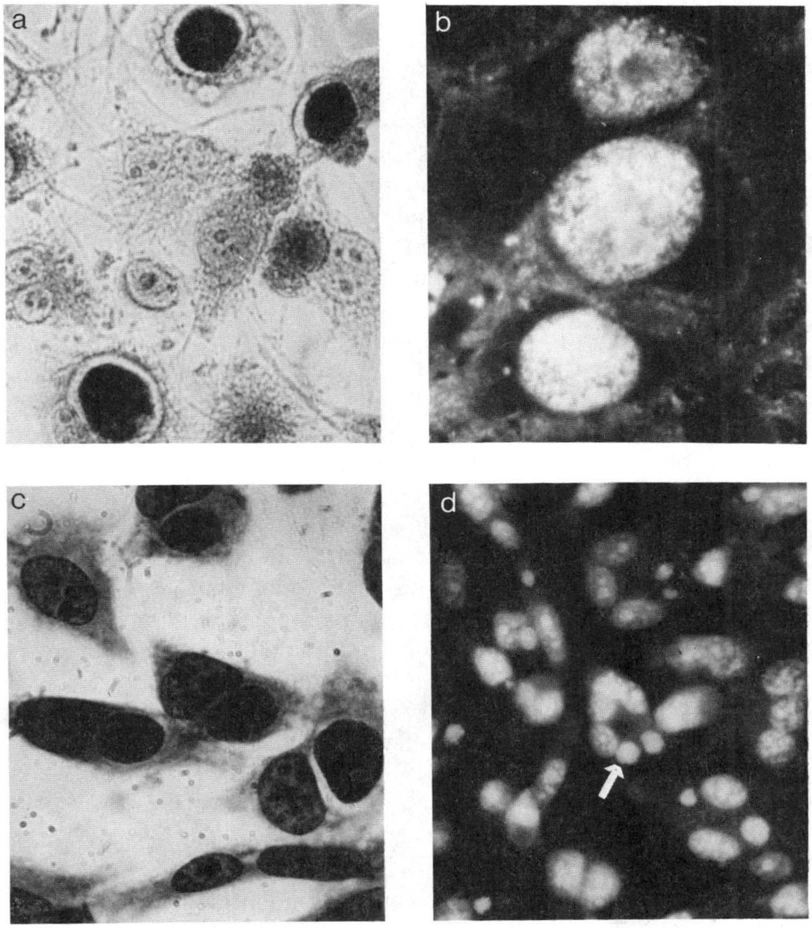

FIG. 4. Chlamydial inclusions in McCoy cells stained by (*a*) iodine (*Chlamydia trachomatis*, 48 h, light microscopy, ×1000); (*b*) Giemsa (*C. trachomatis*, 68 h, DG microscopy, ×1600); (*c*) Giemsa (*C. psittaci*, feline strain, light microscopy, ×1000); and (*d*) acridine orange (*C. trachomatis*, 22 h, fluorescence microscopy, ×1600; one inclusion is indicated by arrow). (Figure 4*c* kindly supplied by Dr. F. W. A. Johnson, Department of Preventive Veterinary Medicine, University of Liverpool.)

other carbohydrate stain (Johnson 1975; Mallinson *et al.* 1981), and (2) staining by Giemsa and then examining the monolayer by dark-ground (DG) microscopy. Iodine staining is quick and simple to perform, but is not entirely reliable, as the mature inclusion does not always stain consistently; moreover, false-positive results may be obtained if the clinical specimen contains iodine-staining particles (e.g. talcum powder) which adhere to the cells (Hipp *et al.* 1980). The Giemsa stain is slightly more time-consuming, but resulting preparations are very quick to scan if examined by DG microscopy with a wide-field DG condenser. This is because Giemsa-stained EBs within mature inclusions stand out as bright greenish white particles, so that each inclusion appears as a bright granular body (Fig. 4*b*). This phenomenon has been referred to as "auto-fluorescence".

Neither iodine nor Giemsa DG staining techniques identify *C. psittaci* inclusions, as this species does not produce a glycogen-containing matrix, and since "auto-fluorescence" of *C. psittaci* EBs does not occur with Giemsa-stained preparations. Methylene blue, however, is a useful substitute for Giemsa for this species, as EBs stained with methylene blue and examined by DG microscopy do auto-fluoresce, and inclusions can be detected quickly and easily (Johnson *et al.* 1978). Both Giemsa and modified Ziehl–Neelsen stains demonstrate inclusion morphology clearly when preparations are examined by bright light (Fig. 4*c*), although neither of these methods are as sensitive for diagnostic work as fluorescence or DG microscopy.

Giemsa stain

The methanol fixative is removed and the monolayer washed once with phosphate buffer (pH 6.8). It is covered with 1 ml of Giemsa stain diluted $\frac{1}{10}$ with phosphate buffer for about 1 hr at room temperature. The stain is then removed and the monolayers washed in phosphate buffer. The coverslip is removed from the tube, dried in air and mounted in DePex. It is examined by DG microscopy with a wide-field DG condenser, at ×250 magnification. Inclusions stand out as bright, white-green granular bodies.

Iodine stain

Monolayers may be fixed in this stain, which contains ethanol. Iodine solution (KI, 5 g; I_2, 5 g; ethanol, 50 ml; water, 50 ml) is applied for 1–2 min. The iodine is then removed and the coverslip mounted in iodine–glycerol mounting medium (I_2, 5 g; glycerol, 50 ml; water, 50 ml). It is examined by bright-light microscopy using ×250 magnification. Inclusions stain reddish brown, and there is often a clear "halo" between the iodine staining material and the clearly demarcated inclusion margin.

Methylene blue stain

The methanol fixative is removed and the monolayer stained with methylene blue, 0.5% w/v in distilled water, for about 30 s. The stain is removed and the cells washed in tap water. They are decolorised in acid water (0.025% H_2SO_4) for 10–15 s and then rinsed in tap water, dehydrated in acetone followed by xylene, and mounted in DePex. Examination by DG microscopy shows inclusions as bright, whitish blue granular bodies.

Modified Ziehl–Neelsen stain

The methanol fixative is removed and the monolayer stained in dilute (about 0.1%) carbol fuchsin for 7–10 min. It is then washed well in tap water, decolorised in 1% acetic acid for about 10 s, washed well in tap water and counterstained with malachite green, 2% w/v freshly prepared in distilled water. The coverslip is washed in water, dried in air, mounted in DePex and examined by bright light microscopy. Inclusions appear as pink, granular bodies.

Acridine orange stain

Of the various nucleic-acid stains which have been described, acridine orange (AO) is more useful than fluorescent DNA-binding dyes, as it stains EBs and RBs differently, and hence differentiates between early and late inclusions (Pollard & Starr 1962; Stirling & Richmond 1977). RBs are stained red, so early inclusions stand out as conspicuous "red balls" against a pale background of pale red cytoplasm and green cell nuclei (Fig. 4*d*). EBs, in contrast, stain a pale greenish brown, and as inclusions mature they become much less conspicuous. Acridine orange is a useful adjunct to the other stains available, particularly for investigating the inclusion morphology and inclusion contents of newly isolated strains at different times in the cycle.

The monolayer is fixed in acid alcohol (ethanol : glacial acetic acid, 2 : 1) for 5 min, rinsed twice in McIlvaine's citrate–phosphate buffer (pH 3.8), and stained with 0.01% acridine orange for 4 min. The stain is made as a 0.1% stock solution in distilled water and diluted with citrate–phosphate buffer before use. After washing twice in the buffer, the coverslip is mounted in it. Early inclusions stain as distinct red balls; as they mature their colour changes from red to orange to greenish brown.

Passage and Storage of New Isolates

Once chlamydia have been isolated in cell culture, infectivity can usually be increased by passaging the cells and supernatants from infected monolayers into fresh cultures, and then repeating the isolation procedure. Each inclusion-forming unit (IFU) inoculated should generate at least 10^2 IFU, so that cell culture-propagated pools of organisms containing $\sim 10^7$ IFU/ml can be produced.

Monolayers infected with chlamydia are scraped from the coverslips with Pasteur pipettes and suspended in medium in a Whirlimixer. The inoculation, centrifugation and incubation procedures previously described are then repeated.

Isolates are stored by adding sorbitol freezing medium to infected cell culture harvests to bring the final concentration of sorbitol to about 10%. The suspensions are then stored at −70 or −160°C.

Sorbitol freezing mixture

The sorbitol freezing mixture contains Earle's saline, 100 ml; sorbitol, 40 g; foetal calf serum, 10 ml; $NaHCO_3$ (4.4% w/v), 8 ml.

Observations

In this chapter we have concentrated on the important factors that should be considered when isolation of any chlamydia in cell culture is attempted. Various additional points should be borne in mind when *C. psittaci* rather than *C. trachomatis* is investigated. Many strains in the *C. psittaci* species complete the growth cycle more quickly than *C. trachomatis*, so incubation times may need to be adjusted accordingly. Also, many *C. psittaci* strains undergo spontaneous secondary growth cycles in *in vitro* cell cultures (Paul 1982), and so amplification of infectivity (i.e. an increase in the number of inclusions in the monolayer) may be achieved by incubation of the cultures beyond the period of one growth cycle. The incubation temperature may also need to be adjusted if the natural host for the particular strain under study has a higher body temperature than man.

Techniques for isolation of both *C. trachomatis* and *C. psittaci* in cell culture now provide a convenient alternative to the earlier methods of diagnosing chlamydial infections either by direct examination of cells from the infected host or by growth in the yolk sac of embryonated eggs. For *in vitro* isolation to be successful, care must be taken to ensure that these fragile organisms reach the laboratory in a viable state. Transport medium in which chlamydia will survive freezing should be used, as specimens must be stored deep-frozen if they cannot be cultured within 24–48 h. Chlamydia studied to date have all grown in CH-treated McCoy cell monolayers, but to achieve maximum uptake of infectious particles by the host cell it is essential to centrifuge the inoculum onto the monolayer after inoculation. *Chlamydia trachomatis* should be incubated for 2–3 days at 35°C to complete one intracellular developmental cycle *in vitro*, but both incubation time and temperature may need to be adjusted to get optimum results with some *C. psittaci* stains. Indirect IF staining of infected monolayers with serum which contains antibodies to the genus antigen will detect all chlamydial inclusions, and this is probably the most sensitive technique for detecting low levels of infectivity. Other stains are useful adjuncts to IF for study of inclusion contents and inclusion morphology at different stages in the developmental cycle.

Cell culture techniques for diagnosis of *C. trachomatis* infections in humans are now widely used in medicine; similar techniques should also prove useful for the study of the *C. psittaci* strains important in veterinary medicine.

References

BEEM, M. O. & SAXON, E. M. 1982 *Chlamydia trachomatis* infections in infants. In *Chlamydial Infections,* ed. Mardh P.-A., Holmes, K. K., Oriel, J. D., Piot, P. & Schachter, J. pp. 199–212. New York: Elsevier Biomedical Press.

DAROUGAR, S. & TREHARNE, J. D. 1982 Cell culture methods for isolation of *C. trachomatis*—a review. In *Chlamydial Infections,* ed. Mardh, P.-A., Holmes, K. K., Oriel, J. D., Piot, P. & Schachter, J., pp. 265–274. New York: Elsevier Biomedical Press.

GORDON, F. B., HARPER, I. A., QUAN, A. L., TREHARNE, J. D., DWYER, R. ST. C. & GARLAND, J. A. 1969 Detection of *Chlamydia* (*Bedsonia*) in certain infections of man. I. Laboratory procedures: comparison of yolk sac and cell culture for detection and isolation. *Journal of Infectious Diseases* **120,** 451–452.

HARRISON, H. R. & RIGGIN, R. T. 1979 Infection of untreated primary human anmion monolayers with *Chlamydia trachomatis. Journal of Infectious Diseases* **140,** 968–971.

HATCH, T. P. 1975 Competition between *Chlamydia psittaci* and L cells for host isoleucine pools: a limiting factor in chlamydial multiplication. *Infection and Immunity* **12,** 211–220.

HIPP, S. S., KIRWOOD, M. & GUMP, D. W. 1980 Artifacts resembling *Chlamydia trachomatis. New England Journal of Medicine* **302,** 1367.

HOBSON, D., JOHNSON, F. W. A. & BYNG, R. E. 1976 Isolation of *Chlamydia trachomatis* in tissue cultures of human thyroid. *Lancet* **ii,** 643.

HOBSON, D., LEE, N., QUAYLE, E. & BECKETT, E. E. 1982 Growth of *Chlamydia trachomatis* in buffalo green monkey cells. *Lancet* **ii,** 872–873.

JOHNSON, F. W. A. 1975 A comparison of staining techniques for demonstrating group A chlamydia in tissue culture. *Medical Laboratory Technology* **32,** 233–238.

JOHNSON, F. W. A. & HOBSON, D. 1976 Factors affecting the sensitivity of replicating McCoy cells in the isolation and growth of chlamydia A (TRIC agents). *Journal of Hygiene* **76,** 441–451.

JOHNSON, F. W. A., CHANCERELLE, L. Y. J. & HOBSON, D. 1978 An improved method for demonstrating the growth of Chlamydiae in tissue culture. *Medical Laboratory Sciences* **35,** 67–74.

KESKI-OJA, J. & PAAVONEN, J. 1982 Isolation of *Chlamydia trachomatis* in untreated MMC-E mouse epithelial cells. *Journal of Clinical Microbiology* **16,** 391–394.

KUO, C.-C., WANG, S.-P., WENTWORTH, B. B. & GRAYSTON, J. T. 1972 Primary isolation of TRIC organisms in Hela 229 cells treated with DEAE-dextran. *Journal of Infectious Diseases* **125,** 665–668.

LA SCOLEA, L. J. & KEDDELL, J. E. 1981 Efficacy of various cell culture procedures for detection of *Chlamydia trachomatis* and applicability to diagnosis of pediatric infections. *Journal of Clinical Microbiology* **13,** 705–708.

MALLINSON, H., SIKOTRA, S. & ARYA, O. P. 1981 Cultural methods for largescale screening for *Chlamydia trachomatis* genital infection. *Journal of Clinical Pathology* **34,** 712–718.

MOULDER, J. W. 1982 A primer for chlamydiae. In *Chlamydial Infections,* ed. Mardh. P.-A., Holmes, K. K., Oriel, J. D., Piot, P. & Schachter, J., pp. 3–14. New York: Elsevier Biomedical Press.

ORIEL D. 1982 Infections of the male genital tract. In *Chlamydial Infections,* ed. Mardh, P.-A.,

Holmes, K. K., Oriel, J. D., Piot, P. & Schachter, J., pp. 93–106. New York: Elsevier Biomedical Press.

PAUL, I. D. 1982 The growth of chlamydia in McCoy cells treated with emetine. *Medical Laboratory Sciences* **39**, 15–32.

PEARCE, J. H. & ALLAN, I. 1982 Differential amino acid requirements of chlamydiae: regulation of growth and relationship with clinical syndrome. In *Chlamydial Infections*, ed. Mardh, P.-A., Holmes, K. K., Oriel, J. D., Piot, P. & Schachter, J., pp. 29–32. New York: Elsevier Biomedical Press.

POLLARD, M. & STARR, T. J. 1962 Study of intracellular virus with acridine orange fluorochrome. *Progress in Medical Virology* **4**, 54–69.

RICHMOND, S. J. 1980 Chlamydial group antigen in McCoy cells infected with *Chlamydia trachomatis* and *Chlamydia psittaci*. *FEMS Microbiology Letters* **8**, 47–50.

RIPA, K. T. & MARDH, P.-A. 1977 Cultivation of *Chlamydia trachomatis* in cycloheximide-treated McCoy cells. *Journal of Clinical Microbiology* **6**, 328–331.

SCHACHTER, J., OSTLER, H. B. & MEYER, K. F. 1969 Human infection with the agent of feline pneumonitis. *Lancet* **i**, 1063–1065.

SOMPOLINSKY, D. & RICHMOND, S. J. 1974 The growth of *Chlamydia trachomatis* in McCoy cells treated with cytochalasin B. *Applied Microbiology* **28**, 912–914.

SPEARS, P. & STORTZ, J. 1979 Biotyping of *Chlamydia psittaci* based on inclusion morphology and response to diethylaminoethyl-dextran and cycloheximide. *Infection and Immunity* **24**, 224–232.

STIRLING, P. & RICHMOND, S. J. 1977 The developmental cycle of *Chlamydia trachomatis* in McCoy cells treated with cytochalasin B. *Journal of General Microbiology* **100**, 31–42.

STORTZ, J. 1971 *Chlamydia and Chlamydia-induced Diseases*. Springfield, Illinois: Charles C Thomas.

WARD, M. E. 1983 Chlamydial classification, development and structure. *British Medical Bulletin* **39**, 109–115.

WENTWORTH, B. B. & ALEXANDER, E. R. 1974 Isolation of *Chlamydia trachomatis* by use of 5-iodo-2 deoxyuridine-treated cells. *Applied Microbiology* **27**, 912–916.

WESTROM, L. & MARDH, P.-A. 1982 Genital chlamydial infection in the female. In *Chlamydia Infections*, ed. Mardh, P.-A., Holmes, K. K., Oriel, J. D., Piot, P. & Schachter, J., pp. 121–139. New York: Elsevier Biomedical Press.

Mycoplasma pneumoniae

P. TAYLOR AND H. GAYA
Microbiology Department, Brompton Hospital, London, UK

The first mycoplasmas to be described (now known as *Mycoplasma mycoides*) were isolated from cattle by Nocard and Roux in 1898. They were then classified as "pleuropneumonia-like organisms", the initials for which, PPLO, are still associated with the names of some of the culture media used for mycoplasmas.

Mycoplasmas have no cell walls. They are therefore resistant to antibiotics such as penicillin, exhibit a high degree of pleomorphism, particularly when viewed by electron microscopy, and are able to pass through filters that retain other bacteria. It was because of this filter-passing property that they were originally thought to be viruses.

There are several genera and about 70 species of mycoplasmas, and this chapter is mainly concerned with the diagnosis of infections of the respiratory tract caused by *M. pneumoniae.*

Mycoplasma pneumoniae is a pathogen associated with infections such as pneumonia, pharyngitis, tracheobronchitis and myringitis. Very rarely, the central nervous system or the skin are involved. Certain other mycoplasmas are commensals of the respiratory tract. These include *M. salivarium, M. orale, M. buccale* and, occasionally, *M. fermentans* and *M. hominis*. As adequate chemotherapy is available, prompt identification or serological evidence of infection by *M. pneumoniae* is important. Cultural methods, although specific and reliable, require at least 14 days, and therefore it is customary to use serological tests as well as, or instead of, culture. The complement fixation test is usually employed, but indirect haemagglutination (IHA) and enzyme-linked immunosorbent assay (ELISA) are becoming important.

Isolation and Identification of
Micro-organisms of Medical and
Veterinary Importance

Copyright © 1985 by the Society for Applied Bacteriology
All rights of reproduction in any form reserved
0-12-181460-2

Isolation

Culture media

Broth

The liquid medium is a beef extract infusion broth containing peptone and NaCl. It may be obtained commercially as mycoplasma broth or PPLO broth (e.g. from Difco and Gibco).

Soft agar

Agar, e.g. Bactoagar (Difco), is added to the aforementioned broth to give a concentration of about 0.8%. Commercial media are available (Difo, Gibco). Broth and agar should be supplemented with horse serum, yeast extract, substrates (glucose, arginine or urea), an indicator, and inhibitors to discourage the growth of other organisms.

Serum. This provides lipid growth factors and cholesterol, necessary for the growth of all mycoplasmas except *Acholeplasma*. Each batch of serum should be tested for its ability to support the growth of particular mycoplasmas. Although human serum has been used in the culture of mycoplasmas affecting man, specific antibody may be present; horse and pig sera are therefore used by most workers. The serum should be stored at −20°C.

Yeast extract. Some species, including *M. pneumoniae* and *M. orale*, grow poorly in the absence of yeast extract. This extract provides a heat-stable, low molecular weight supplement containing vitamins, purines, pyrimidines and other essential nutrients. Several commercial preparations are available, but freshly prepared extract is recommended as its use allows greater control over preparation and quality control.

Fresh bakers' yeast, 250 g, is suspended in 1 litre of deionised water and autoclaved at 115°C for 10 min. The suspension is then cooled and clarified by centrifugation at 1500 rev/min for 10 min. The supernatant fluid is dispensed in 20-ml volumes, autoclaved again, and stored at −20°C.

To avoid contamination of other media and cell cultures with this yeast, the extract should be prepared in a separate room.

Substrates. Most mycoplasmas may be placed in one of three groups according to whether they ferment glucose, produce ammonia from arginine, and hydrolyse urea. The substrates are prepared as 10% stock solutions and are sterilised by filtration.

Indicator. Phenol red is prepared as a stock stolution by dissolving 1 g of the powder (Gurr, BDH Ltd) in 2.5 ml of normal NaOH and making up to 100 ml with deionised water. The solution is filtered and sterilised by passing it through ordinary filter paper and then through a membrane filter.

Inhibitors. These are (a) thallous acetate, 10% w/v in water, stored at 4°C, (b) penicillin, 100,000 i.u./ml, stored at −20°C and (c) amphotericin B, 0.5 mg/ml, stored at −20°C. These are sterilised by filtration.

Final medium

Agar plates are prepared by melting 70 ml of the base, cooling it to 54°C and adding the following: serum, 20 ml; yeast extract, 10 ml; substrate, 10 ml; phenol red, 0.2 ml; thallous acetate, 0.2 ml; penicillin, 0.2 ml; amphotericin B, 0.1 ml. The final pH of glucose medium should be 7.2; of arginine medium 6.5; and of urea medium 6.8. Plates may be stored at 4°C for 4 weeks (Hayflick 1965).

Broth media are prepared by adding the same amounts of the supplements to 70 ml of broth and dispensing the mixture in 5-ml amounts in 7-ml screw-capped (bijou) bottles. They may be stored at 4°C for up to 4 weeks.

Diphasic medium, selective

Mycoplasma base is dispensed in 1.5-ml amounts in 7-ml screw-capped (bijou) bottles and allowed to set. This is then overlayed with 2.5 ml of a broth medium containing mycoplasma broth base, 30 ml; phenol red (1%), 0.6 ml; thallous acetate (10%), 0.4 ml; glucose (10%), 15 ml; horse serum, 30 ml; yeast extract (25%), 15 ml; methylene blue (0.1%), 1.5 ml; penicillin (100,000 i.u./ml), 0.65 ml.

The colour of the final medium should be adjusted to grey or purple with NaOH. Occasionally it may be necessary to add HCl if the colour goes beyond purple. It may be stored at 4°C for up to 4 weeks (Smith *et al.* 1967).

Diphasic medium, SP4

The solid phase contains the following, dissolved in 615 ml of deionised water (Tully *et al.* 1979): mycoplasma broth base (Difco), 3.5 g; tryptone (Difco), 10.0 g; peptone (Difco), 5.3 g; glucose, 5.0 g; agar (Noble), 8 g. This is held at 56°C and these supplements are added: CMRL 1066 medium, 10× (Gibco Europe), 50 ml; yeast extract (25%), 35 ml; Yeastolate (Difco), 100 ml; fetal calf serum (Flow Laboratories; inactivated), 170 ml; penicillin (100,000 i.u./ml), 10 ml; phenol red (1%), 2.0 ml.

One milliliter of the SP4 agar is placed in a 7-ml screw-capped (bijou) bottle and allowed to set. This is then overlayed with 2 ml of complete SP4 broth (i.e. SP4 medium without the agar).

Methylene blue inhibits the growth of *M. pneumoniae* in this medium.

Choice of specimens

Suitable specimens include sputum, nose and throat swabs, pleural fluid, lung biopsies and, if indicated, cerebrospinal fluid. Sputum is homogenised by shaking it in a suitable container with glass beads and an equal volume of either virus transport medium (VTM: Hanks basal salts plus 0.2% bovine serum albumin but no antibiotics) or in complete mycoplasma broth. Virus transport medium and complete mycoplasma broth are also used as transport media for nose and throat swabs. The swabs are squeezed into the fluid, which is used to inoculate the media.

Inoculation and incubation

Two prepared plates, two tubes of mycoplasma broth, and two tubes of either selective or SP4 broth are each inoculated with 0.1 ml of the prepared specimen.

All cultures are incubated at 37°C, one set aerobically sealed in a plastic bag or tin to prevent dessication. The other is incubated in an atmosphere of hydrogen or nitrogen containing 5–10% carbon dioxide. An anaerobic jar with the GasPak (BBL) system is suitable. *Mycoplasma pneumoniae* and *M. hominis* will grow under both conditions, but some other mycoplasmas will not.

Examination of cultures

Broth and plate cultures are examined twice weekly for up to 3 weeks. Occasionally, for example during epidemiological surveys, diphasic broths are reincubated for up to 6 weeks; this increases the chances of detecting the mycoplasmas.

Acid is produced when mycoplasma grow in the horse serum broth, and the indicator changes colour to orange and finally to yellow. When this change is noted, two mycoplasma agar plates are each inoculated with 0.2 ml of the broth culture. They are incubated aerobically and anaerobically as described above.

Mycoplasma pneumoniae usually grows in the selective diphasic broth within 8–12 days. Acid, produced from glucose, changes the colour of the medium from blue to clear light green. The colour of the agar plug changes from blue to yellow. Bacterial and fungal contamination produces a cloudy green or yellow discoloration of the broth within 24 h, and these tubes should be discarded. Broths showing a distinctive colour change should be subcultured (0.2 ml) onto mycoplasma agar plates.

Although some mycoplasmas form tiny pin-point colonies that can be seen with the naked eye, a low-power microscope is essential to see the finer detail and to distinguish between colonies and artefacts. The plates are examined unopened at a magnification of ×80 or ×100. The surface of the

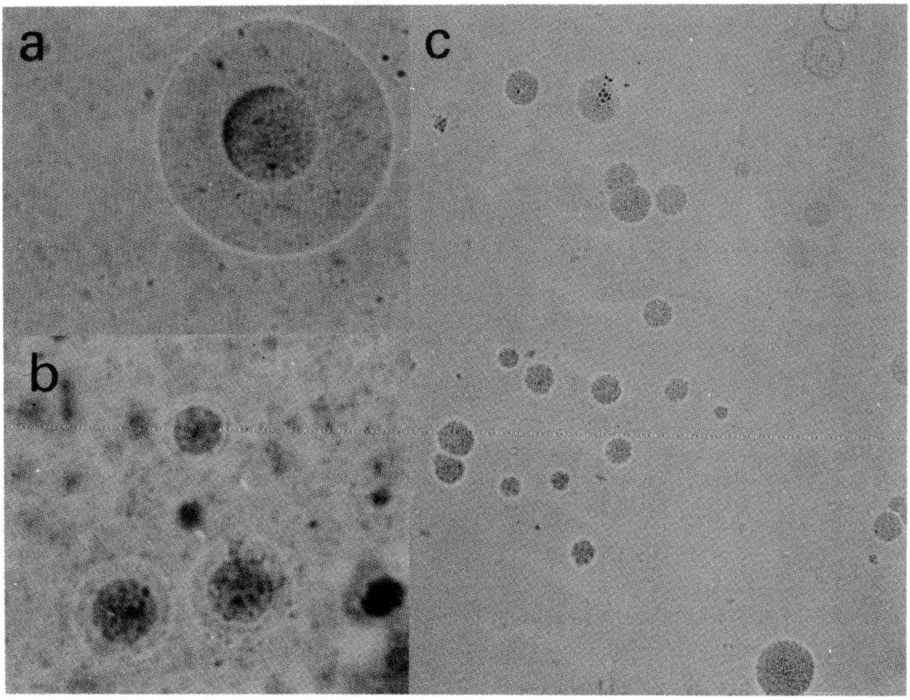

FIG. 1. Colonial morphology of some *Mycoplasma* species growing on agar surfaces: (a) *Acholeplasma laidlawii*, (b) *M. hominis*, (c) *M. pneumoniae (×150).*

medium is scanned for what have been appropriately described as "fried egg" colonies (Fig. 1). There is a dark central zone, usually surrounded by a lighter, peripheral zone. A cross section of a colony is shown in Fig. 2. It will be seen that the central zone corresponds to growth of the colony down into the medium, whereas the peripheral zone is surface growth. Colonies of *M. pneumoniae* growing in agar media in primary culture lack the peripheral zone, but this is formed after several agar to agar subcultures. Colony size varies from 10 to 500 μm.

Artefacts such as serum precipitates, fat globules and cell nuclei (the so-called pseudocolonies) are distinguished from mycoplasma colonies by focussing in different planes. Careful attention to media preparation will avoid the appearance of at least some of these artefacts.

Mycoplasma colonies cannot be removed with a loop because they grow into the agar (Fig. 2). The following method is suitable. A block of agar 5–10 mm square and containing as many colonies as possible is cut from the

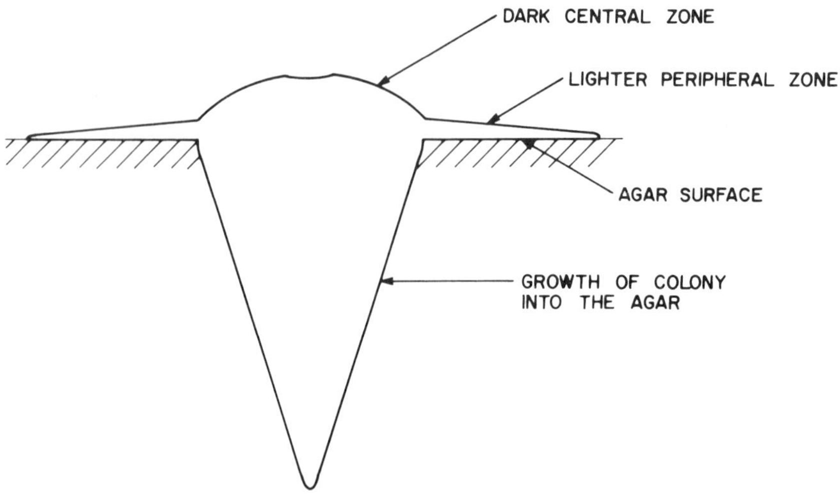

FIG. 2. Section of a mycoplasma colony growing on and into an agar surface.

original plate with a sterile scalpel. The block is inverted on to the surface of a fresh mycoplasma agar plate which is incubated at 37°C for 3–5 days.

The block is then moved around the agar surface and the plate is reincubated. After a further 3–5 days, young colonies of mycoplasmas should appear along the track of the agar block.

Identification

Staining

This method is useful for distinguishing between mycoplasma colonies and artefacts or bacterial colonies.

Dienes stain contains (g/100 ml in distilled water): methylene blue, 2.5; azure II, 1.25; maltose, 10; Na_2CO_3, 0.25; benzoic acid, 0.2.

A block of agar containing the colonies and about 5 mm square is cut from the plate and transferred, colony side up, to a microscope slide. A drop of the stain is placed on a coverslip and this is inverted over the agar block. The preparation is examined with a low-power microscope.

Mycoplasma colonies stain dark blue and the stain is retained for over 24 h. Ureaplasma colonies stain a greenish blue. The stain is eliminated from bacterial colonies within 30 min.

Biochemical and Biological Reactions

Carbohydrates, amino acids and urea are metabolised by certain mycoplasmas of medical interest (Table 1). This provides a convenient method of separating them into groups. Young colonies, prepared by agar block transfer, are seeded into broths containing the substrates. Fermentation of glucose and hydrolysis of urea are indicated by the colour changes of the indicators. Arginine hydrolysis is accompanied by a change to an alkaline pH.

Other properties which may assist in identification include growth rates, aerobic reduction of tetrazolium, inhibition by 0.01% methylene blue, and the presence of various receptors which permit haemadsorption of erythrocytes and cell cultures, which is described below.

Erythrocyte haemadsorption

A 10% suspension of washed erythrocytes in PBS is poured over a plate which shows about 100 young colonies. The plate is incubated at room temperature for 30 min and the suspension of cells is removed. The agar surface is washed with PBS and examined under a low-power microscope for haemadsorption of erythrocytes to the colonies.

Haemolysis of sheep erythrocytes

An agar overlay consisting of 1 ml of 25% washed sheep or guinea pig erythrocytes and 2 ml of mycoplasma agar base (at 45°C) is poured carefully over the plate on which the colonies are growing. After the agar has set, the plate is incubated aerobically at 37°C. Before examination at 24 and 48 h, it should be refrigerated for 30 min.

Clear (β-) haemolytic plaques are produced by *M. pneumoniae*. Other respiratory mycoplasmas give incomplete (α) plaques which are smaller and take longer to develop.

These biochemical and biological tests may aid in the identification of colonies of mycoplasmas but serological methods are necessary for specific identification.

Growth inhibition

A 3- to 7-day-old culture of the isolate is prepared by seeding it with young colonies obtained by the block transfer method. The growth should be dense. Paper discs, about 5 mm in diameter, are impregnated with hyperim-

TABLE 1

Biochemical and biological properties of mycoplasmas commonly isolated from man

Organism	Common site of isolation*	Substrates required for growth†	Colonial morphology‡	Growth rate	Aerobic tetrazolium reduction	Inhibition by 0.01% methylene blue	Haemadsorption of erythrocytes (E) and HeLa cell cultures (H)
Mycoplasma pneumoniae	R	YE+G	G	Slow	+	(reduction)	H + E (sheep, guinea pig, rat or chicken)
M. hominis	G (B, R)	A	FE	Medium	−	+	H
M. orale							
Types 1 and 3	R	YE + A	FE	Medium	−	+	E (chicken)
Type 2	(R)	A		Slow/medium	−	+	—
M. salivarium	R	A	FE	Medium	−	+	—
M. fermentans	(R)	G+A	G	Medium	−	+	—

* R, Respiratory; B, blood; G, genital tract.
† YE, Yeast extract; A, arginine; G, glucose.
‡ Granular (no peripheral zone), G; "fried egg," FE.

mune specific anti-sera (CF titre not less than 1 : 320). Dried discs give similar results.

A dry mycoplasma agar plate is flooded with the broth culture and the excess fluid is removed. The small amount of broth remaining is allowed to absorb for 15–30 min. The discs (up to 5) are applied around the periphery of the plate which is then incubated at 37°C, under appropriate atmospheric conditions, for 5–7 days.

A zone of inhibition appears round the disc containing the anti-serum homologous to the strain of mycoplasma (Clyde 1964; Dighero *et al.* 1970).

Immunofluorescent staining

This is the most reliable method for the identification of mycoplasmas. It has the advantage over the growth inhibition tests in that pure cultures (and therefore sub-cultures) are not required. In addition, individual species in mixed cultures may be identified. The results are available within a few hours and correlate well with those of other identification tests.

A number of methods have been proposed for the initial preparation of the colonies. These include the staining of colonies growing on agar blocks or on agar-coated slides and colony impressions obtained by hot-water fixation. The problems associated with these techniques are background fluorescence, necessity to sub-culture, and loss of specificity as a result of heat inactivation. They have been overcome by the use of rubber bungs to transfer colonies to microscope slides (Bradbury *et al.* 1976).

Rubber or neoprene bungs, 10 mm in diameter, are boiled in distilled water and dried before use. The bottom of a bung is placed gently on the top of the chosen colony. It is then lifted carefully and pressed on to a microscope slide. A number of replicates may be made from the same bung by pressing harder on each successive slide. The bung is discarded into disinfectant.

The impression is fixed by passing the slide through the Bunsen flame and is stained by the indirect immunofluorescent method using specific sera from man or animals and anti-species FITC conjugates. Both sera and FITC may require adsorption with pig liver powder and washed yeast cells (Stewart 1967) if media constituents cause nonspecific fluorescence. Ultraviolet microscopy should show bright green fluorescence with homologous sera.

Serological Methods

The isolation of *M. pneumoniae* takes too long for the prompt treatment of the patient. Most infections are therefore diagnosed by serological tests, usually by the complement fixation (CF) test, using the method of Bradstreet & Taylor (1962) and reagents supplied by the Department of Microbiological Reagents and Quality Control, Central Public Health Laboratory, Colindale. A disadvantage of this method is that both acute- and convalescent-phase sera must be tested to arrive at a definite diagnosis. The method also requires overnight incubation.

Two new tests have recently been introduced: the IHA test by Dowdle & Robinson (1964) and the ELISA by Raisanan *et al.* (1980) and Busulo *et al.* (1980). These tests are rapid, taking 1 h (IHA) or 5 h (ELISA), and sensitive, because they detect IgM class antibody. Furthermore, they may be done on single serum specimens.

Modifications to the IHA method were made by Lind (1968), who used formalised erythrocytes, which are more stable than fresh cells. This enabled larger batches of reagents to be prepared, giving more reproducible results. Ghyka *et al.* (1977) produced IHA reagents that were stable for 1 year by coupling sheep erythrocytes with glutaraldehyde rather than tannic acid. The selection of optimum coupling conditions and correct erythrocyte donor offered difficulties when this technique was first used. Fortunately, however, a commercial preparation of chicken erythrocytes coated with the Mac strain of *M. pneumoniae* by tannic acid became available in 1977 (Fujirebio, Diamed Diagnostics, Derby Road, Merseyside). This preparation is freeze-dried and is stable for at least 1 year at 4°C.

The indirect haemagglutination test

Microtitre plates (e.g. Sterilin), with U-shaped wells are used. The kit contains a phosphate-buffered saline (PBS) diluent which is used for the reconstitution of both the test (sensitised) and control (unsensitised) chicken erythrocytes as well as for the dilution of patients' sera. Test cells are used within 2 days of reconstitution, although they are stable for 7 days at 4°C. Control cells are used within 4 weeks of reconstitution. Instead of the quantitative assay described by the manufacturer, all sera are screened at an initial dilution of 1 : 10 and titrated if positive. The technique used is as follows:

(1) A 1 : 5 dilution of serum (not inactivated) is prepared by adding 25 μl of the test serum to 100 μl of diluent in the first well of a Microtitre plate.
(2) After mixing, 25 μl of this is removed and added to 25 μl of PBS in the second well to give a 1 : 10 dilution.
(3) Then 25 μl of this is removed to the third well.
(4) Next 75 μl of test cells are added to the second well.
(5) Finally 75 μl of control cells are added to the third well.

(6) The contents of the wells are mixed by tapping the plates and the cells are allowed to settle for 1 h at room temperature.

The test is positive if there is agglutination of test cells but no agglutination of control cells. If such specific agglutination occurs at 1 : 10, serial dilutions of the sera to at least 1 : 80 are tested. If non-specific agglutination occurs, the sera are titrated over at least four wells. A two-well (i.e. fourfold) or greater titre between control and test cells indicates a true positive. A titre of 1 : 40 or greater is taken as evidence of a recent infection.

This IHA test was compared with the CF test in a retrospective study by Taylor (1979) and later by a prospective study. Positive results by one or both methods were obtained from 402 patients. Of these, 264 (66%) were positive by both methods, 115 (28%) were positive by CF but negative by IHA, and 23 (6%) were negative by CF but positive by IHA.

The "false positive" IHA results probably reflect a greater sensitivity of this method for IgM class antibodies. There are a number of reasons for the "false negative" IHA results. Firstly, non-specific agglutination occurred in sera taken from 8% of the patients; these sera had been stored at −20°C for over 12 months. No non-specific agglutination occurred with fresh sera. Secondly, sera from 38% of the patients in this group gave relatively low CF titres (1 : 160–1 : 320), which may suggest past, rather than current, infection with *M. pneumoniae*. Finally, sera from 22% of the patients gave IHA titres of 1 : 10 to 1 : 20, which suggest that they had been collected too early or too late in the course of the infection. It was concluded that the IHA test provides a rapid and relatively sensitive method of detecting recent infections with *M. pneumoniae*, especially when combined with the CF test.

Enzyme-linked immunosorbent assay (ELISA)

The tests are done in flat-bottomed Microtitre plates (Sterilin). Each batch of plates should be screened for use in ELISA tests, as the degree of binding may vary.

Reagents

All reagents are made in glass-distilled water.

Carbonate–bicarbonate buffer (0.05 mole/litre; pH 9.6). This contains (g/litre): NaN_3, 0.02; Na_2CO_3, 1.59; $NaHCO_3$, 2.9. It may be stored at 4°C for no more than 4 weeks.

Wash buffer. Ten PBS tablets (Oxoid) are dissolved in 1 litre of water containing NaN_3, 0.2 g, and Tween 80 (Koch Light), 0.5 ml. This is stored at 4°C.

Diluent. Gelatin, 0.1 g, is dissolved in 100 ml of the wash buffer.

Diethanolamine buffer (pH 9.8). This contains diethanolamine (BDH), 9.7 ml;

1 mole/litre HCl, 9.7 ml; $MgCl_2 \cdot 6H_2O$, 0.01 g; NaN_3, 0.02 g, dissolved in 84.3 ml of water. This is stored at 4°C.

Alkaline phosphate substrate. Four phosphate tablets (5 mg; Sigma) are dissolved in 20 ml of diethanolamine buffer. This is prepared immediately before use. The phosphate tablets should be stored at −20°C.

Antigen. This is produced by growing *M. pneumoniae* on a glass surface by the method of Somerson *et al.* (1967).

Enzyme conjugate. Anti-human IgG and anti-human IgM alkaline phosphate conjugates are supplied by Miles Laboratories.

Optimum concentrations. A chessboard titration is done to determine the optimum concentrations of antigen and enzyme conjugate. A typical result is shown in Fig. 3, where the optimum concentration of the antigen is 1 : 40.

Indirect ELISA for antibody measurement

(1) Antigen is diluted in coating buffer to its optimum concentration and 200 μl added to each well of a Microtitre plate. The plate is incubated at 4°C overnight.

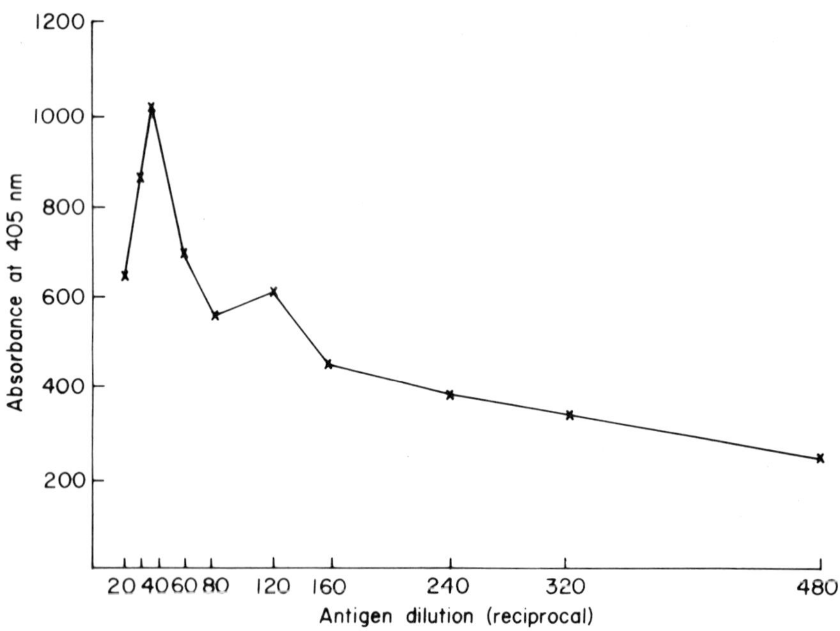

FIG. 3. Titration to establish the optimum concentration of *M. pneumoniae* antigen in the indirect ELISA system.

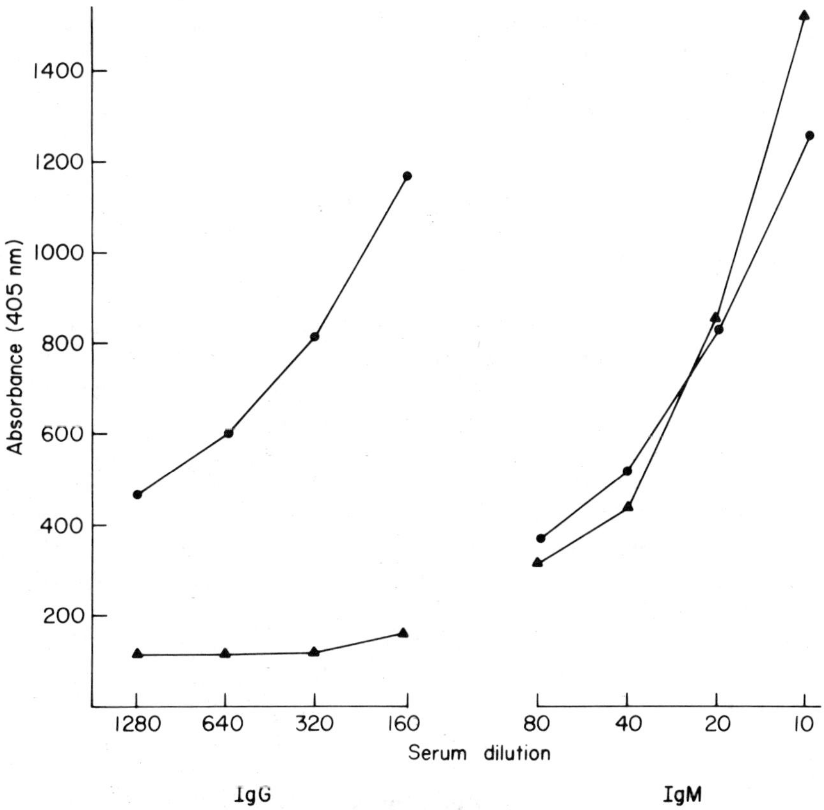

FIG. 4. The detection of specific IgG and IgM antibodies to *M. pneumoniae* by the indirect ELISA test. Acute sample (▲) at 1 week after onset, CFT titre <1 : 10, IHA titre <1 : 10. Convalescent sample (●) at 3 weeks after onset, CFT titre >1 : 320, IHA titre >1 : 1280.

(2) The plates are rinsed twice in wash buffer, allowing 5–10 min for each rinse.
(3) The patient's serum is diluted serially in the gelatin diluent and 200 μl of each dilution is added to a row of wells. A positive control serum is treated in the same way. The plates are incubated at room temperature for 2 h.
(3) Step (2) above is repeated.
(4) Alkaline phosphatase antiglobulin is diluted to its optimum concentration and 200 μl is added to each well. The plates are incubated at room temperature for 2 h.
(5) Step (2) above is repeated.
(6) Alkaline phosphatase substrate, 200 μl, is added to each well and the

reaction is allowed to proceed either until the colour compares with that of the positive control, or for 30 min.

(7) The reaction is stopped by adding 50 μl of 0.3 mole/litre NaOH.

(8) The colour in each well is read visually or in a spectrophotometer at 405 nm.

The results may be expressed as absorbance units or as a ratio of test results to a range of known negative findings. In the latter case, a ratio of 3 or more indicates a significant level of specific antibody.

Figure 4 shows typical results, obtained with an antigen dilution of 1 : 40 on the serum of a young adult. The test was done 1 week after the onset of the disease; this was probably a primary infection and hence produced high levels of specific IgM. This response is usually weaker in secondary infections with *M. pneumoniae,* especially in older patients.

Future developments with ELISA include the use of IgM-coated beads to capture specific antibody. The quantity of IgM antibody bound is detected by the sequential addition of specific antigen, rabbit anti-*M. pneumoniae* antibody, anti-rabbit alkaline phosphatase conjugate, and phosphate substrate. The ELISA test has yet to be established for the detection of *M. pneumoniae* antigen in respiratory secretions.

Whichever serological approach is chosen for the diagnosis of *M. pneumoniae* infections, the cross reaction between this organism and *M. genitalium* must always be evaluated (Taylor-Robinson *et al.* 1983).

References

BRADBURY, J. M., ORIEL, C. A. & JORDAN, F. T. W. 1976 Simple method for immunofluorescent identification of mycoplasma colonies. *Journal of Clinical Microbiology* **3**, 449–452.

BRADSTREET, C. M. P. & TAYLOR, C. E. D. 1962 Technique of complement fixation applicable to the diagnosis of virus diseases. *Monthly Bulletin of the Ministry of Health and Public Health Laboratory Service (Great Britain)* **21**, 96–104.

BUSULO, F., TONIN, E. & CONVENTI, L. 1980 Enzyme linked immunosorbent assay for detection of *M. pneumoniae* antibodies. *Journal of Clinical Microbiology* **12**, 69–73.

CLYDE, W. A. 1964 Mycoplasma species identification based upon growth inhibition by specific antisera. *Journal of Immunology* **92**, 958–965.

DIGHERO, M. W., BRADSTREET, C. M. P. & ANDREWS, B. E. 1970 Dried paper discs for serological identification of human mycoplasmas. *Journal of Applied Bacteriology* **33**, 750–757.

DOWDLE, W. R. & ROBINSON, R. Q. 1964 An indirect haemagglutination test for diagnosis of *mycoplasma pneumoniae* infections. *Proceedings of the Society for Experimental Biology and Medicine* **116**, 947–950.

GHYKA, G. R., STOIAN, N., SORODOC, G., PEIULESCU, P., OSMAN, J. & DUCA, D. 1977 An antigen for the diagnosis of *M. pneumoniae* infections by passive haemagglutination. *Revue Roumaine de Medecine, Virologie* **28**, 111–115.

HAYFLICK, L. 1965 Cell cultures and mycoplasma. *Texas Reports on Biology and Medicine* **23**, Suppl. 1, 285–303.

LIND, K. 1968 An indirect haemagglutination test for serum antibodies against *mycoplasma pneumoniae* using formalised tanned sheep erythrocytes. *Acta Pathologica Scandinavica* **73,** 459–472.

NOCARD, E. I. E. & ROUX, P. P. B. 1898. Le microbe de la péripneumonie. *Annals Institute Pasteur* **12,** 240–262.

RAISANEN, S. M., SUNI, J. & LEINIKKI, P. O. 1980 Serological diagnosis of *Mycoplasma pneumoniae* infections by enzyme immunoassay. *Journal of Clinical Pathology* **33,** 836–840.

SMITH, C. B., CHANOCK, R. M., FRIEDWALD, W. T. & ALFORD, R. H. 1967 *Mycoplasma pneumoniae* infections in volunteers. *Annals of the New York Academy of Sciences* **143,** 471–483.

SOMERSON, N. L., JAMES, W. D., WALLS, B. E. & CHANOCK, R. M. 1967 Growth of *Mycoplasma pneumoniae* on a glass surface. *Annals of the New York Academy of Sciences* **143,** 384–389.

STEWART, S. M. 1967 The use of fluorescent antibody technique for the identification of cultures of *Mycoplasma. Immunology* **13,** 513–516.

TAYLOR, P. 1979 Evaluation of an indirect haemagglutination kit for the rapid serological diagnosis of *Mycoplasma pneumoniae* infections. *Journal of Clinical Pathology* **32,** 280–283.

TAYLOR-ROBINSON, D., FURR, P. M. & TULLY, J. G. 1983 Serological cross-reactions between *Mycoplasma genitalium* and *Mycoplasma pneumoniae. Lancet* **i,** 527.

TULLY, J. G., ROSE, D. L., WHITCOMB, R. F. & WENZEL, R. P. 1979 Enhanced isolation of *Mycoplasma pneumoniae* from throat washings with a newly modified culture medium. *Journal of Infectious Diseases* **139,** 478–482.

Detection and Isolation of Pathogenic Fungi

C. K. CAMPBELL, C. DAVIS AND D. W. R. MACKENZIE

Public Health Laboratory Service, Mycological Reference Laboratory, London School of Hygiene and Tropical Medicine, London, UK

The significance of fungi in causing serious infections in immunocompromised patients is well established. Patients with severely debilitating diseases often succumb to infections by otherwise harmless fungi, because of failure of the cellular immune system, either mediated by steroid therapy or arising as a direct consequence of the primary condition. In addition, the use of broad-spectrum antibiotics and the presence of in-dwelling catheters and other lines constitute pre-disposing factors.

In this chapter we describe the methods currently in use at the Mycological Reference Laboratory for the detection of the commoner fungal pathogens of immuno-compromised patients and for interpretation of their significance.

Detection by Microscopy

In the diagnosis of active infections by organisms common in the patient's environment, direct visual proof of the presence of the pathogen in an intimate association with the affected tissue is particularly important. Fortunately, owing to the relatively large size of fungal cells, this is often not difficult by low power microscopy. Certainly the microscopical proof of presence of hyphae or other fungal structures is unequivocal and is often of more value to the clinician than evidence offered by culture or serology. Histological sections play a very significant role in the diagnosis of deep fungal infection.

It is well recognised that not all infections are amenable to direct microscopy, as removal of deep tissue for study is often precluded by the poor

Isolation and Identification of Micro-organisms of Medical and Veterinary Importance

Copyright © 1985 by the Society for Applied Bacteriology
All rights of reproduction in any form reserved
0-12-181460-2

medical state of the patient. Where possible, however, material from biopsy or aspirations are examined by classical histo-pathological techniques and as wet preparations in potassium hydroxide (KOH). These preparations have several advantages over histology in that they are quick, easy and reveal more complete mycelial fragments. Using KOH wet preparations alone, it is often possible within a few minutes to give the clinician sufficient information to influence the choice of therapy.

For detection of fungi by histo-pathology, the most suitable stain is probably the silver method of Grocott, counter-stained with Cole's haematoxylin and eosin (H and E). Rolph & Austwick (1973) showed that Cole's method distinguished actively growing from old hyphae. All fungal cell walls stain black and are easily seen against the background of the familiar H and E section. In addition to this method there are two more specific staining procedures, the mucicarmine stain, useful only in differentially staining the capsular material of *Cryptococcus neoformans*, and the cresol fast violet stain, a procedure which differentiates hyphae of Zygomycetes (*Mucor*, etc.) from those of *Aspergillus* (Rolph & Austwick 1973). These and the procedure for immunofluorescent staining of fungi in smears and tissue sections are given next.

Grocott silver technique

This stain is made as follows: 5 ml of a 5% solution of silver nitrate in water is added to 100 ml of methenamine. The mixture is shaken to dissolve the precipitate. For use it is diluted with an equal part of water and pre-heated in a water bath at 48–50°C. Immediately before use, 1 ml of 5% borax solution is added and the mixture stirred thoroughly.

(1) The specimens are brought to water and placed in 5% chromic acid for 1 hr.
(2) They are washed in running tap water for 10 min, placed in 1% sodium bisulphite for 1 min, washed in running tap water for 5 min and then in three changes of distilled water.
(3) They are then stained with silver methenamine (hexamine) at 58–60°C for 1½ h until a golden colour develops. A positive control specimen should be included and examined after 30 mil.
(4) The specimen is then rinsed in three changes of distilled water, 2 min each, and toned in 1% gold chloride for 5 min.
(5) It is then rinsed in distilled water and in 2% sodium thiosulphate, washed in running tap water for 15 min, and then counter-stained with Cole's haematoxylin and eosin (H and E).

All fungal cell walls stain black or brown; host tissues stain as expected with H and E.

Mucicarmine stain

(1) The section is taken to water and stained with haematoxylin for 15–30 min.
(2) It is rinsed in tap water and briefly in acid alcohol (0.5% HCl in 70% ethanol) and washed in running water for 10 min or more.
(3) It is placed in Southgate's mucicarmine stain for 45 min, rinsed in water, dehydrated and mounted.

Cryptococcal capsules and epithelial mucin stain red.

Cresol fast violet stain

(1) The sections are taken to water and stained in cresol fast violet for 5 min.
(2) They are washed in running tap water for 5 min, drained well, dehydrated in three changes of absolute alcohol and mounted.

Zygomycete (mucoraceous) hyphal walls stain orange. *Aspergillus* and *Candida* hyphal walls are not stained (contents are metachromatic).

Immunofluorescent staining

(1) The sections are brought to water. Specific anti-serum (*A. fumigatus*, *C. albicans*, etc.) is spread over them and they are incubated at 37°C for 30 min.
(2) The excess serum is then shaken off and the sections are washed in three changes each of phosphate-buffered saline and distilled water.
(3) They are then stained in fluorescein-conjugated homologous antiglobulin (diluted 1 : 16) for 30 min at 37°C, washed in water, mounted and examined by fluorescence microscopy.

Key to Fungi in KOH Preparations and Histological Sections

Table 1 shows a key to the appearance of fungal elements in deep tissues and body fluids. No mention is made of the classical systemic or sub-

TABLE 1

Key to appearance of fungal elements in deep tissues and body fluids

1a	Hyphae present.	2
1b	Hyphae not present.	6
2a	Hyphae mixed with yeast cells.	3
2b	Hyphae not mixed with yeast cells.	4
3a	Hyphae septate, often constricted at the septa (pseudohyphae). Oval budding cells present, often in small numbers. (*Note:* It is wise to establish that the budding cells are growing from the hyphae, to exclude the possibility of a mixture of two organisms.)	*Candida* spp.
3b	Hyphae septate, often breaking at the septa to form square-ended arthrospores which may become rounded to resemble budding cells.	*Trichosporon* spp.
4a	Unstained hyphae (KOH or H and E only), brown walled.	*Alternaria* and *Cladosporium* etc.
4b	Unstained hyphae (KOH or H and E only), not brown walled.	5
5a	Hyphae non-septate, branching at wide angles, thin walled and with great variation in diameter, appearing knobbly or irregularly swollen; cresol fast violet stains them orange.	*Zygomycetes* (*Rhizopus, Mucor, Absidia* spp.)
5b	Hyphae regularly septate, uniform in diameter, branching acute and often dichotomous; walls not stained by cresol fast violet. (*Note:* A range of less common moulds may have similar hyphae, notably *Pseudallescheria* and *Fusarium* spp. Also, *Candida* and *Trichosporon* may "key out" here if no yeast cells are seen.)	*Aspergillus* spp.
6a	Cells mostly 8–15 μm diam., sub-spherical or broadly ovoid, budding. A capsule may be seen as a clear halo in stained sections, though it is frequently absent. Confirmatory staining with mucicarmine is desirable. (*Note:* Other cryptococci and species of *Rhodotorula* look similar, but are rare.)	*Cryptococcus neoformans*
6b	Cells mostly 5–10 μm, ovoid, budding. A halo due to cytoplasmic shrinkage may occur around the cell. This will fail to stain with mucicarmine.	Some infections with *Candida* spp.
6c	Cells mostly 2–5 μm, ovoid, budding, sometimes clearly intracellular. (*Note:* Check for histoplasmosis.)	*Candids glabrata*

cutaneous pathogens (*Histoplasma, Coccidioides,* etc.), any of which may also invade patients with immune deficiencies. The aim of the key is to reflect systemic mycoses as they commonly appear in Europe.

Detection by Culture

All the common pathogenic fungi grow well on media generally employed for their isolation, and culture is often less of a problem than is subsequent identification. It is beyond the scope of this chapter to describe methods of identification in detail, but several points of relevance to primary isolation may be made.

For primary isolation, either 2% malt agar or Sabouraud's glucose peptone agar with supplementary antibiotics such as chloramphenicol (0.05%) are ideal. The anti-fungal compound cycloheximide, often employed in isolation of dermatophytes and of the classical systemic fungal pathogens, is inhibitory to most of the fungi causing disease in immuno-suppressed patients and should not be used.

Fungi will grow on normal blood agar, and many laboratories rely on this medium for fungal as well as bacterial isolation. There is no doubt, however, that the moulds grow more luxuriantly on mycological media, and that they are easily inhibited by the excessive bacterial growth possible on blood agar.

Patients with yeast infections frequently yield positive blood cultures, whereas those infected by mycelial fungi do not. This may be explained by the relative ease with which budding cells of yeasts may be caught up and transported from the site of infection. Yeast cells are much larger than most bacteria, however, and if only a few are present in the blood there is a risk that they may be caught in capillary beds of the body organs, rendering venous blood an unlikely source from which to isolate them. In such cases, and particularly where infection is suspected in heart or lungs, culture of arterial blood may be useful.

One method of blood culture is the Castaneda type of biphasic medium, using Sabouraud's agar and broth. The solid phase should be sloped against the wide face of a flat bottle for easy examination of the colonies. The automated systems for blood culture using, for instance, detection of radio-labelled CO_2 appear to be satisfactory for yeasts.

Finally, whilst the culture of a fungus from an immuno-suppressed patient will need to be interpreted in the light of the findings of direct microscopy, serology and the clinical syndrome, its identity may give some clue about its probable involvement. A brief summary of the pathogenic potentials of the species most commonly encountered is presented here.

(1) Recognised destructive invaders of visceral tissues: *Aspergillus fumigatus*, *A. niger*, *A. flavus*, *A. terreus*, *Candida albicans*, *C. parapsilosis* (especially endocarditis), *C. krusei*, *C. tropicalis*, *Pseudallescheria boydii*, *Rhizopus oryzae*, *Absidia corymbifera*, *Rhizomucor pusillus*, *Cryptococcus neoformans*.

(2) Largely restricted to peritonitis, wound contamination, but capable of spread to solid tissues: *Candida guilliermondii*, *C. pseudotropicalis*, *C. glabrata*, *Trichosporon beigelii*, *T. capitatum*.

(3) Rare causes of restricted abscesses or subcutaneous granulomata in immuno-suppressed patients: *Alternaria* spp., *Cladosporium* spp.

(4) Commonly isolated, but rarely, if ever implicated in disease: *Saccharomyces* spp., *Rhodotorula* spp., *Penicillium* spp.

Detection by Serology

Serological tests can often provide information which complements other types of investigation in helping to establish a diagnosis. Serological results on their own are generally an indication of possible disease by fungi, and seldom provide confirmatory evidence.

The types of serological tests currently available can be split into two broad categories: detection of antibody and detection of antigenic cell constituents. The latter category has many potential advantages over antibody detection but, with the exception of the latex agglutination test for *Cryptococcus neoformans*, the methods are still in the developmental stage. Detection of antibody to an organism may provide important information to the clinician, but has several drawbacks.

(1) The results may only be of retrospective significance because of the time lag between onset of infection and production of antibodies.
(2) One of the groups of patients most vulnerable to opportunistic fungal infections (immuno-suppressed and leukaemic patients) will not generally produce an antibody response.
(3) Because of the widespread occurrence of many fungal organisms, detectable levels of antibodies are often found in normal, healthy people.

Detection of circulating fungal antigens may overcome some of these difficulties. Antigen should be present in the early stages of the infection, and should be detectable in immuno-suppressed and leukaemic patients. As fungal organisms are capable of producing a rapidly fatal outcome, there is currently great interest in the development of techniques for estimating fungal antigen. The levels of fungal antigen may enable the monitoring of the disease process to be more accurate, as the titre should fall and eventually disappear on recovery, whilst a rise in titre should provide an indication of worsening infection.

Aspergillosis

Aspergillosis can present as three different types of disease: allergic, invasive and "aspergilloma". With the exception of the invasive form, diagnosis of aspergillosis is generally made on the clinical picture with serological tests providing supporting evidence.

Allergic aspergillosis is the commonest form of the disease in the UK, and results from an allergic reaction to inhaled spores of *Aspergillus fumigatus*. Antibodies to aspergilli can be detected by counterimmunoelectrophoresis (CIE) or by double diffusion (DD). At the Mycological Reference Laboratory the former method is used, because of its greater sensitivity and rapidity. The commercially available kits mostly rely on double diffusion tests. The precipitin reactions in allergic disease are generally weak, and the serum may require concentrating before detection by DD is possible.

Invasive aspergillosis is the most difficult form of the disease to diagnose, and one which is particularly associated with immuno-suppressed and leukaemic patients. There are few diagnostic symptoms to the disease, and both cultural and serological results may be unhelpful. Because of the type of patient involved, antibodies are rarely produced, and serological tests are usually weak or negative. This is a disease where detection of circulating antigen could be of great benefit. Various techniques have been tried for this, including radioimmunoassay (Weiner & Coates-Stephen 1979a) and enzyme-linked immunosorbent assay, but, although promising results have been obtained, these methods are still in a developmental stage.

Aspergilloma occurs in patients with a history of lung disease, such as tuberculosis, which has resulted in cavities remaining in the lung. Growth of aspergilli in a cavity results in the production of a fungus ball. *Aspergillus fumigatus* is the most usual causative agent, although it is recommended to look for antibodies to *A. flavus*, *A. terreus*, *A. niger* and *Pseudallescheria boydii*. Strong precipitin lines are usually produced both in CIE and DD in cases of aspergilloma. A certain amount of cross reactivity may occur between antibodies to *A. fumigatus* and the other organisms mentioned.

Candidosis

An invasive infection by a *Candida* sp. is probably one of the most difficult fungal infections to diagnose, and one in which detection of antibody has a distinct role to play. High antibody titres are not generally produced in cases of localised colonisation or in superficial infections, such as thrush, and they may therefore indicate invasive disease. The normal population may have antibodies to *Candidas* spp., however, and it may be difficult to interpret moderate levels. False negative reactions may be obtained with immunocompromised patients who often fail to produce antibodies. As in invasive aspergillosis, the eventual development of antigen detection techniques may provide more valuable information than tests for the presence of antibody.

A range of serological procedures has been used for the detection of antibody to *Candida* spp. These include CIE and DD for detection of precipitating antibodies (mostly IgG), whole-cell agglutination and haemagglutination for detection of agglutinating antibodies (generally IgM), and the indirect fluorescent antibody test. At the Mycological Reference Laboratory, the methods routinely used are CIE and whole-cell agglutination (WCA). Because healthy people often have low levels of antibody to *Candida* spp., it is important to establish a cut-off point between normal healthy and disease state titres. A general guideline to titres regarded as indicative of probable infection is as follows:

CIE, DD	Weak or strong precipitins (>1 : 2)
WCA	>1 : 32
IFA	>1 : 40

As with other antibody level determinations, demonstration of a rise in titre is more significant than ascertaining a single level.

Several methods for determination of circulating antigenic cell constituents have been examined by various workers. Serological methods include radioimmunoassay (RIA: Weiner & Coates-Stephen 1979b), ELISA (Scheld *et al.* 1980; Araj *et al.* 1982), and latex agglutination (Gentry *et al.* 1983). All these methods claim to produce positive reactions in patients with disseminated candida infection, but not in patients with colonisation, superficial infections, non-candida infections, or in healthy subjects. The antigens detected vary with the methods, and include mannan as well as cytoplasmic antigens of unknown chemistry. The latex agglutination method is available commercially in kit form.

Non-serological methods for determination of the presence of candida cell constituents have also been recently evaluated. Gas-liquid chromatography has been used to determine the presence of arabinitol (Kiehn *et al.* 1979) and of mannose (Monson & Wilkinson 1981).

At the present time, these methods are still under evaluation. The latex agglutination method appears to have the advantage that it is very simple to perform and does not require special apparatus.

Cryptococcosis

Serological diagnosis of cryptococcal infection is unusual in that the most valuable test relies on the detection of antigen rather than antibody. Antibodies are often not detectable in patients with cryptococcosis, either because of overwhelming quantities of antigen or of failure of the patient to produce antibodies. Antigen may be detected both in serum and cerebrospinal fluid (CSF) by latex agglutination, and this provides a highly specific and useful test. The method first published by Bloomfield *et al.* (1963) depends on the fact that cells of *Cryptococcus neoformans* are surrounded by a thick polysaccharide capsule. In an infected patient, large amounts of this capsular material circulate as free antigen in the serum and CSF. The test has been shown to be positive in almost 90% of cases with proven cryptococcal meningitis. This figure is lower in patients with pulmonary or skin infections in the absence of involvement of the central nervous system. False positive results may occur, but these may be removed by treatment of the specimen. They are mainly due to the presence of rheumatoid factor. The test is very rapid; a result may be obtained within 2 h of the specimen reaching the laboratory. Another major advantage of the test to the clinician is that the course of infection can be followed by monitoring the antigen titre over a period of weeks, thereby providing valuable information on progress.

Antibodies to *C. neoformans* can be detected serologically by WCA or by indirect fluorescent antibody (IFA). Unlike the test for *C. albicans*, WCA titres of more than 1 : 2 are classed as significant in cases of cryptococcosis. Titres of more than 1 : 8 are rare.

Zygomycosis

Zygomycosis (mucormycosis) is another disease which occasionally occurs in immuno-suppressed and also in diabetic patients. The commonest causative agent is *Rhizopus oryzae*, but other organisms are less commonly implicated. Diagnosis is generally made from the clinical picture and from culture and smears of biopsy specimens. Serological tests performed by CIE can sometimes be of help in diagnosis. As a rule, only weak precipitin lines are produced, and in some cases there appears to be no response. Cross-reactions between the various species involved are frequent, and identification of the causative organism according to the type of antibody present is difficult. As prognosis and choice of therapy are apparently not affected by species type, however, this is of little significance.

Production of antigens

The production of fungal antigens for use in serological tests at the Mycological Reference Laboratory involves three main stages: (1) growth of the organism, (2) rupture of the cells to release soluble components, and (3) concentration of the antigen-containing fraction of the resultant liquor.

The organism is grown in two stages. The first involves the preparation of a "starter culture" for inoculation of a second, larger culture. For aspergilli and similar organisms, including the agents of zygomycosis, the initial stage is carried out in a broth culture, whereas yeast organisms are best grown on agar plates. In both cases a glucose peptone medium is used. To prepare a broth culture, 300 ml of broth is autoclaved in a 500-ml flask containing a magnetic stirrer. The cooled medium is then inoculated with a fragment of mycelium or a few spores and maintained at 30°C for 2–3 days, until a dense growth is obtained. During this time the culture is stirred (slowly at first, until the growth is established) to promote good aeration.

To produce large quantities of the organism, the starter culture may either be used to inoculate a chemostat or to inoculate 3000 ml of glucose peptone broth in a 5000-ml flask. This is then incubated under the same conditions of temperature and aeration. Adequate growth should be obtained in 2–3 days; if left longer, the mycelium may begin to autolyse. The growth is harvested by filtration through nylon gauze, washed once with water, and centrifuged.

Large quantities of *Candida* spp. can be grown on a 9-cm petri dish of glucose peptone agar, a method more convenient than the one employed for moulds. A suspension is made from the initial growth on the first petri dish, and 0.5 ml of this is run over the surface of the plate. The plates are incubated at 30°C for 2 nights and 1 day, when the growth is washed off with saline and centrifuged. The harvested cells are weighed in a tared beaker before further treatment.

There are several ways of rupturing fungal cells. Overnight storage at −20°C immediately after harvest appears to help the process. The organisms can be left at this temperature for several weeks, if required, before being treated in one of the following ways:

(1) The fungus is mixed with a volatile buffer in an equal volume of sand, frozen at −20°C and beaten in a pestle and mortar until liquid.
(2) The fungus is ruptured by ultrasonic process in suspension.
(3) The cells are ground by high speed fracture with glass ballotini beads (Dyno-mill).

After disintegration, the suspension is again frozen (−20°C) overnight to aggregate insoluble material, which may be removed later by centrifugation. The supernatant resulting contains the soluble antigen. It is concentrated, with removal of low molecular weight impurities, by dialysis at 4°C against 10% polyethylene glycol 6000 in volatile buffer. The latter is changed daily until sufficient concentration has been obtained. The concentrated solution is again frozen overnight, centrifuged for 1 h at 15000 rev/min. The resultant supernatant constitutes the antigen and can be freeze-dried for keeping in pre-weighed ampoules.

Standardisation of antigens

Before a new batch of antigen is used it must be standardised against test anti-sera. Each freeze-dried ampoule should contain a known fixed weight of antigen. This is reconstituted in 1 ml of antigen diluent (veronal buffer, containing 1 : 5000 merthiolate). It is then titrated against a reference anti-serum to obtain the concentration producing optimum precipitin lines for use. It is recommended that the antigen be used at two dilutions, one producing several fairly strong precipitin lines and one which gives fewer, weaker lines. In this way, there is an increased chance of detecting a range of antibody concentrations.

Counter-immunoelectrophoresis

Preparation of gels

Clean glass slides (3 in. × 2 in.) are precoated with a very thin layer of buffered 1% agarose gel and allowed to dry. These coated slides may be stored for several months. For use, 5.5 ml of the gel is poured onto each coated slide and allowed to dry.

Buffered gel. This contains (g/100 ml in water): agar purified, 1.0; and agarose, 1.0. It is steamed to dissolve, cooled to 50°C, and 100 ml of double-strength veronal buffer is added.

Veronal buffer. This is prepared by dissolving 12.04 g of barbitone; 26.295 g of sodium barbitone and 1.75 g of sodium azide in 3.5 litres of water.

Addition of reagents and sera

A plastic template is used to cut wells into the gel with a stainless-steel cutter in the pattern as shown below. The serum wells are 4.5 mm in diameter and hold 20 μl; the antigen wells are 2.5 mm in diameter and hold 8 μl.

The wells are then filled with the appropriate antigens and sera. In this laboratory, each antigen is used at two concentrations, in an attempt to eliminate false negative reactions which may be due to very large and very small amounts of antibody. It is important to include positive controls using reference anti-sera for every antigen.

The slides are placed in an electrophoretic tank and a potential of 4 V/cm applied across the gels. The buffer placed in both sides of the tank is veronal buffer (pH 8.2).

After completion of the electrophoresis, slides are placed at 45°C in a borax bath to wash for 48 h.

Borax bath. This contains (g/litre in distilled water): borax ($Na_2B_4O_7 \cdot 10H_2O$), 4; NaCl, 4.0. The slides are then removed from the bath, immersed in 2% glutaraldehyde for 10 min and washed in running water for a further 10 min. The gels are then dried onto the slides for 60 min at 60°C before staining.

Stain. This is prepared by dissolving 10.5 g of Coomassie brilliant blue R in 900 ml of ethanol, 200 ml of glacial acetic acid and 900 ml of distilled water. The slides are stained for 10 min, rinsed briefly in methylated spirits, and then washed in two changes of the ethanol–acetic acid mixture without Coomassie blue for 2 min and 1 min, respectively. The slides are then dried and examined.

The tests are read as positive or negative, depending on the presence of

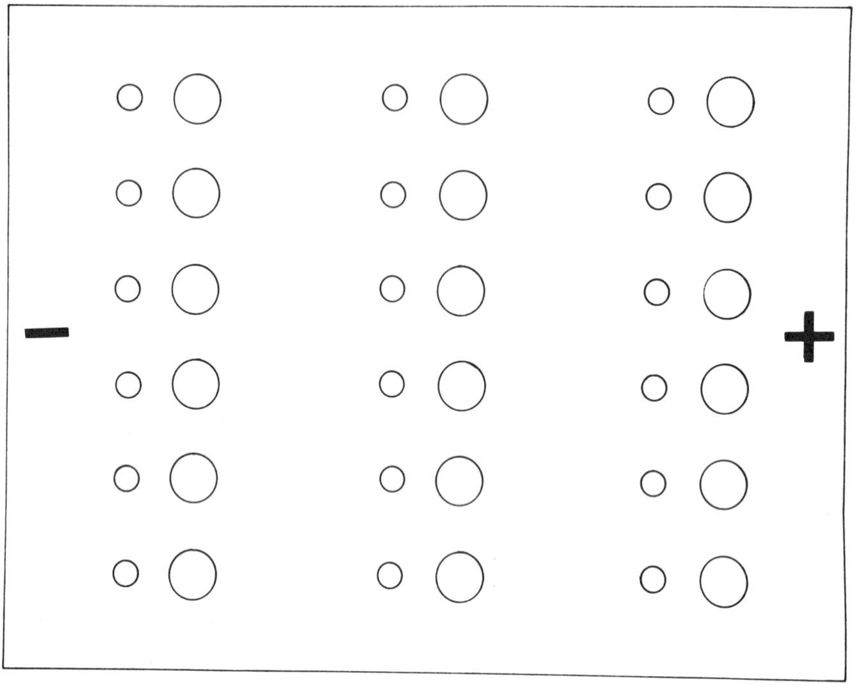

FIG. 1. Counter-immunoelectrophoresis: pattern for template.

precipitin lines. The results are not quantitative, although a weak or strong reaction can be differentiated with experience for each antigen system. For detection of aspergillosis such a qualitative result is generally sufficient as the normal population does not usually possess detectable antibodies; with candidosis patients, however, a more quantitative result is desirable. This can be obtained by repeating the CIE on sera which show relatively strong precipitin lines and using doubling dilutions of the patient's sera. Again, the titration should be performed against two different concentrations of antigen.

It is possible, when very rapid results are required, to read the gels wet as soon as the electrophoresis has finished. Such results should be given with reservation, as both false positive and false negative results occur. The former is due to non-specific lines which may wash out and the latter is due to very weak lines which can only be seen after staining.

Whole-cell agglutination

Whole-cell agglutination (WCA) is carried out in plastic WHO trays by the addition of whole killed cells to doubling dilutions of the patients' sera. The organisms which are used in this system in our laboratory are *Candida albicans*, *C. parapsilosis* and *Cryptococcus neoformans*.

Preparation of cells. *Candida albicans* cells are prepared by growth on the surface of glucose peptone agar plates for 24 h. Prolonged incubation is not recommended, as it may result in development of filamentous cells. The growth is harvested and washed three times in sterile saline by centrifugation. The packed cells are then suspended in saline to a concentration of about 20% (the exact concentration can be determined using a haematocrit tube). For use, the 20% suspension is adjusted to 2% by the addition of phosphate-buffered saline (PBS) containing 1% sodium azide. The stock suspensions are maintained at −20°C. The in-use suspension will keep for 2–3 weeks in the refrigerator.

Cells from *C. neoformans* are prepared in a similar manner. To discourage florid capsule formation, a strain of *C. neoformans* with a small capsule should be used. A medium which also helps to discourage too much capsule production contains (g/litre in distilled water): glucose, 10; Bacto peptone, 10; $NaH_2PO_4 \cdot 2H_2O$, 5; and agar, 15.

The test

Serial twofold dilutions of the test sera are made as follows: for each test, 0.6 ml of PBS is placed in the first well of a plastic WHO tray and 0.4 ml PBS in each well of the row. Serum, 0.2 ml, is added to the first well, mixed, and 0.4 ml double-diluted across the row, leaving the final well with PBS alone to test for autoagglutination of the cell suspension. One drop of yeast cell suspension is then added to each well with a pasteur pipette. A positive control of known agglutinin titre is included on each plate for each type of yeast cell suspension used.

This method gives a starting dilution for the test serum of 1 : 4. For *Cryptococcus neoformans* a starting dilution of 1 : 2 should be used, as positive results at this dilution can be significant (0.4 ml PBS + 0.4 ml test sera in the first well).

After careful but thorough mixing, the trays are covered with a plastic lid and left at room temperature for 24 h. To read the results, the trays are shaken gently to resuspend the deposit and agglutination observed by dark-ground illumination. Very fine agglutination may occur due to the serum alone and should be ignored.

Latex agglutination test for cryptococcal antigen

The materials for this test are available commercially (Conalco Diagnostics, Rockville, Md.), but the method here describes the procedure as carried out at the Mycological Reference Laboratory.

Preparation of sensitised latex

The sensitised latex is prepared by the addition of latex to *C. neoformans* globulin raised in rabbits. The latex (Difco) is used at a dilution of 1 : 10, whilst the optimum dilution of globulin is determined by titration with latex suspensions diluted in glycine-buffered saline (GBS), which contains (per litre in distilled water): glycine, 7.31 g; NaCl, 10.0 g; normal NaOH, 3.5 ml.

Equal quantities of the diluted latex and globulin are prepared and mixed together very quickly. The mixture is then left overnight at 4°C before use.

The test

For routine screening of serum and CSF, the test is performed on the undiluted fluid as well as at 1 : 5 and 1 : 25 dilutions (made in GBS with the addition of 1% bovine serum albumin, BSA). This enables weak reactions to be detected, but also allows for the presence of prozone effect at low serum dilutions.

Two drops of each dilution of test serum or CSF are placed on separate sections of a grooved slide (microscope slides can be used, but are less convenient). One drop of sensitised latex suspension is added to each. A known positive control and a negative control (the diluent is used for this) are also tested on the same slide. The slide is then rocked for exactly 3 min, and agglutination observed against dark background illumination. The test is read as positive at the greatest dilution which shows definite agglutination as compared with the negative control. If strongly positive at a 1 : 25 dilution, the test is repeated using further dilutions until an endpoint is reached.

False positive reactions do occur and must be excluded. Most of them occur in sera and are due to the presence of rheumatoid factor (RF). This can be tested for, using a commercially available kit (Difco). Occasionally, false positives occur in CSF (where RF is thought not to occur) and in RF-negative sera. All false positive reactions can be removed by treating the specimens in the following manner.

Serum specimens are heated at 56°C for 30 min, 0.003 *M* dithiothreitol is added, and the mixture is incubated at 37°C for 30 min.

CSF specimens are placed in a boiling water bath for 10 min.

It is recommended that all positive specimens be so treated and that the test be repeated on the treated specimen before a positive result is reported.

References

ARAJ, G. F., HOPFER, R. L., CHESNUT, S., FAINSTEIN, V. & BODEY, G. P. 1982 Diagnostic value of the enzyme-linked immunosorbent assay for detection of *Candida albicans* cytoplasmic antigen in sera of cancer patients. *Journal of Clinical Microbiology* **16**, 46–52.

Bloomfield, N., Gordon, M. A. & Elmendorf, D. F. 1963 Detection of *Cryptococcus neoformans* antigen in body fluids by latex particle agglutination. *Proceedings of the Society for Experimental Biology and Medicine* **114**, 64–67.

Gentry, L. O., Wilkinson, A. S., Lea, A. S. & Price, M. F. 1983 Latex agglutination test for detection of *Candida* antigen in patients with disseminated disease. *European Journal of Clinical Microbiology* **2**, 122–128.

Kiehn, T. E., Bernard, E. M., Gold, J. W. M. & Armstrong, D. 1979 Candidiasis: detection by gas–liquid chromatography of D-arabinitol, a fungal metabolite, in human serum. *Science (Washington, D.C.)* **206**, 577–580.

Monson, T. P. & Wilkinson, K. P. 1981 Mannose in body fluids as an indicator of invasive candidiasis. *Journal of Clinical Microbiology* **14**, 557–562.

Rolph, L. & Austwick, P. K. C. 1973 Differential staining of fungi in tissues. *Journal of Science Technology* **17**, 22–26.

Scheld, W. M., Brown, R. S., Jr., Harding, S. A. & Sande, M. A. 1980 Detection of circulating antigen in experimental *Candida albicans* endocarditis by an enzyme-linked immunosorbent assay. *Journal of Clinical Microbiology* **12**, 679–683.

Weiner, M. H. & Coats-Stephen, M. 1979*a* Immunodiagnosis of systemic aspergillosis. I. Antigenaemia detected by radioimmunoassay in experimental infection. *Journal of Laboratory and Clinical Medicine* **93**, 111–119.

Weiner, M. H. & Coats-Stephen, M. 1979*b* Immunodiagnosis of systemic candidiasis: Mannan antigenaemia detected by radioimmunoassay in experimental and human infections. *Journal of Infectious Diseases* **140**, 989–993.

Bovine Mastitis: Examination of Milk in Control Schemes

P. G. FRANCIS

Ministry of Agriculture, Fisheries and Food
Central Veterinary Laboratory
Bacteriology Department
New Haw, Weybridge, Surrey, UK

Mastitis in dairy cows causes a loss in milk production. Clinical mastitis cases require treatment and additional attention. Milk produced by such animals must be withheld from sale for a prescribed period to allow drug residues to be cleared. Data from the Ministry of Agriculture's Mastitis Surveillance Scheme suggest that 43 clinical cases occur in every 100 cows each year and 1% of cows exhibiting signs of clinical mastitis die from the condition. Clinical cases are common during the early months of lactation, whilst sub-clinical infection levels increase as lactation proceeds. The herds involved in the Surveillance Scheme are considered to be "better than average" herds, but extrapolation of the results from these to the National Dairy Herd suggests that 3500 new clinical cases occur each day. Sub-clinical mastitis causes an insidious loss of milk production and a reduction in the productive life of the cow. A national survey (Wilson & Richards 1980) showed that 32% of cows and 14.1% of quarters were sub-clinically infected with major mastitis pathogens. These sub-clinically infected quarters constitute sources from which infection is easily spread, usually at milking times, to other uninfected quarters. The investigations of Pearson *et al.* (1972) showed that annual milk sales from herds with high levels of sub-clinical udder infection were 20% less than similar herds with low levels of sub-clinical infection.

During the last 15 years the predominant pathogens involved in bovine mastitis have shown some changes. The most marked has been the substantial reduction in sub-clinical infection due to *Streptococcus agalactiae* as a result of the extensive adoption of a mastitis control programme involving the use

Isolation and Identification of
Micro-organisms of Medical and
Veterinary Importance

Copyright © 1985 by the Society for Applied Bacteriology
All rights of reproduction in any form reserved
0-12-181460-2

of teat disinfection and dry cow therapy on all cows. *Staphylococcus aureus* remains the predominant pathogen involved in sub-clinical infection. Clinical cases present a different picture because increased intensification in dairy farming coupled with the concentration of calvings into autumn and early winter has increased the bacterial challenge due to organisms from the cow's environment. This increased challenge at housing coincides with the early lactation period. The Mastitis Surveillance Scheme data show that *Escherichia coli* and *Streptococcus uberis* account for nearly one-third of clinical cases during the housing period.

Bacteriological examinations of milk samples from a proportion of cows in a herd or from clinical cases occurring during a period are carried out to determine the predominant pathogens affecting udders. Control methods can then be concentrated on the reservoirs of these organisms and on the means by which they are spread to unaffected quarters.

The value of the results of laboratory examinations is dependent upon the quality of the sample. Training or instruction on sample collection is given to those persons supplying samples to a mastitis laboratory.

Collection of Samples

Teats are washed and dried before sample collection only if they are obviously soiled. The teat end is thoroughly cleaned with a cotton wool pledget soaked in 70% alcohol, and the first few streams of milk (the strict foremilk) are discarded on the assumption that this will wash extraneous bacteria from the teat canal and reduce false positive bacteriological results. The sample of foremilk is collected into a sterile container as quickly as practicable, holding the container at an angle of at least 45° to reduce the likelihood of dust and debris entering. The aim is to draw 7–10 ml of milk into the tube with the minimum of teat handling.

Preliminary Examination

The appearance of the milk sample is noted and then one or more of the following tests is carried out.

Whiteside test

This is performed on a glass plate or black tile by mixing five drops of milk with one drop of normal NaOH and stirring briskly with a glass rod for 20 s

(Murphy & Hanson 1941). Milk from infected quarters (which will contain many polymorphs) gives a viscous mass when mixed with sodium hydroxide because the sodium salt of nucleic acid is sticky. Attempts have been made to grade reactions to the Whiteside test to indicate various cell count levels, but the subjectivity of the test is difficult to overcome.

Newman's stain

The Levowitz–Weber modification of the Newman–Lampert stain de-fats the milk film and stains both inflammatory cells and the bacteria blue during a single 2-min operation.

The stain is prepared by heating 54 ml of industrial methylated spirit (95% ethanol) with 40 ml of trichlorethane on a water bath at no higher than 70°C. Methylene blue, 1 g, is then added and dissolved. The solution is cooled and 6 ml of glacial acetic acid added slowly. The mixture is then filtered.

Examination of milk films by this method assists in the differentiation between samples which have been accidentally contaminated with bacteria and those from truly sub-clinically infected quarters.

Gram's stain

Air-dried films are de-fatted for 10 min by immersion in one of these mixtures: (1) equal parts of methanol and ether, or (2) ethanol, 52%; xylene, 44%; glacial acetic acid, 4%. They are then stained by Gram's method.

Culture Methods

Culture media

Sheep blood aesculin medium

This is blood agar base (Oxoid CM55) containing 0.5% NaCl and is bottled in 200-ml amounts. For use it is melted, and cooled to 56°C. Sheep blood, 15 ml, and aesculin (5% w/v in water), 5 ml, are added. One bottle provides 25 poured plates.

Edwards TA medium

This medium is supplied in dehydrated form by Oxoid. Nutrient agar is melted, cooled to 56°C and these ingredients added: aesulin, 1%; crystal violet, to 1 : 500,000; aesculin to 1 : 3000; defibrinated sheep blood, 5%.

Mycoplasma medium

Mycoplasma medium is a modification of the medium of Eaton & Lowe (1967). It contains PPLO broth base (Difco), 10.5 g; yeast extract*, 50 ml; unheated horse serum (e.g. Wellcome no. 3), 100 ml; glucose (5%), 20 ml; DNA (0.2%), 5 ml; ampicillin, 50 mg; phenol red (0.2%), 6.25 ml; distilled water, to 495 ml. The pH is adjusted to 7.6–7.8 with normal NaOH (usually about 5 ml) and the medium sterilized by filtration. If solid medium is required the filtered medium is added to previous sterilized agar solution. It is dispensed in 3-ml amounts in small screw-capped bottles.

Leptospire medium

EMJH medium is used. This is a modification by Johnson & Harris (1967) of the medium of Ellinghausen & McCullough (1965) and is obtained from Difco.

Choice and inoculation of media

When milk samples are collected by persons other than trained samplers, it is advisable to culture milk on 5% sheep blood agar, MacConkey's medium and on Edwards medium supplemented by the addition of thallium acetate. Brown & Gulbransen (1918) showed that streptococci were resistant to thallium acetate up to a concentration of 0.1%, and McKenzie (1941) specifically recommended a crystal violet–thallium acetate agar for mastitis bacteriology.

When the quality of the sample is expected to be high a 7.5% sheep blood aesculin agar is used alone. No more than 8 ml of medium is used per standard 9.5-cm diameter petri dish, as the thin layer of medium enables the haemolysis to be clearly seen. Increasing the blood content from the customary 5% to 7.5% promotes the growth of streptococci (Wilson, personal communication).

The milk samples are shaken manually to disperse the cream layer and 0.05 ml of milk is dispensed on to one-half of the plate. This drop is spread with a metal spreader, four of which are kept in boiling water and used in rotation. The milk is not spread over the entire area of the half plate, but a small area is left at the margin so that there is ample room between bacterial growth and the edge of the plate for the haemolytic zones to be visible. The plates are incubated at 37°C for 48 h and the bacteria are identified mainly by colony recognition.

The culture of mycoplasmas and leptospires is discussed later.

*The yeast extract is prepared by suspending 17 g of dried yeast in 100 ml of 0.2 *M* KH_2PO_4 and heating at 80–85°C for 20 min. The suspension is then centrifuged, sterilized by filtration and stored at −20°C.

The Visual Identification of Bacteria on Thin Blood Aesculin Agar Plates

Staphylococci

α-*Staphylococci* produce an α-haemolysin causing a clear β-haemolytic zone which has a hazy edge.

β-*Staphylococci* produce a β-haemolysin which causes a very wide zone of incomplete lysis. The zone of β lysis has a sharply defined margin.

γ-β-*Staphylococci* produce a combination of haemolysis with the clear α-haemolysin zone proximal to the colony and the β-haemolysin zone distal to the colony.

γ-*Staphylococci* produce no visible haemolysis, but these organisms are almost always coagulase positive and therefore likely to be pathogenic.

Δ-*Lysin staphylococci* produce a narrow zone of β-hemolysis with a clearly defined margin. These organisms are coagulase negative and not a major pathogen. Coagulase tests should be done on all staphylococci except those producing Δ-lysin. Coagulase-positive staphylococci are usually golden yellow in colour, but white variants are not uncommon. Small colony variants are also seen where prior antibiotic treatment has occurred; these are white and often have a narrow zone of β-hemolysis.

Streptococci

Streptococcus agalactiae appears as a slightly bluc colony with a narrow β-hemolytic zone. It does not split aesculin.

Streptococcus dysgalactiae shows α-haemolysis with a distinct green circle around each colony.

Streptococcus uberis forms a non-haemolytic grey colony which splits the aesulin producing a diffuse zone of browning of the agar. Occasionally *Strep. uberis* colonies are mucoid in appearance.

Streptococcus faecalis is non-haemolytic but sometimes greening. It splits aesculin, usually in a narrow distinctive black zone immediately underneath and around the colony. This distinguishes it from *Strep. uberis* on primary isolation. Aesculin fluoresces in UV light, so that when incubated plates are viewed by UV light, a dark zone is seen around colonies which split aesculin.

Micrococci

These organisms produce colonies similar in size to staphylococci. They are usually off-white in colour, but yellow colonies occasionally occur. They are coagulase negative. (Δ-Lysin staphylococci are also reported as micrococci.)

Coliforms

Coliform organisms appear as 2–4 mm diameter grey colonies.

Corynebacteria

Corynebacterium pyogenes is a strongly β-haemolytic organism somewhat similar in appearance to *Strep. agalactiae*, but it has no blue colour and produces a smaller colony.

Corynebacterium ulcerans appears similar to micrococci but is very white in colour, has a dry appearance and can be removed whole from the plate on a loop.

Corynebacterium bovis produces a small, very dry colony which is often restricted to areas of the plate containing most milk fat.

Pseudomonads

Pseudomonas aeruginosa produces large (2–4 mm diameter) blue/green opaque colonies which are strongly β-haemolytic and emit a distinct pungent smell.

Bacillus *species*

These are seen as large (up to 10 mm diameter), irregular, dry and usually crusty colonies which sometimes show haemolysis. These are usually considered to be contaminants.

Yeasts

Yeasts appear as 1–2 mm diameter white or grey colonies.

Rapid-growing acid-fast organisms

These organisms are unlikely to be seen after 48 h incubation, as they rarely appear until after 3–4 days. They usually form 2 mm diameter white or cream glistening colonies. The usual isolate from cases of mycobacterial mastitis is *Mycobacterium smegmatis*, and this can be identified by carrying out the presumptive test described by Naylor (1972).

Further Tests

Coagulase test

The slide coagulase test, should be carried out on all staphylococci. Tube tests should be done on all strains that give a negative slide test, as about 5% of coagulase-positive staphylococci will give a negative slide reaction. One drop of 0.85% saline is placed on a clean microscope slide and a colony is emulsified with the minimum amount of spreading. One drop of reconstituted rabbit plasma (Wellcome) is stirred into the bacterial suspension with a loop. Clumping becomes visible to the naked eye in 5–10 s if positive. A slower reaction is negative.

The tube test is carried out by mixing 0.5 ml of an 18 to 24-h broth culture of the organism with 0.5 ml of reconstituted rabbit plasma. This mixture is incubated for 4 h at 37°C and examined after 1 and 4 h for definite clot formation. Negative tubes should be left at room temperature overnight and then re-examined.

Modified CAMP test

This test is carried out on a sheep blood agar plate and is useful for distinguishing *Strep. agalactiae* from other streptococci. A β-haemolytic staphylococcus is streaked with a loop down the middle of a blood agar plate. The streptococcus under test is streaked at 90° to the staphylococcus. A known *Strep. agalactiae* is also streaked in a similar way to serve as a positive control. The plate is examined after overnight incubation at 37°C when the characteristic "arrow head" of complete clearing of the β-haemolytic zone caused by *Strep. agalactiae* can be seen. *Streptococcus dysgalactiae* produces no arrow head but causes greening of the blood agar.

Biochemical tests

Streptococci may be conclusively identified if required, by means of standard biochemical procedures.

Antibiotic sensitivity testing

Sensitivity tests are carried out on sensitivity test agar, although streptococci do not grow well in this medium. In the unlikely event of sensitivity tests being required on streptococci, blood agar should be used. Several identical colonies are emulsified in 4 ml of peptone water or normal saline and three loops full of this suspension are placed equidistantly on the sensitivity plate. The material is spread by loop technique, revolving the plate constantly so that all the area is covered. The aim is to produce a near-confluent growth with colonies just nudging each other after incubation. Antibiotic sensitivity discs (Difco, Mast, Oxoid) are applied to the surface of these inoculated plates and incubated for 18 h. A zone of growth inhibition 2 mm or more around any antibiotic disc indicates sensitivity. Sensitivity discs must be stored with a desiccant in sealed containers at 4°C. To prevent condensation the container should be allowed to come to room temperature before opening.

Presumptive test for M. smegmatis

Two drops of milk or part of the culture is subcultured on the oxalate organic acid medium (Naylor 1972) and incubated at 37°C. Cultures developing a deep red colour are considered to be positive. The caps of the bottles of those cultures which do not develop a colour after 48 h are loosened, and those not developing a deep red colour 4 h after loosening the caps are recorded as negative.

Oxalate medium

This is a modification of that of Gordon & Mihm (1959) containing (g/litre in distilled water): sodium oxalate, 2.0; NaCl, 1.0; $MgSO_4 \cdot 7H_2O$, 0.2; Na_2HPO_4, 1.1; KH_2PO_4, 0.5; agar, 15.0. The ingredients are dissolved by heat, the pH adjusted to 6.7 and 20 ml of phenol red (0.4%) added. It is dispensed in 5-ml amounts, autoclaved at 115°C for 20 min and cooled to set slopes.

Mycoplasmas

Mycoplasma bovis, *M. canadensae* and *M. californicum* have been recovered from cases of clinical mastitis in the UK. A history of multiple clinical cases which fail to respond to intramammary antibiotics should give an indication that mycoplasma may be involved. Further cultures on specific media, usually a modified form of Eaton's medium (Thorns & Boughton 1978), are then carried out. See also Taylor & Gaya, this volume.

Leptospires

Leptospira interrogans serotype hardjo causes the "milk drop syndrome" in dairy herds. Many cows are affected at the same time, usually with pyrexia and anorexia accompanied by significantly lowered milk yields. Affected cows yield a colostrum-like substance from all four quarters, and the udders are flaccid. Culture of the organism requires samples to be collected at the early stage of the disease, and three drops of each sample should be inoculated within 15 min of collection into 8 ml of EMJH medium to which has been added 5-fluorouracil and 2% rabbit serum (Hathaway & Little 1983). Further four 10-fold serial dilutions into the same media are made and all bottles are incubated at 30°C for 12 weekly examinations by dark-ground microscopy. See also Waitkins, this volume.

Observations

The aim of this scheme for examining milk samples is to identify the common causal organisms 48 h after receipt of the sample. By this time the animal will have received antibacterial therapy so the result is of historic interest. Nevertheless, a number of samples taken from any herd will give an indication of the predominant pathogens causing mastitis at that time.

The methods assume:

(1) That mastitis of economic importance is caused by bacteria which can be isolated by relatively simple methods.

(2) Most of the common primary pathogens are normally shed in sufficient numbers even in sub-clinically infected quarters for them to be detected when a small volume of foremilk is examined.

(3) The pathogens are regularly shed from infected quarters, so that it is unusual to fail to recover pathogens from foremilk samples from infected quarters.

(4) These pathogens are rarely found in foremilk samples from uninfected quarters.

References

Brown, C. H. & Gulbransen, R. 1918 *Applied Bacteriology* pp. 147–149. London: Oxford University Press.

Eaton, M. D. & Lowe, I. E. 1967 Propagation of *Mycoplasma pneumoniae* and other fastidious strains of PPLO. *Annals of the New York Academy of Sciences* **143**, 375–383.

Ellinghausen, H. C. & McCullough, W. G. 1965 Nutrition of *Leptospira pomona* and 13 other serotypes. A serum-free medium employing oleic albumin complex and polysorbate 80. *American Journal of Veterinary Research* **26**, 39–51.

Gordon, R. E. & Mihm, J. M. 1959 A comparison of four species of mycobacteria. *Journal of General Microbiology* **21**, 736–740.

Hathaway, S. C. & Little, T. W. A. 1983 Epidemiological study of *Leptospira* hardjo infection in second calf dairy cows. *Veterinary Record* **112**, 215–218.

JOHNSON, R. C. & HARRIS, V. G. 1967 Differentiation of pathogenic and saprophytic leptospires. *Journal of Bacteriology* **94**, 27–31.

MCKENZIE, D. A. 1941 The use of thallium acetate broth in the diagnosis of streptococcal mastitis. *Veterinary Record* **53**, 473–480.

MURPHY, J. M. & HANSON, J. H. 1941 A modified Whiteside test for the detection of chronic bovine mastitis. *Cornell Veterinarian* **31**, 47–55.

NAYLOR, P. F. 1972 A presumptive test for the detection of *Mycobacterium smegmatis* in milk. *Research in Veterinary Science* **13**, 93–94.

PEARSON, J. K. L., GREER, D. O., SPENCE, D. K., MCPARLAND, P. J., MCKINLEY, D. L., DUNLOP, W. L. & ACHESON, A. W. 1972 Factors involved in mastitis control—A comparative study between high and low incidence herds. *Veterinary Record* **91**, 615–624.

THORNS, C. J. & BOUGHTON, E. 1978 Studies on film production and its specific inhibition with special reference to *Mycoplasma bovis* (*M. agalactiae* var. *bovis*). *Zentralblatt für Veterinärmedizin, Reihe* B **25**, 657–667.

WILSON, C. D. & RICHARDS, M. S. 1980 A survey of mastitis in the British dairy herd. *Veterinary Record* **106**, 431–435.

Bacteria Associated with Diseases of Fish

G. N. FRERICHS

Institute of Aquaculture, University of Stirling, Stirling, UK

MARGARET S. HENDRIE

Torry Research Station, Aberdeen, UK

The principles followed in the diagnosis of bacterial fish diseases and the isolation and identification of the causal organisms are essentially the same as those applied to man and other animals. Rapid diagnostic methods such as the examination of wet preparations or stained smears from affected tissues can be useful where prompt treatment or prophylactic measures are required and there is a continuing demand for the extension and refinement of such techniques. The specific identification of fish bacteria, however, still requires the isolation of the organisms on bacteriological media and the application of physical, biochemical and, if necessary, serological tests to characterise the isolates.

Bacterial Disease Agents

Most of the bacteria associated with fish diseases are naturally saprophytic organisms, widely distributed in the aquatic environment; only a comparatively few species are classified as true obligate pathogens. Both classes of organism, however, may be present on the external body surface or within the tissues of apparently healthy fish and only assume a pathogenic role following the imposition of "stress" factors such as poor water quality, temperature changes, nutritional deficiencies, overcrowding, trauma, parasitism or primary viral infections. It is important that this be borne in mind when

Isolation and Identification of
Micro-organisms of Medical and
Veterinary Importance

Copyright © 1985 by the Society for Applied Bacteriology
All rights of reproduction in any form reserved
0-12-181460-2

TABLE 1

Bacterial pathogens of fish

Gram negative
Flexibacter
F. columnaris
*F. psychrophila**
*F. marinus**
Unclassified myxobacteria
Edwardsiella
E. tarda
*E. ictaluri**
*Yersinia ruckeri**
Pseudomonas
P. fluorescens
*P. anguilliseptica**
Aeromonas
A. hydrophila
A. salmonicida
Vibrio
V. anguillarum
Vibrio spp.
Flavobacterium spp.
Haemophilus piscium
*Pasteurella piscicida**
Gram positive
*Renibacterium salmoninarum**
Streptococcus
S. faecalis
Streptococcus spp.
Clostridium botulinum
Acid-fast
Mycobacterium
M. fortuitum
M. marinum
Nocardia
N. asteroides
*N. kampachi**

* Not described in Bergey's *Manual*, 8th ed. (1974).

assessing the significance of any bacterial isolates made from fish. Those organisms currently recognised as potential pathogens are listed in Table 1. Many other species of bacteria have been isolated from fish from time to time, but their role as pathogens remains uncertain and they have been omitted from this listing.

Examination of Fish

Fish presented for bacteriological examination fall into one of two categories, although the dividing line between one and the other is not always immediately obvious.

Diseased fish

These are specimens whose appearance or behavior indicates the existence of a disease process in which bacteria may be implicated. Whenever possible, a number of affected fish should be examined, ranging from those showing only mild clinical signs to moribund specimens. The isolation of the same organism from a number of fish greatly facilitates an assessment of its significance in the disease process. Fish that have been dead for some time are unsatisfactory, as bacterial invasion by commensal organisms occurs readily and rapidly in the dead fish and the isolation of even large numbers of any particular organism from such material may be of no clinical significance.

Healthy fish

These are usually either fish taken from the same population as diseased stock and submitted for simultaneous examination or a sample from a clinically healthy stock which is the subject of a disease monitoring or health certification programme.

Irrespective of category, a complete bacteriological investigation can be carried out only on dead specimens, and the usual procedure is to kill the fish either by decapitation or anaesthetic overdose immediately before examination. Benzocaine (0.2 g dissolved in 5 ml acetone and added to 8 litres of water) is an effective and relatively inexpensive anaesthetic agent for the purpose (Bullock 1978). It may sometimes be necessary to avoid killing the fish, in which case an examination will be restricted to the external surface, biopsy material or tissue fluid obtained from the anaesthetized specimen.

Selection of material

Diseased fish

The history and appearance of the fish will normally indicate the appropriate tissue(s) for examination. Many disease processes, including some forms of septicaemic conditions, give rise to visible lesions, particularly ulcers, on the external surface of the body and in these cases material from obviously affected areas is taken for examination. Similarly, if fins, tail or gills show pathological changes material from these sites is selected. An examination of the internal organs may reveal diffuse or discrete lesions and these should be examined as appropriate. When a septicaemic condition is suspected, the kidney is invariably selected for examination, although other internal organs are often affected and may also be used if considered appropriate.

Healthy fish

In the absence of any visible pathology or specific health certificate testing requirements, kidney tissue should be taken for further examination.

Sampling

Samples of material are conveniently obtained by inserting a sterile inoculating loop into the selected lesion or tissue. When external lesions are to be sampled, the surface of the lesion should be first seared with a hot scalpel blade and the inoculating loop inserted through the seared surface into the actively diseased edge of the lesion. Disinfectants should not be applied to external lesions before sampling but may be used to treat the body surface before opening the fish along the ventral surface to expose the internal organs.

Examination of material

Material taken from fish may be examined in one or more of the following ways.

Wet preparations

These are useful for the rapid presumptive diagnosis of myxobacterial infections, including columnaris, cold water (peduncle) or bacterial gill diseases and some forms of fin and tail rot. Myxobacteria can be seen as long, slender rods often exhibiting a characteristic flexing motility. Identification should be confirmed by culture.

Stained smears

These may indicate the presence of bacteria in affected tissues. Gram and Ziehl–Neelsen stains are appropriate. The morphology and size of the organisms can be noted and this may indicate a likely identity, which will, however, require confirmation by culture.

Stained smears alone are often used for the diagnosis of bacterial kidney disease (BKD) in salmonid fish. The presence of small, strongly Gram positive bacilli in kidney smears is considered diagnostic for infection with *Renibacterium salmoninarum* (Fig. 1). This procedure is usually satisfactory in cases of clinical disease where large numbers of bacilli may be found in affected tissue. The presence of melanin granules or other bacteria of similar morphology such as lactobacilli may, however, confuse the diagnosis (Ross &

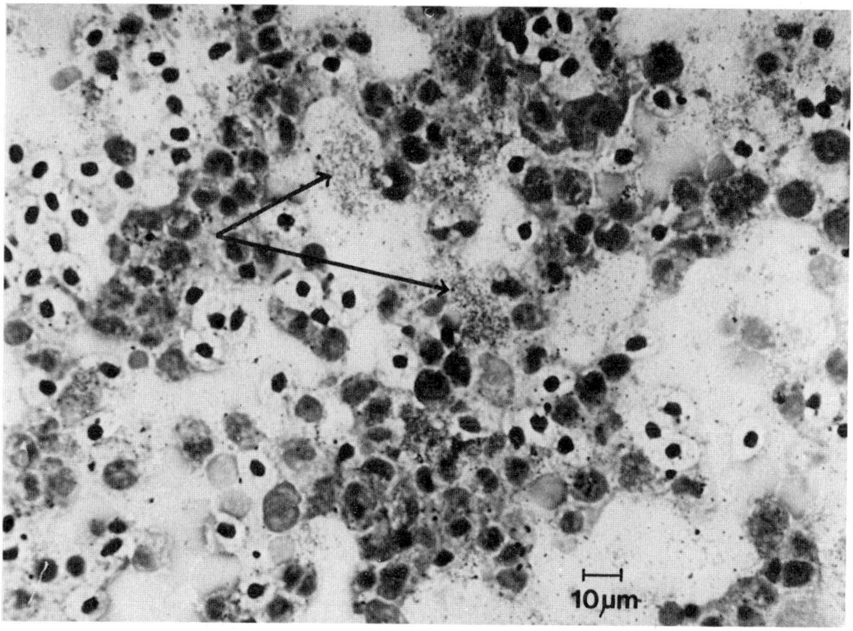

FIG. 1. Large numbers of *Renibacterium salmoninarum* organisms (arrow) in kidney smear of trout with macroscopic lesions of bacterial kidney disease.

Toth 1974), particularly in carrier fish or those in the early stages of infection when only small numbers of organisms may be present. In such cases more sensitive and specific procedures such as fluorescent antibody tests are indicated (Bullock *et al.* 1980). Whenever possible, the diagnosis should be confirmed by culture of the causal organism.

Culture

Apart from the possible exception of *R. salmoninarum* infection already noted, cultural techniques are required to identify bacterial infections of fish. This is necessary because different bacteria may give rise to similar clinical signs (e.g. septicaemia, granulomas, ulcers), and it is also not uncommon to encounter mixed infections which can be recognised and evaluated only by cultural methods.

Isolation of Bacteria

Two general approaches to the isolation of fish pathogens may be followed: selective and non-selective. In the selective approach, media designed to isolate a particular family or genus of organisms are used and are of value where specimens are being examined for the presence of specific organisms. This has the advantage of avoiding time spent identifying a multiplicity of organisms and lends itself to the application of rapid diagnostic procedures using a few identification tests. In the non-selective approach media which will support the growth of a wide range of organisms are used. Although this may result in a considerable amount of time and materials being spent in identification procedures, it has the advantage of providing both bacteriologists and clinicians with a comprehensive picture of the bacteriological status of the fish. It also lessens the possibility of an organism of significance remaining undetected which might happen with the selective approach.

The methods described here are orientated towards the non-selective approach, as this provides a generally applicable procedure from which a selective approach can be constructed according to requirements. An example of a selective procedure for the identification of Gram negative pathogens has been described by Shotts & Bullock (1975).

Culture media

The direct inoculation of solid media with tissue samples is satisfactory for the primary isolation of fish pathogens. The majority of these organisms are non-fastidious in their nutrient requirements and many general purpose media (e.g. nutrient agar) may be used, although Tryptone Soya agar (TSA; Oxoid) has become the medium of choice in most laboratories. This will support the growth of all enterobacteria, pseudomonads, vibrios, flavobacteria, pasteurellae, streptococci, micrococci and nocardias likely to be encountered in freshwater fish and has the particular advantage of clearly demonstrating the formation of the diffusible brown pigment characteristically produced by the majority of *Aeromonas salmonicida* isolates (Fig. 2). It is necessary to supplement the medium with sodium chloride to a final concentration of 2% when fish from marine or other salt-water environments are examined, as some strains of *Vibrio anguillarum* in particular have a salt requirement for initial isolation. Marine agar (Difco), which contains a supplement which duplicates the major mineral components of seawater, is a good alternative medium for this purpose. Once isolated, the salt requirement is often lost and sub-culturing can usually be made on standard TSA medium (see also Lee & Donovan, this volume).

The most commonly encountered group of pathogens which are not recoverable or grow only poorly on TSA medium are the myxobacteria. These organisms require comparatively low levels of both nutrients and agar, and *Cytophaga* agar (Anacker & Ordal 1959) [composed of (g/litre in distilled water): tryptone, 0.5; sodium acetate, 0.2; beef extract, 0.2; and agar, 0.9]

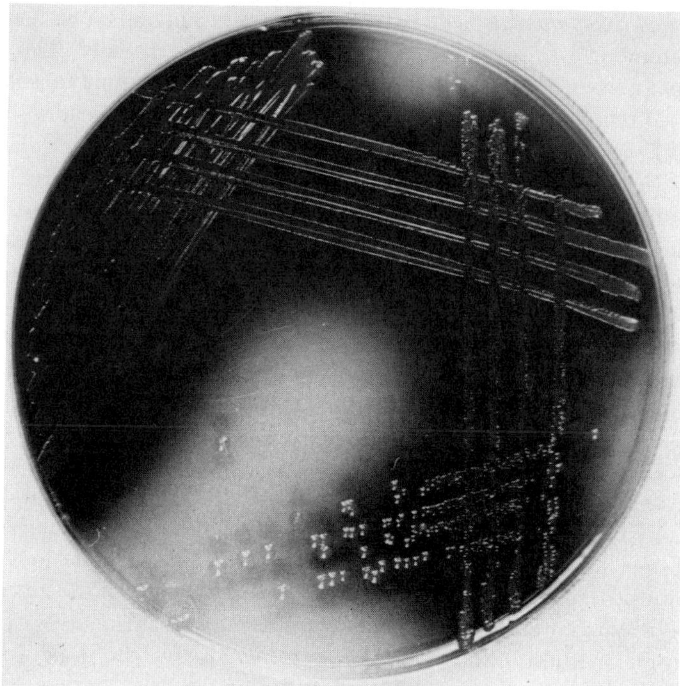

FIG. 2. A 72-h culture of *Aeromonas salmonicida* on TSA medium showing development of diffusible brown pigment.

provides a suitable medium for the isolation and maintenance of freshwater strains. Other general purpose media diluted to contain about one-tenth of nutrients and 1% agar are satisfactory alternatives. Few marine strains of myxobacteria tolerate transfer to fresh water or vice versa (Lewin & Lounsbery 1969), and isolation from marine fish should be attempted using medium containing 50–70% seawater (Sawyer 1976; Hikida *et al.* 1979) or the commercially available marine agar (Campbell & Buswell 1982).

Enriched media are essential for the isolation of *Haemophilus piscium*, the causal agent of ulcer disease in trout, the BKD agent (*Renibacterium salmoninarum*) and *Mycobacteria* spp., although some of the latter organisms are capable of growth on TSA medium. As with myxobacterial diseases, infection with these agents is usually suspected on specific clinical or pathological evidence or following examination of stained smears, and isolation is not normally attempted in the absence of such indications. *Haemophilus piscium* has a requirement for adenosine triphosphate or co-carboxylase (diphosphothiamine) which may be provided using Mueller–Hinton agar (Difco) containing 2 μg/ml of commercially available co-carboxylase (Bullock 1965*a*).

Renibacterium salmoninarum is an extremely fastidious organism with an absolute requirement for L-cysteine. Ordal & Earp (1956) initially developed a medium containing 20% human blood and 0.1% w/v L-cysteine to isolate the bacterium directly from fish tissue, and a number of modifications of this medium have since been described, of which the KDM-2 formulation of Evelyn (1977) has become the most widely used. This contains (g/litre in distilled water): peptone, 10.0; yeast extract, 0.5; cysteine HCl, 1.0; and agar, 15.0. It is supplemented with fetal calf serum (10% v/v). In addition to its fastidious nutritional requirements, *R. salmoninarum* grows slowly and on initial isolation may require 6 weeks of incubation before visible colonies appear. Overgrowth of plates with faster growing contaminants or other pathogens may thus be a problem, and the incorporation of selective inhibitors such as cyclohexamide (50 mg/litre), D-cycloserine (12.5 mg/litre), oxolinic acid (2.5 mg/litre) and polymyxin B sulphate (25 mg/litre) may be indicated (Austin *et al.* 1983). Serum-free media such as cysteine-supplemented Mueller–Hinton broth (Wolf & Dunbar 1959) and the semi-defined medium of Embley *et al.* (1982) have been used successfully for the maintenance of stock cultures, but the presence of serum is advisable for primary isolation purposes. *Mycobacteria* spp. can be isolated on any of the glycerol- or egg-based media commonly used for this group of organisms (e.g. Lowenstein–Jensen medium) and *Clostridium botulinum* by the inoculation of Robertson's cooked meat broth (Cann & Taylor 1982).

Incubation

The temperature range over which the different pathogens will grow and the optimum temperature for each species varies considerably, but, as a matter of practical expediency, a single incubation temperature within the range 18–23°C is generally used for isolation purposes. Incubation temperatures outside this range are required for certain subsequent identification tests.

All organisms, apart from *C. botulinum*, which is a strict anaerobe, can be cultured aerobically.

Visible growth may be expected to appear within 48 h in most cultures, but, as already noted, *R. salmoninarum* may require 6 weeks and some of the mycobacteria require at least 3 weeks of incubation before colonies are seen.

Identification of Bacteria

The importance of obtaining pure cultures before carrying out identification tests cannot be over-emphasized. Mixed infections, giving rise to mixed cultures on primary isolation, particularly when non-selective media are used, are commonly encountered, and purification by sub-culture is essential to obtain an accurate identification of isolates.

As with almost all other identification procedures used in medical and veterinary bacteriology, the Gram reaction is the first characterisation test applied to isolates obtained from fish. Hucker's modification of the method (Hucker & Conn 1923) using crystal violet/ammonium oxalate to stain the preparation and alcohol/acetone for decolorization is most commonly used.

Gram negative bacteria

Flexibacter

Although the term myxobacteria is commonly used to describe the group of gliding bacteria associated with a number of fish diseases, all the recognised pathogens are taxonomically members of the family Cytophagaceae, genus *Flexibacter* (Leadbetter 1974). These organisms can usually be recognised at the primary isolation stage of the identification procedure, as they form characteristic yellow, spreading colonies and show gliding motility on low nutrient media. Gliding is determined by direct microscopic observation of the swarming edge of a colony (Hayes *et al.* 1979), and in stained smears the organisms are seen as long, slender rods (0.5–0.8 μm × 2–100 μm). Further identification to the species level is seldom required for diagnostic purposes, as *F. columnaris* is pathogenic only at water temperatures above 15°C, unlike *F. psychrophila* which causes disease at temperatures below 12°C (Richards & Roberts 1978). Differentiation between the two species on the basis of growth at 25 and 37°C, degradation of tributyrin, H_2S production and tyrosine decomposition has also been described (Pacha & Porter 1968.) It has been proposed that those strains of *Flexibacter* with an obligate growth requirement for seawater, which cannot be replaced by NaCl alone, should be designated *F. marinus* (Hikida *et al.* 1979).

Aeromonas, Vibrio, Pseudomonas, Flavobacterium, Edwardsiella, Yersina *and* Pasteurella

Potential pathogenic species from all these genera may be isolated on TSA medium. In stained smears they may appear almost coccobacillary in form or as short to medium, straight or curved rods (0.5–0.8 μm × 1–3 μm).

Although the flavobacteria, some pseudomonads and the majority of *A. salmonicida* strains produce characteristically pigmented colonies, this is not adequate for diagnostic purposes, and additional tests are required to establish the identities of both pigmented and non-pigmented isolates. Determination of oxidase activity using the method of Kovács (1956) or commercially available impregnated paper strips (Pathotec, General Diagnostics, New Jersey) and oxidative–fermentative (O–F) attack on glucose, using O–F

TABLE 2

*Primary tests for Gram negative bacteria**

Bacteria	Oxidase	O/F test†
Aeromonas	+	F
Vibrio	+	F
Pasteurella	+	F
Enterobacteriaceae	−	F
Pseudomonas	+	O/−
Flavobacterium	+	O/−

* These are the results to be expected from at least 90% of strains within the genera.
† Oxidation–fermentation test: F, fermentative; O/−, oxidative/no reaction.

basal medium (Difco) or NaCl-supplemented basal medium as appropriate, will initially divide the organisms into groups as shown in Table 2.

A motility test using broth cultures will provide further differentiation within the initial groupings. *Aeromonas salmonicida* is the only non-motile member of the Vibrionaceae, and this characteristic together with the production of a brown, diffusible pigment on solid media is virtually diagnostic for the species. A macroscopic slide agglutination test which may be used for both confirmatory and rapid identification purposes has also been developed for *A. salmonicida* (Rabb *et al.* 1964). A small amount of bacterial growth taken from a solid medium is emulsified with a drop of saline on a microscope slide and mixed with a similar volume of anti-serum. Rapid clumping of the organisms within a minute or so constitutes a positive agglutination reaction. The vibrios and motile aeromonads can be distinguished on the basis of sensitivity to novobiocin, 10 μg/ml, and the O/129 vibriostatic compound (BDH), 150 μg/ml. Almost all *Vibrio* spp. are sensitive to these two agents, whereas *Aeromonas* spp. are typically resistant.

In the oxidase-positive, non-fermentative group of organisms, all Pseudomonadaceae are motile whereas the flavobacteria, with rare exceptions, are non-motile (Hayes *et al.* 1979). Flavobacteria also form yellow, orange or red pigmented colonies on solid media (Weeks 1974), but as similarly pigmented pseudomonads may be isolated, pigmentation alone cannot be relied upon to distinguish between the genera.

Although this simplified test procedure will enable the majority of fish pathogens to be identified at least to the genus level, it should be remembered that, as with all bacteria, species variants may be encountered. For example, both non-pigmented strains of *A. salmonicida* and brown-pigmented strains of *A. hydrophila* have been isolated. Similarly, pseudomonads

Table 3

Secondary tests for oxidase-positive, fermentative, Gram negative rods*

Test	*Aeromonas hydrophila*	*Aeromonas salmonicida*	*Vibrio anguillarum*	*Vibrio ordalii*	*Vibrio tubiashii*	*Vibrio vulnificus* biogroup 2
Motility	+	–	+	+	+	+
Growth at 5°C	+	+	+	–	–	–
Growth at 37°C	+	–	d	–	d	+
Growth without added NaCl	+	+	–	–	–	–
Growth in 5.0% NaCl	–	–	+		+	+
Growth in 7.0% NaCl	–	–	+	–	+	–
Brown diffusible pigment	–	+	–	–	–	–
Sensitivity to novobiocin	–	–	+	+	+	+
Sensitivity to O/129	–	–	+	+	+	+
Indol	+	–	d	–	+	–
Methyl red	d	–	d	–	–	+
Voges–Proskauer	d	–	+	–	–	–
Citrate	d	–	d	–	+	+
Gluconate	d	–	–	–	–	+
Arginine dihydrolase	+	+	+	–	+	–
Lysine decarboxylase	–	–	–	–	–	+
Hydrolysis of starch	+	+	+	–	+	+
Hydrolysis of casein	+		+	+	+	+
β-Galactosidase (ONPG)	+	–	+	–	–	+
Xanthine degradation	–	–	–	–	–	+
Gas from glucose	d	d	–	–	–	–
Acid from arabinose	+	d	d	–	–	–
sucrose	+	d	+	+	+	–
trehalose	+	d	+	–		+
glycerol	d	+	d	–	–	d
mannitol	+	d	+	+	–	–
sorbitol	d	–	+	–	–	–
salicin	+	–	–	–	+	+
DNA base ratio (mole % G+C)	57–63	56.5–58.5	43.5–45			45.7–46.7

* Features common to all species: Gram negative rods; motile species have polar flagella; fermentative; oxidase (Kovács) and catalase positive; gelatin hydrolysed; urease, ornithine decarboxylase and malonate negative. Acid produced from glucose, galactose and maltose; no acid in raffinose. +, 90% or more strains positive; –, 90% or more strains negative; d, 11–89% strains positive.

which are non-oxidizing or give an alkaline reaction in the O–F test may be recovered, and in order to confirm the identity of the recognised pathogens a range of secondary tests should be carried out where possible. Such additional testing is in any case necessary to identify both vibrios and pseudomonads at the species level and also those species not specifically considered here but which may also be isolated quite frequently. A selected range of secondary tests is given in Tables 3–5.

Haemophilus piscium

This organism is specifically associated with ulcer disease in salmonids and can be isolated only on a suitably enriched medium. It is a non-motile rod (0.8–1 μm × 1–3 μm) which may show bipolar staining with Giemsa stain. Nitrates are not reduced and indole is not formed. The Voges–Proskauer test is negative but the methyl red test is positive (Snieszko *et al.* 1950).

Identification tables

The properties of the Gram negative organisms are given in Tables 3, 4 and 5, which have been constructed from the papers of the following workers, and also from the personal experience of M.S.H.: Austin *et al.* (1981); Bullock (1965*b*); Bullock *et al.* (1978); Egidius & Andersen (1977); Evelyn (1971); Ewing *et al.* (1978); Ezura *et al.* (1980); Hada *et al.* (1984); Hansen *et al.* (1965); Håstein *et al.* (1978); Hawke (1979); Hawke *et al.* (1981); Hodgkiss & Shewan (1950); Jeffries (1982); Kimura (1969); Mandel *et al.* (1965); Nishibuchi *et al.* (1980); Nishibuchi *et al.* (1979); O'Leary *et al.* (1979); Ross (1962); Ross *et al.* (1966); Schiewe *et al.* (1981); Smith (1963); Steward *et al.* (1983); Tison *et al.* (1982); Wakabayashi & Egusa (1979).

Gram positive bacteria

Renibacterium salmoninarum

The presence of small Gram positive bacilli (0.1–1 μm × 0.3–1.5 μm) in kidney smears is considered diagnostic for *R. salmoninarum* infection (Fig. 1). This should, whenever possible, be confirmed by isolation of the organism on an appropriately enriched medium. Incubation for at least a week, and sometimes considerably longer, is required before visible growth of white to creamy yellow colonies of varying size is observed (Fryer & Sanders 1981).

Streptococcus, Micrococcus *and* Nocardia

Streptococci, micrococci and nocardias associated with disease in fish may be isolated on TSA medium. They can be recognised in stained smears

TABLE 4

Secondary tests for oxidase-positive, non-fermentative, motile, Gram negative rods*

Test	*Alcaligenes faecalis* sp. *homari*	*Alteromonas piscicida*	*Pseudomonas anguilliseptica*	*Pseudomonas fluorescens*	*Pseudomonas putida*
Diffusible pigment	−	−	−	Green, fluorescent	Green, fluorescent
Non-diffusible pigment	−	Yellow	−	−	−
Growth at 5°C	+	−	+	+	+
Growth at 37°C	+	+	−	d	d
Growth without added NaCl	−	−	−	+	+
Growth in 5.0% NaCl	+	+	−	+	+
Growth in 7.0% NaCl	+	+	−	−	−
Growth in 10.0% NaCl	+	+	−	−	−
Sensitivity to penicillin	+	−	−	−	−
Sensitivity to novobiocin	−	+	+	−	−
O-F medium (glucose)	Alkaline	Oxidative	−	Oxidative	Oxidative
Nitrate reduction	+	+	−	+	+
Citrate	+	−	+	+	I
Gluconate		+	−	+	+
Arginine dihydrolase	−	−	−	+	+
Hydrolysis of gelatin	−	+	+	+	−
Hydrolysis of casein	−	+	+	+	−
Hydrolysis of starch	+(Slow)	+	−	−	−
DNAase		+	−	−	−
Flagella	Peritrichous	Polar	Polar	Polar	Polar
DNA base ratio (mole % G + C)	56.6	43.5–45	56.5–57.4	59.4–61.3	60–63

* Features common to all species: Gram negative, motile rods; non-fermentative; oxidase (Kovács) positive; insensitive to O/129 (2,4-diamino-6,7-di*iso*propyl pteridine); indol, methyl red, and Voges–Proskauer tests all negative. +, 90% or over positive; −, 90% or over negative; d, 11–89% positive.

TABLE 5

Secondary identification tests for oxidase-negative, fermentative, Gram negative rods*

Test	*Edwardsiella ictaluri*	*Edwardsiella tarda*	*Yersinia ruckeri*
Indol	−	+	−
H_2S production	−	+	−
Citrate	−	−	+
Gelatin hydrolysis	−	−	+
Gas from glucose	+	+	−
Acid from mannitol	−	−	+
Acid from trehalose	−	−	+
β-Galactosidase (ONPG)	+	−	+
DNA base ratio (mole % G+C)	53	56–59	47.5–48.5

* Features common to all species: Gram negative rods, motile by polar flagella; non-pigmented; oxidase (Kovács) negative; fermentative; positive in methyl red, lysine and ornithine decarboxylase tests; negative in Voges–Proskauer, arginine dihydrolase, urease, gluconate and malonate tests. Acid produced from glucose and maltose; no acid from adonitol, arabinose, dulcitol, lactose, salicin, sorbitol, sucrose or xylose.

either as cocci in short chains (*Streptococcus* spp.) or irregular clumps (*Micrococcus* spp.) or as branching rods (*Nocardia* spp.). The latter organisms are also weakly acid-fast by Ziehl–Neelsen staining. Reference should be made to the *Manual for the Identification of Medical Bacteria* (2nd ed.) (Cowan 1974) if identification of streptococci and micrococci at the species level is required. *Streptococcus faecalis* is the only member of the genus which has so far been specifically identified as a fish pathogen, although organisms belonging to Lancefield's group B and a number of untyped isolates have also been incriminated. Species identification of nocardias is seldom required. *Nocardia asteroides* and *N. kampachi* are the two species for which a pathogenic role has been established (Snieszko *et al.* 1964; Kusuda & Taki 1973).

Mycobacteria

These organisms are not readily stained by Gram's method but are acid/alcohol-fast with the Ziehl–Neelsen stain. *Mycobacterium fortuitum* and some other saprophytic strains will grow within 3–4 days on simple media and may be mistaken for *Nocardia* spp., but *M. marinum* may take 3 weeks to grow and requires an egg- or glycerol-based medium. Identification of *Mycobacteria* spp. is best carried out by a specialist laboratory with appropriate facilities for handling these potential human pathogens.

Clostridium botulinum

As fish often harbour *C. botulinum* as a commensal in their intestines, the isolation of the organism in suspected cases of botulism is not sufficient to confirm the diagnosis. Definitive diagnosis is dependent upon the demonstration of the toxin in tissues from affected fish (Cann *et al.* 1965).

References

ANACKER, R. L. & ORDAL, E. J. 1959 Studies on the myxobacterium *Chondrococcus columnaris* I. Serological typing. *Journal of Bacteriology* **78**, 25–32.

AUSTIN, B., RODGERS, C. J., FORNS, J. M. & COLWELL, R. R. 1981 *Alcaligenes faecalis* subsp. *homari* subsp. nov., a new group of bacteria isolated from moribund lobsters. *International Journal of Systematic Bacteriology* **31**, 72–76.

AUSTIN, B., EMBLEY, T. M. & GOODFELLOW, M. 1983 Selective isolation of *Renibacterium salmoninarum*. *FEMS Microbiology Letters* **17**, 111–114.

BULLOCK, G. L. 1965*a* Simple enrichment of commercial media for growth of *Haemophilus piscium*. *Progressive Fish-Culturist* **27**, 163–164.

BULLOCK, G. L. 1965*b* Characteristics and pathogenicity of a capsulated *Pseudomonas* isolated from goldfish. *Applied Microbiology* **13**, 89–92.

BULLOCK, A. M. 1978 Laboratory methods. In *Fish Pathology*, ed. Roberts, R. J., pp. 235–267. London: Baillière Tindall.

BULLOCK, G. L., STUCKEY, H. M. & SHOTTS, E. B., JR. 1978 Enteric redmouth bacterium: comparison of isolates from different geographic areas. *Journal of Fish Diseases* **1**, 351–356.

BULLOCK, G. L., GRIFFIN, B. R. & STUCKEY, H. M. 1980 Detection of *Corynebacterium salmoninus* by direct fluorescent antibody test. *Canadian Journal of Fisheries and Aquatic Sciences* **37**, 719–721.

CAMPBELL, A. C. & BUSWELL, J. A. 1982 An investigation into the bacterial aetiology of 'black patch necrosis' in Dover Sole, *Solea solea* L. *Journal of Fish Diseases* **5**, 495–508.

CANN, D. C. & TAYLOR, L. Y. 1982 An outbreak of botulism in rainbow trout, *Salmo gairdneri* Richardson, farmed in Britain. *Journal of Fish Diseases* **5**, 393–399.

CANN, D. C., WILSON, B. B., HOBBS, G., SHEWAN, J. M. & JOHANNSEN, A. 1965 The incidence of *Clostridium botulinum* Type E in fish and bottom deposits in the North Sea and off the coast of Scandinavia. *Journal of Applied Bacteriology* **28**, 426–430.

COWAN, S. T. 1974 *Cowan and Steel's Manual for the Identification of Medical Bacteria* 2nd edn. London: Cambridge University Press.

EGIDIUS, E. & ANDERSEN, K. 1977 Norwegian reference strains of *Vibrio anguillarum*. *Aquaculture* **10**, 215–219.

EMBLEY, T. M., GOODFELLOW, M. & AUSTIN, B. 1982 A semi-defined growth medium for *Renibacterium salmoninarum*. *FEMS Microbiology Letters* **14**, 299–301.

EVELYN, T. P. T. 1971 First records of vibriosis in Pacific salmon cultured in Canada and taxonomy of the responsible bacterium *Vibrio anguillarum*. *Journal of the Fisheries Research Board of Canada* **28**, 517–525.

EVELYN, T. P. T. 1977 An improved growth medium for the Kidney Disease bacterium and some notes on using the medium. *Bulletin de l'Office International des Epizooties* **87**, 511–513.

EWING, W. H., ROSS, A. J., BRENNER, D. J. & FANNING, G. R. 1978 *Yersinia ruckeri* sp. nov., the redmouth (RM) bacterium. *International Journal of Systematic Bacteriology* **28**, 37–44.

EZURA, Y., TAJIMA, K., YOSHIMIZU, M. & KIMURA, T. 1980 Studies on the taxonomy and serology of causative organisms of fish vibriosis. *Fish Pathology* **14**, 167–179.

FRYER, J. L. & SANDERS, J. E. 1981 Bacterial kidney disease of salmonid fish. *Annual Review of Microbiology* **35**, 273–298.

HADA, H. S., WEST, P. A., LEE, J. V., STEMMLER, R. J. & COLWELL, R. R. 1984 *Vibrio tubiashii* sp. nov., a pathogen of bivalve mollusks. *International Journal of Systematic Bacteriology* **34**, 1–4.

HANSEN, A. J., WEEKS, O. B. & COLWELL, R. R. 1965 Taxonomy of *Pseudomonas piscicida* (Bein) Buck, Meyers & Leifson. *Journal of Bacteriology* **89**, 752–761.

HÅSTEIN, T., SALTVEIT, S. J. & ROBERTS, R. J. 1978 Mass mortality among minnows, *Phoxinus phoxinus* (L.) in Lake Tveitevatn, Norway, due to an aberrant strain of *Aeromonas salmonicida*. *Journal of Fish Diseases* **1**, 241–249.

HAWKE, J. P. 1979 A bacterium associated with disease of pond cultured channel catfish, *Ictalurus punctatus*. *Journal of the Fisheries Research Board of Canada* **36**, 1508–1512.

HAWKE, J. P., MCWHORTER, A. C., STEIGERWALT, A. G. & BRENNER, D. J. 1981 *Edwardsiella ictaluri* sp. nov., the causative agent of enteric septicemia of catfish. *International Journal of Systematic Bacteriology* **31**, 396–400.

HAYES, P. R., MCMEEKIN, T. A. & SHEWAN, J. M. 1979 The identification of gram negative, yellow pigmented rods. In *Identification Methods for Microbiologists* 2nd edn, ed. Skinner, F. A. & Lovelock, D. W., pp. 177–187. London & New York: Academic Press.

HIKIDA, M., WAKABAYASHI, H., EGUSA, S. & MASUMURA, K. 1979 *Flexibacter* sp., a gliding bacterium pathogenic to some marine fishes in Japan. *Nippon Suisan Gakkaishi* **45**, 421–428.

HODGKISS, W. & SHEWAN, J. M. 1950 *Pseudomonas* infection in a plaice. *Journal of Pathology and Bacteriology* **62**, 655–657.

HUCKER, G. J. & CONN, H. J. 1923 Methods of Gram staining. *New York, Agricultural Experimental Station, Geneva, Technical Bulletin* No. 93.

JEFFRIES, V. E. 1982 Three *Vibrio* strains pathogenic to larvae of *Crassostrea gigas* and *Ostrea edulis*. *Aquaculture* **29**, 201–226.

KIMURA, T. 1969 A new subspecies of *Aeromonas salmonicida* as an etiological agent of furunuculosis on "Sakuramasu" (*Oncorhynchus masou*) and pink salmon (*O. gorbuscha*) rearing for maturity. Part 1. On the morphological and physiological properties. *Fish Pathology* **3**, 34–44.

KOVACS, N. 1956 Identification of *Pseudomonas pyocyanea* by the oxidase reaction. *Nature (London)* **178**, 703.

KUSUDA, R. & TAKI, H. 1973 Studies on a nocardial infection of cultured yellowtail—I. Morphological and biochemical characteristics of *Nocardia* isolated from diseased fishes. *Nippon Suisan Gakkaishi* **39**, 937–943.

LEADBETTER, E. R. 1974 Cytophagales. In *Bergey's Manual of Determinative Bacteriology* 8th edn, ed. Buchanan, R. E. & Gibbons, N. E., pp. 99–107. Baltimore, Maryland: Williams & Wilkins.

LEWIN, R. A. & LOUNSBERY, D. M. 1969 Isolation, cultivation and characterisation of flexibacteria. *Journal of General Microbiology* **58**, 145–170.

MANDEL, M., WEEKS, O. B. & COLWELL, R. R. 1965 Deoxyribonucleic acid base of *Pseudomonas piscicida*. *Journal of Bacteriology* **90**, 1492–1493.

NISHIBUCHI, M., MUROGA, M., SEIDLER, R. J. & FRYER, J. L. 1979 Pathogenic *Vibrio* isolated from cultured eels. IV. Deoxyribonucleic acid studies. *Nippon Suisan Gakkaishi* **45**, 1469–1473.

NISHIBUCHI, M., MUROGA, K. & JO, Y. 1980 Pathogenic *Vibrio* isolated from cultured eels. V. Diagnostic tests for the disease due to the present bacterium. *Fish Pathology* **14**, 125–130.

O'Leary, P. J., Rohovec, J. S. & Fryer, J. L. 1979 A further characterization of *Yersinia ruckeri* (Enteric Redmouth Bacterium). *Fish Pathology* **14,** 71–78.

Ordal, E. J. & Earp, B. J. 1956 Cultivation and transmission of etiological agent of kidney disease in salmonid fishes. *Proceedings of the Society for Experimental Biology and Medicine* **92,** 85–88.

Pacha, R. E. & Porter, S. 1968 Characteristics of myxobacteria isolated from the surface of freshwater fish. *Applied Microbiology* **16,** 1901–1906.

Rabb, L., Cornick, J. W. & McDermott, L. A. 1964 A macroscopic-slide agglutination test for the presumptive diagnosis of furunculosis in fish. *Progressive Fish-Culturist* **26,** 118–120.

Richards, R. H. & Roberts, R. J. 1978 The bacteriology of teleosts. In *Fish Pathology*, ed. Roberts, R. J., pp. 183–204. London: Baillière Tindall.

Ross, A. J. 1962 Isolation of a pigment-producing strain of *Aeromonas liquefaciens* from silver salmon (*Oncorhynchus kisutch*). *Journal of Bacteriology* **84,** 590–591.

Ross, A. J. & Toth, R. J. 1974 *Lactobacillus*—A new fish pathogen? *Progressive Fish-Culturist* **36,** 191.

Ross, A. J., Rucker, R. R. & Ewing, W. H. 1966 Description of a bacterium associated with redmouth disease of rainbow trout (*Salmo gairdneri*). *Canadian Journal of Microbiology* **12,** 763–770.

Sawyer, E. S. 1976 An outbreak of myxobacterial disease in coho salmon (*Oncorhynchus kisutch*) reared in a Maine estuary. *Journal of Wildlife Diseases* **12,** 575–578.

Schiewe, M. H., Trust, T. J. & Crosa, J. H. 1981 *Vibrio ordalii* sp.nov.: a causative agent of vibriosis in fish. *Current Microbiology* **6,** 343–348.

Shotts, E. B. & Bullock, G. L. 1975 Bacterial diseases of fishes: Diagnostic procedures for gram-negative pathogens. *Journal of the Fisheries Research Board of Canada* **32,** 1243–1247.

Smith, I. W. 1963 The classification of '*Bacterium salmonicida*'. *Journal of General Microbiology* **33,** 263–274.

Snieszko, S. F., Griffin, P. J. & Friddle, S. B. 1950 A new bacterium (*Haemophilus piscium* n. sp.) from ulcer disease of trout. *Journal of Bacteriology* **59,** 699–710.

Snieszko, S. F., Bullock, G. L., Dunbar, C. E. & Pettijohn, L. L. 1964 Nocardial infection in hatchery-reared fingerling rainbow trout (*Salmo gairdneri*). *Journal of Bacteriology* **88,** 1809–1810.

Stewart, D. J., Woldermariam, K., Dear, G. & Mochaba, F. M. 1983 An outbreak of 'Sekiten-byo' among cultured European eels, *Anguilla anguilla* L., in Scotland. *Journal of Fish Diseases* **6,** 75–76.

Tison, D. L., Nishibuchi, M., Greenwood, J. D. & Seidler, R. J. 1982 *Vibrio vulnificus* biogroup 2: new group pathogenic for eels. *Applied and Environmental Microbiology* **44,** 640–646.

Wakabayashi, H. & Egusa, S. 1979 Characteristics of a *Pseudomonas* sp. from an epizootic of pond-cultured eels (*Anguilla japonica*). *Nippon Suisan Gakkaishi* **38,** 577–587.

Weeks, O. B. 1974 Genus *Flavobacterium* Bergey et al 1923. In *Bergey's Manual of Determinative Bacteriology* 8th edn, ed. Buchanan, R. E. & Gibbons, N. E., pp. 357–364. Baltimore, Maryland: Williams & Wilkins.

Wolf, K. & Dunbar, C. E. 1959 Test of 34 therapeutic agents for control of Kidney Disease in trout. *Transactions of the American Fisheries Society* **88,** 117–124.

Subject Index

THE SOCIETY FOR APPLIED BACTERIOLOGY TECHNICAL SERIES

General Editor: F. A. Skinner

1 Identification Methods for Microbiologists Part A *1966*
Edited by B. M. Gibbs and F. A. Skinner
Out of print

2 Identification Methods for Microbiologists Part B *1968*
Edited by B. M. Gibbs and D. A. Shapton
Out of print

3 Isolation Methods for Microbiologists *1969*
Edited by D. A. Shapton and G. W. Gould

4 Automation, Mechanization and Data Handling in Microbiology *1970*
Edited by Ann Baillie and R. J. Gilbert

5 Isolation of Anaerobes *1971*
Edited by D. A. Shapton and R. G. Board

6 Safety in Microbiology *1972*
Edited by D. A. Shapton and R. G. Board

7 Sampling—Microbiological Monitoring of Environments *1973*
Edited by R. G. Board and D. W. Lovelock

8 Some Methods for Microbiological Assay *1975*
Edited by R. G. Board and D. W. Lovelock

9 Microbial Aspects of the Deterioration of Materials *1975*
Edited by R. J. Gilbert and D. W. Lovelock

10 Microbial Ultrastructure: The Use of the Electron Microscope *1976*
Edited by R. Fuller and D. W. Lovelock

11 Techniques for the Study of Mixed Populations *1978*
Edited by D. W. Lovelock and R. Davies

12 Plant Pathogens *1979*
Edited by D. W. Lovelock

13 Cold Tolerant Microbes in Spoilage and the Environment *1979*
Edited by A. D. Russell and R. Fuller

14 Identification Methods for Microbiologists (2nd Edn) *1979*
Edited by F. A. Skinner and D. W. Lovelock

15 Microbial Growth and Survival in Extremes of Environment *1980*
Edited by G. W. Gould and Janet E. L. Corry

16 Disinfectants: Their Use and Evaluation of Effectiveness *1981*
Edited by C. H. Collins, M. C. Allwood, Sally F. Bloomfield and A. Fox

17 Isolation and Identification Methods for Food Poisoning Organisms *1982*
Edited by Janet E. L. Corry, Diane Roberts and F. A. Skinner

18 Antibiotics: Assessment of Antimicrobial Activity and Resistance *1983*
Edited by A. D. Russell and L. B. Quesnel

THE SOCIETY FOR APPLIED BACTERIOLOGY TECHNICAL SERIES

19 Microbiological Methods for Environmental Biotechnology *1984*
Edited by J. J. Grainger and J. M. Lynch

20 Chemical Methods in Bacterial Systematics *1985*
Edited by M. Goodfellow and D. E. Minnikin

21 Isolation and Identification of Micro-organisms of Medical and Veterinary Importance *1985*
Edited by C. H. Collins and J. M. Grange